# THE WORLD OF
# MATHEMATICS

TEMPUS™

# THE WORLD OF
# MATHEMATICS

### A Small Library of the Literature of Mathematics from A'h-mosé the Scribe to Albert Einstein

*Presented with Commentaries and Notes by James R. Newman*

TEMPUS™

PUBLISHED BY

Tempus Books of Microsoft Press
A Division of Microsoft Corporation
16011 NE 36th Way, Box 97017, Redmond, Washington 98073-9717

Library of Congress Cataloging in Publication Data

The world of mathematics.
    Reprint. Originally published: New York: Simon and Schuster, 1956.
    Includes index.
    1. Mathematics. I. Newman, James Roy, 1907-1966.
QA3.W67      1988          510               88-20040
ISBN 1-55615-149-7 (cloth)
ISBN 1-55615-148-9 (paper)

THE EDITOR wishes to express his gratitude for permission to reprint material from the following sources:

Cambridge University Press for "Mathematics of Music," from *Science and Music*, by Sir James Jeans; and for excerpt from *A Mathematician's Apology*, by G. H. Hardy. Reprinted by permission of the publisher.

Chatto & Windus, Ltd., and Mr. Victor Butler for "Geometry in the South Pacific," from *Mr. Fortune's Maggot*, by Sylvia Townsend Warner. © 1927 by The Viking Press, Inc. Reprinted by permission of the publisher.

Constable and Co., Ltd., for "Mathematics in Warfare," from *Aircraft in Warfare*, by Frederick William Lanchester. Reprinted by permission of the publisher.

Ernest D. Courant for "The Lever of Mahomet," from *What Is Mathematics?*, by Richard Courant and Herbert Robbins. © 1941 by Richard Courant. Reprinted by permission of Ernest D. Courant.

Harper & Row, Publishers, Inc., for "Young Archimedes," by Aldous Huxley. © 1924, 1952 by Aldous Huxley. Reprinted by permission of the publisher.

Harvard University Press for "Mathematics of Aesthetics," from *Aesthetic Measure*, by George David Birkhoff. © 1933 by The President and Fellows of Harvard College. Reprinted by permission of the publisher.

Alfred A. Knopf, Inc., for "Meaning of Numbers," from *The Decline of the West*, by Oswald Spengler. © 1927 by Alfred A. Knopf, Inc. Reprinted by permission of the publisher.

The Macmillan Publishing Company, Inc., for "Arithmetical Restorations," from *Mathematical Recreations and Essays*, by W. W. Rouse Ball. Reprinted by permission of the publisher.

McClelland & Stewart and the Estate of Stephen Leacock for "Common Sense and the Universe," by Stephen Leacock, from *The Atlantic Monthly*; and for "Mathematics for Golfers."

M.I.T. Press for "How to Hunt a Submarine," from *Methods of Operations Research*, by Phillip M. Morse and George E. Kimball. Reprinted by permission of the publisher.

Estate of John von Neumann for "The General and Logical Theory of Automata," by John von Neumann, from *Cerebral Mechanisms in Behavior*. Reprinted with the permission of the Hixon Foundation and Dr. Lloyd A. Jeffress.

*The New Yorker* for "Inflexible Logic," by Russell Maloney, © 1940, 1968 by The New Yorker Magazine, Inc.; and for "The Law," by Robert M. Coates, © 1947, 1975 by The New Yorker Magazine, Inc. Reprinted by permission of the publisher.

Ruth G. Newman for "Pastimes of Past and Present Times," from *Mathematics and the Imagination*. © 1940 by Edward Kasner and James R. Newman, renewed by Ruth G. Newman.

Oxford University Press, Inc., for "Can a Machine Think?," by A. M. Turing, from *Mind*, 1950. Reprinted by permission of the publisher.

The Science Press for "Mathematical Creation," from *Foundations of Science*, by Henri Poincaré, translated by George Bruce Halsted.

*Scientific American* for "A Chess-Playing Machine," by Claude Shannon. © 1950 by Scientific American, Inc. Reprinted by permission of the publisher.

University of Chicago Press for "The Mathematician," by John von Neumann, from *The Works of the Mind*, edited by Heywood and Nef. © 1950 by University of Chicago Press Reprinted by permission of the publisher.

Unwin Hyman, Ltd., for "Easy Mathematics and Lawn Tennis," by T. J. I'A. Bromwich, from *Mathematical Gazette XIV*, October 1928. Reprinted by permission of the publisher.

Williams & Wilkins, Co., for "The Locus of Mathematical Reality: An Anthropological Footnote," by Leslie A. White, from *Philosophy of Science*, October, 1947. Reprinted by permission of the publisher.

Printed and bound in the United States of America

1 2 3 4 5 6 7 8 9 HCHC 3 2 1 0 9 8

Distributed to the book trade in the United States by Harper & Row.
Distributed to the book trade in Canada by General Publishing Company, Ltd.

Tempus Books and the Tempus Books logo are trademarks of Microsoft Press.
Tempus Books is an imprint of Microsoft Press.

Cover artwork © Richard Kehl 1988

Publisher's Note:

This Tempus Books edition is a faithful reproduction of the 1956 edition of James R. Newman's THE WORLD OF MATHEMATICS. Biographical material has been updated where appropriate and available, and the original index has been updated and included in all four volumes. The Publisher has made every effort to credit the appropriate parties for their ongoing permission to use material that appears in this edition of THE WORLD OF MATHEMATICS. Any questions or information regarding credits should be directed to Tempus Books of Microsoft Press.

# CONTENTS

# PART XVIII

# THE MATHEMATICIAN

# COMMENTARY ON
# G. H. HARDY

G. H. Hardy was a pure mathematician. The boundaries of this subject cannot be precisely defined but for Hardy the word "pure" as applied to mathematics had a clear, though negative, meaning. To qualify as pure, Hardy said, a mathematical topic had to be useless; if useless, it was not only pure, but beautiful. If useful—which is to say impure—it was ugly, and the more useful, the more ugly. These opinions were not always well received. The noted chemist Frederick Soddy, reviewing the book from which the following excerpts are taken, pronounced as scandalous Hardy's expressed contempt for useful mathematics or indeed for any applied science. "From such cloistral clowning," wrote Soddy, "the world sickens."[1] Hardy was a strange, original and enigmatic man. He was also a fine mathematician and a charming writer.

Godfrey Harold Hardy was born in Surrey in February 1877. His parents were teachers and "mathematically minded." He was educated first at Winchester—which he hated—and then at Cambridge, where he taught the greater part of his life. From 1919 to 1931 he held the Savilian chair of geometry at Oxford; in 1931 he was elected to the Sadlerian chair of pure mathematics at Cambridge and resumed the Fellowship at Trinity College which he had held from 1898 to 1919.

Hardy's main work was in analysis and arithmetic. He is known to students for his classic text, *A Course of Pure Mathematics,* which set a new standard for English mathematical education. But his reputation as the leader of pure mathematicians in Great Britain rests on his original and advanced researches. He wrote profound and masterly papers on such topics as the convergence and summability of series, inequalities and the analytic theory of numbers. The problems of number theory are often very easily stated (e.g., to prove that every even number is the sum of two prime numbers) "but all the resources of analysis are required to make any impression on them."[2] The problem quoted, and others of equally innocent appearance, are still unsolved "but they are not now—as they were in 1910—unapproachable."[3] This advance is due mainly to the joint work of Hardy and the British mathematician J. E. Littlewood. Their collaboration was exceptionally long and immensely fruitful; it is considered the most remarkable of all mathematical partnerships. An equally brilliant but unhappily brief partnership existed between Hardy and the self-taught Indian genius Ramanujan (see p. 360). It is hard to imagine two men further apart in training and background, yet Hardy was one of the first to discern what he termed Ramanujan's "profound and invincible originality." Ramanujan "called out Hardy's equal but quite different powers." "I owe more to him," Hardy said, "than to anyone else in the world with one exception, and my association with him is the one romantic incident of my life."[4]

I once encountered Hardy in the early 1930s at the subway entrance near Columbia University in New York City. It was a raw, wet winter day, but he was bareheaded, had no overcoat and wore a white cable-stitched turtle-necked sweater and a baggy pair of tennis slacks. I recall his delicately cut but strong features, his high coloring and the hair that fell in irregular bangs over his forehead. He was a strikingly handsome and graceful man who would have drawn attention even in more conventional dress. Hardy had strong opinions and vehement prejudices; some were admirable, some merely eccentric, and, I cannot help thinking, deliberately assumed. In political opinion as well as in his mathematical philosophy, he shared Bertrand Russell's views. His hatred of war was one reason why he regarded applied mathematics (ballistics or aerodynamics, for example) as "repulsively ugly and intolerably dull."[5] Hardy "always referred to God as his personal enemy. This was of course a joke but there was something real behind it. . . . He would not enter a religious building, even for such purpose as the election of a Warden of New College."[6] A special exemption clause had to be written into the by-laws of the college to enable him to discharge certain duties by proxy which otherwise would have required him to attend Chapel.

His love of mathematics was almost equaled by his passion for ball games: cricket, tennis and even baseball.[7] Justice Frankfurter tells the story of Hardy's visit to Boston in 1936 when he delivered his Ramanujan lectures at the Harvard Tercentenary. He was to be the house guest of a well-known lawyer, later a United States senator, and was terrified that he would find little to talk about with his host. The host was similarly alarmed, but the visit turned out to be easy and pleasant for both. For while the lawyer was no better prepared to discuss zeta functions than the mathematician to comment upon the rule in Shelley's case, they discovered a common enthusiasm for baseball. The Red Sox were playing a home stand at the time and Hardy could barely spare the time for his lectures.

"I have never done anything 'useful.' No discovery of mine has made, or is likely to make, directly or indirectly, for good or ill, the least difference to the amenity of the world." These lines appear in Hardy's half-defiant, half-ironical apology for his misspent life as a pure mathematician. The statement is nonsense. Hardy, I do not doubt, knew it was nonsense. Contributions such as his are certain to be useful; unexpectedly, and considering the world of today, perhaps even disagreeably useful. To make matters worse by his standards, it appears that Hardy once made a contribution to genetics. Writing a letter to *Science* in 1908 on a problem involving the transmission of dominant and recessive Mendelian characters in a mixed population, he established a principle known as Hardy's Law. This law (though he attached "little weight to it") turns out to be of "central importance in the study of Rh-blood groups and the treatment of haemolytic disease of the newborn."[8]

Hardy received many degrees and honors, including of course election to a Fellowship in the Royal Society in 1910. He died on December 1, 1947, the day the Copley Medal of the Royal Society, its highest award, was to have been presented to him.

## ENDNOTES

[1]*Nature*, Vol. 147, January 4, 1941.

[2]Obituary of G. H. Hardy, *Nature*, Vol. 161, May 22, 1948, pp. 797–98.

[3]*Ibid.*

[4]Obituary of G. H. Hardy by E. C. Titchmarsh, *The Journal of the London Mathematical Society*, April 1950, pp. 81–88. One of the joint papers of Hardy and Ramanujan is worth noting briefly. "Denote by $p(n)$ the number of ways of denoting the positive integer n as the sum of integers. For example, 5 can be expressed as $1 + 1 + 1 + 1 + 1$ or $1 + 1 + 1 + 2$ or $1 + 2 + 2$ or $1 + 1 + 3$ or $2 + 3$ or $1 + 4$ or 5, and so $p(5) = 7$. It is plain that $p(n)$ increases rapidly with $n$; and $p(200) = 3972999029388$ (as was shown by a computation which took a month). Hardy and Ramanujan's achievement was to establish an explicit formula for $p(n)$, of which the leading term is:

$$\frac{1}{2\pi\sqrt{2}} \frac{d}{dn} \frac{\exp\left\{\frac{2\pi}{\sqrt{6}}\sqrt{(n - \frac{1}{24})}\right\}}{\sqrt{(n - \frac{1}{24})}}$$

Five terms of the formula give the correct value of $p(200)$.

[5]Titchmarsh, *op. cit.,* p. 84.

[6]*Ibid.,* p. 86.

[7]Hardy frequently enlivened his discussions of philosophy or mathematics by illustrations taken from cricket. One of his papers "A maximal theorem with function-theoretic applications" contains the sentence "The problem is most easily grasped when stated in the language of cricket. . . . Suppose that a batsman plays, in a given season, a given 'stock' of innings." This paper, published in *Acta Mathematica* (54) and "presumably addressed to European mathematicians in general" must not have been very helpful to the Hungarians, say, who may not have appreciated all the fine points of the example.

[8]Titchmarsh, *op. cit.,* p, 83. J. B. S. Haldane gives another example of the useful, if unintentional, consequences of Hardy's work. There is a function called Riemann's Zeta function "which was devised, and its properties investigated, to find an expression for the number of prime numbers less than a given number. Hardy loved it. But it has been used in the theory of pyrometry, that is to say the investigation of the temperature of furnaces." *Everything Has a History,* London, 1951, p. 240.

**1**

*Mark all Mathematical heads which be wholly and only bent on these sciences, how solitary they be themselves, how unfit to live with others, how unapt to serve the world.*

—Roger Ascham (*ca. 1550*) (*Quoted in E. G. R. Taylor, "The Mathematical Practitioners of Tudor and Stuart England"*)

*I admit that mathematical science is a good thing. But excessive devotion to it is a bad thing.*

—Aldous Huxley (*Interview, J. W. N. Sullivan*)

# A Mathematician's
# Apology

## By G. H. Hardy

A mathematician, like a painter or a poet, is a maker of patterns. If his patterns are more permanent than theirs, it is because they are made with *ideas*. A painter makes patterns with shapes and colours, a poet with words. A painting may embody an 'idea,' but the idea is usually commonplace and unimportant. In poetry, ideas count for a good deal more; but, as Housman insisted, the importance of ideas in poetry is habitually exaggerated: 'I cannot satisfy myself that there are any such things as poetical ideas. . . . Poetry is not the thing said but a way of saying it.'

> Not all the water in the rough rude sea
> Can wash the balm from an anointed King.

Could lines be better, and could ideas be at once more trite and more false? The poverty of the ideas seems hardly to affect the beauty of the verbal pattern. A mathematician, on the other hand, has no material to work with but ideas, and so his patterns are likely to last longer, since ideas wear less with time than words.

The mathematician's patterns, like the painter's or the poet's, must be *beautiful*; the ideas, like the colours or the words, must fit together in a harmonious way. Beauty is the first test: there is no permanent place in the world for ugly mathematics. And here I must deal with a misconception which is still widespread (though probably much less so now than it was twenty years ago), what Whitehead has called the 'literary superstition' that love of and aesthetic appreciation of mathematics is 'a monomania confined to a few eccentrics in each generation.'

It would be difficult now to find an educated man quite insensitive to the aesthetic appeal of mathematics. It may be very hard to *define* mathematical beauty, but that is just as true of beauty of any kind—we may not know quite what we mean by a beautiful poem, but that does not prevent us from recognizing one when we read it. Even Professor Hogben, who is out to minimize at all costs the importance of the aesthetic element in mathematics, does not venture to deny its reality. 'There are, to be sure, individuals for whom mathematics exercises a coldly impersonal attraction. . . . The aesthetic appeal of mathematics may be very real for a chosen few.' But they are 'few,' he suggests, and they feel 'coldly' (and are really rather ridiculous people, who live in silly little university towns sheltered from the fresh breezes of the wide open spaces). In this he is merely echoing Whitehead's 'literary superstition.'

The fact is that there are few more 'popular' subjects than mathematics. Most people have some appreciation of mathematics, just as most people can enjoy a pleasant tune; and there are probably more people really interested in mathematics than in music. Appearances may suggest the contrary, but there are easy explanations. Music can be used to stimulate mass emotion, while mathematics cannot; and musical incapacity is recognized (no doubt rightly) as mildly discreditable, whereas most people are so frightened of the name of mathematics that they are ready, quite unaffectedly, to exaggerate their own mathematical stupidity.

A very little reflection is enough to expose the absurdity of the 'literary superstition.' There are masses of chess-players in every civilized country —in Russia, almost the whole educated population; and every chess-player can recognize and appreciate a 'beautiful' game or problem. Yet a chess problem is *simply* an exercise in pure mathematics (a game not entirely, since psychology also plays a part), and everyone who calls a problem 'beautiful' is applauding mathematical beauty, even if it is beauty of a comparatively lowly kind. Chess problems are the hymn-tunes of mathematics.

We may learn the same lesson, at a lower level but for a wider public, from bridge, or descending further, from the puzzle columns of the popular newspapers. Nearly all their immense popularity is a tribute to the drawing power of rudimentary mathematics, and the better makers of puzzles, such as Dudeney or 'Caliban,' use very little else. They know their business: what the public wants is a little intellectual 'kick,' and nothing else has quite the kick of mathematics.

I might add that there is nothing in the world which pleases even famous men (and men who have used disparaging language about mathematics) quite so much as to discover, or rediscover, a genuine mathematical theorem. Herbert Spencer republished in his autobiography a theorem about circles which he proved when he was twenty (not knowing that it had been proved over two thousand years before by Plato). Professor Soddy is a more recent and a more striking example (but *his* theorem really is his own).[1]

A chess problem is genuine mathematics, but it is in some way 'trivial' mathematics. However ingenious and intricate, however original and surprising the moves, there is something essential lacking. Chess problems are *unimportant*. The best mathematics is *serious* as well as beautiful—'important' if you like, but the word is very ambiguous, and 'serious' expresses what I mean much better.

I am not thinking of the 'practical' consequences of mathematics. I have to return to that point later: at present I will say only that if a chess problem is, in the crude sense, 'useless,' then that is equally true of most of the best mathematics; that very little of mathematics is useful practically, and that that little is comparatively dull. The 'seriousness' of a mathematical theorem lies, not in its practical consequences, which are usually negligible, but in the *significance* of the mathematical ideas which it connects. We may say, roughly, that a mathematical idea is 'significant' if it can be connected, in a natural and illuminating way, with a large complex of other mathematical ideas. Thus a serious mathematical theorem, a theorem which connects significant ideas, is likely to lead to important advances in mathematics itself and even in other sciences. No chess problem has ever affected the general development of scientific thought: Pythagoras, Newton, Einstein have in their times changed its whole direction.

The seriousness of a theorem, of course, does not *lie in* its consequences, which are merely the *evidence* for its seriousness. Shakespeare had an enormous influence on the development of the English language, Otway next to none, but that is not why Shakespeare was the better poet. He was the better poet because he wrote much better poetry. The inferiority of the chess problem, like that of Otway's poetry, lies not in its consequences but in its content.

There is one more point which I shall dismiss very shortly, not because it is uninteresting but because it is difficult, and because I have no qualifications for any serious discussion in aesthetics. The beauty of a mathematical theorem *depends* a great deal on its seriousness, as even in poetry the beauty of a line may depend to some extent on the significance of the ideas which it contains. I quoted two lines of Shakespeare as an example of the sheer beauty of a verbal pattern; but

After life's fitful fever he sleeps well

seems still more beautiful. The pattern is just as fine, and in this case the ideas have significance and the thesis is sound, so that our emotions are stirred much more deeply. The ideas do matter to the pattern, even in poetry, and much more, naturally, in mathematics; but I must not try to argue the question seriously.

It will be clear by now that, if we are to have any chance of making progress, I must produce examples of 'real' mathematical theorems, theorems which every mathematician will admit to be first-rate. And here I am very heavily handicapped by the restrictions under which I am writing. On the one hand my examples must be very simple, and intelligible to a reader who has no specialized mathematical knowledge; no elaborate preliminary explanations must be needed; and a reader must be able to follow the proofs as well as the enunciations. These conditions exclude, for instance, many of the most beautiful theorems of the theory of numbers, such as Fermat's 'two square' theorem or the law of quadratic reciprocity. And on the other hand my examples should be drawn from 'pukka' mathematics, the mathematics of the working professional mathematician; and this condition excludes a good deal which it would be comparatively easy to make intelligible but which trespasses on logic and mathematical philosophy.

I can hardly do better than go back to the Greeks. I will state and prove two of the famous theorems of Greek mathematics. They are 'simple' theorems, simple both in idea and in execution, but there is no doubt at all about their being theorems of the highest class. Each is as fresh and significant as when it was discovered—two thousand years have not written a wrinkle on either of them. Finally, both the statements and the proofs can be mastered in an hour by any intelligent reader, however slender his mathematical equipment.

1. The first is Euclid's[2] proof of the existence of an infinity of prime numbers.

The *prime numbers* or *primes* are the numbers

(A)                    2, 3, 5, 7, 11, 13, 17, 19, 23, 29, . . .

which cannot be resolved into smaller factors.[3] Thus 37 and 317 are prime. The primes are the material out of which all numbers are built up by multiplication: thus $666 = 2 \cdot 3 \cdot 3 \cdot 37$. Every number which is not prime itself is divisible by at least one prime (usually, of course, by several). We have to prove that there are infinitely many primes, i.e., that the series (A) never comes to an end.

Let us suppose that it does, and that

$$2, 3, 5, \ldots, P$$

is the complete series (so that $P$ is the largest prime); and let us, on this hypothesis, consider the number $Q$ defined by the formula

$$Q = (2 \cdot 3 \cdot 5 \cdot \ldots \cdot P) + 1.$$

It is plain that $Q$ is not divisible by any of 2, 3, 5, . . ., $P$; for it leaves the remainder 1 when divided by any one of these numbers. But, if not itself prime, it is divisible by *some* prime, and therefore there is a prime (which may be $Q$ itself) greater than any of them. This contradicts our hypothesis, that there is no prime greater than $P$; and therefore this hypothesis is false.

The proof is by *reductio ad absurdum,* and *reductio ad absurdum,* which Euclid loved so much, is one of a mathematician's finest weapons.[4] It is a far finer gambit than any chess gambit: a chess player may offer the sacrifice of a pawn or even a piece, but a mathematician offers *the game.*

2. My second example is Pythagoras's[5] proof of the 'irrationality' of $\sqrt{2}$.

A 'rational number' is a fraction $a/b$, where $a$ and $b$ are integers: we may suppose that $a$ and $b$ have no common factor, since if they had we could remove it. To say that '$\sqrt{2}$ is irrational' is merely another way of saying that 2 cannot be expressed in the form $(a/b)^2$; and this is the same thing as saying that the equation

(B) $$a^2 = 2b^2$$

cannot be satisfied by integral values of $a$ and $b$ which have no common factor. This is a theorem of pure arithmetic, which does not demand any knowledge of 'irrational numbers' or depend on any theory about their nature.

We argue again by *reductio ad absurdum*; we suppose that (B) is true, $a$ and $b$ being integers without any common factor. It follows from (B) that $a^2$ is even (since $2b^2$ is divisible by 2), and therefore that $a$ is even (since the square of an odd number is odd). If $a$ is even then

(C) $$a = 2c$$

for some integral value of $c$; and therefore

$$2b^2 = a^2 = (2c)^2 = 4c^2$$

or

(D) $$b^2 = 2c^2.$$

Hence $b^2$ is even, and therefore (for the same reason as before) $b$ is even. That is to say, $a$ and $b$ are both even, and so have the common factor 2. This contradicts our hypothesis, and therefore the hypothesis is false.

It follows from Pythagoras's theorem that the diagonal of a square is incommensurable with the side (that their ratio is not a rational number, that there is no unit of which both are integral multiples). For if we take the side as our unit of length, and the length of the diagonal is $d$, then, by a very familiar theorem also ascribed to Pythagoras,[6]

$$d^2 = 1^2 + 1^2 = 2,$$

so that $d$ cannot be a rational number.

I could quote any number of fine theorems from the theory of numbers whose *meaning* anyone can understand. For example, there is what is called 'the fundamental theorem of arithmetic,' that any integer can be resolved, *in one way only,* into a product of primes. Thus $666 = 2 \cdot 3 \cdot 3 \cdot 37$, and there is no other decomposition; it is impossible that $666 = 2 \cdot 11 \cdot 29$ or that $13 \cdot 89 = 17 \cdot 73$ (and we can see so without working out the products). This theorem is, as its name implies, the foundation of higher arithmetic; but the proof, although not 'difficult,' requires a certain amount of preface and might be found tedious by an unmathematical reader.

Another famous and beautiful theorem is Fermat's 'two square' theorem. The primes may (if we ignore the special prime 2) be arranged in two classes; the primes

$$5, 13, 17, 29, 37, 41, \ldots$$

which leave remainder 1 when divided by 4, and the primes

$$3, 7, 11, 19, 23, 31, \ldots$$

which leave remainder 3. All the primes of the first class, and none of the second, can be expressed as the sum of two integral squares: thus

$$5 = 1^2 + 2^2, \; 13 = 2^2 + 3^2,$$
$$17 = 1^2 + 4^4, \; 29 = 2^2 + 5^2;$$

but 3, 7, 11, and 19 are not expressible in this way (as the reader may check by trial). This is Fermat's theorem, which is ranked, very justly, as one of the finest of arithmetic. Unfortunately there is no proof within the comprehension of anybody but a fairly expert mathematician.

There are also beautiful theorems in the 'theory of aggregates' (*Mengenlehre*), such as Cantor's theorem of the 'non-enumerability' of the continuum. Here there is just the opposite difficulty. The proof is easy enough, when once the language has been mastered, but considerable explanation is necessary before the *meaning* of the theorem becomes clear. So I will not try to give more examples. Those which I have given are test cases, and a reader who cannot appreciate them is unlikely to appreciate anything in mathematics.

I said that a mathematician was a maker of patterns of ideas, and that beauty and seriousness were the criteria by which his patterns should be judged. I can hardly believe that anyone who has understood the two theorems will dispute that they pass these tests. If we compare them with Dudeney's most ingenious puzzles, or the finest chess problems that masters of that art have composed, their superiority in both respects stands out: there is an unmistakable difference of class. They are much more serious, and also much more beautiful: can we define, a little more closely, where their superiority lies?

In the first place, the superiority of the mathematical theorems in *seriousness* is obvious and overwhelming. The chess problem is the product of an ingenious but very limited complex of ideas, which do not differ from one another very fundamentally and have no external repercussions. We should think in the same way if chess had never been invented, whereas the theorems of Euclid and Pythagoras have influenced thought profoundly, even outside mathematics.

Thus Euclid's theorem is vital for the whole structure of arithmetic. The primes are the raw material out of which we have to build arithmetic, and Euclid's theorem assures us that we have plenty of material for the task. But the theorem of Pythagoras has wider applications and provides a better text.

We should observe first that Pythagoras's argument is capable of far-reaching extension, and can be applied, with little change of principle, to very wide classes of 'irrationals.' We can prove very similarly (as Theaetetus seems to have done) that

$$\sqrt{3}, \sqrt{5}, \sqrt{7}, \sqrt{11}, \sqrt{13}, \sqrt{17}$$

are irrational, or (going beyond Theatetus) that $\sqrt[3]{2}$ and $\sqrt[3]{17}$ are irrational.[7]

Euclid's theorem tells us that we have a good supply of material for the construction of a coherent arithmetic of the integers. Pythagoras's theorem and its extensions tells us that, when we have constructed this arithmetic, it will not prove sufficient for our needs, since there will be many magnitudes which obtrude themselves upon our attention and which it will be unable to measure: the diagonal of the square is merely the most obvious example. The profound importance of this discovery was recognized at once by the Greek mathematicians. They had begun by assuming (in accordance, I suppose, with the 'natural' dictates of 'common sense') that all magnitudes of the same kind are commensurable, that any two lengths, for example, are multiples of some common unit, and they had constructed a theory of proportion based on this assumption. Pythagoras's discovery exposed the unsoundness of this foundation, and led to the construction of the much more profound theory of Eudoxus which is set out in the fifth book of the *Elements,* and which is regarded by many modern mathematicians as the finest achievement of Greek mathematics. This theory is astonishingly modern in spirit, and may be regarded as the beginning of the modern theory of irrational number, which has revolutionized mathematical analysis and had much influence on recent philosophy.

There is no doubt at all, then, of the 'seriousness' of either theorem. It is therefore the better worth remarking that neither theorem has the slightest 'practical' importance. In practical applications we are concerned only with comparatively small numbers; only stellar astronomy and atomic physics deal with 'large' numbers, and they have very little more practical importance, as yet, than the most abstract pure mathematics. I do not know what is the highest degree of accuracy which is ever useful to an engineer—we shall be very generous if we say ten significant figures. Then

$$3.141592654$$

(the value of $\pi$ to nine places of decimals) is the ratio

$$\frac{3141592654}{1000000000}$$

of two numbers of ten digits. The number of primes less than 1,000,000,000 is 50,847,478: that is enough for an engineer, and he can be perfectly happy without the rest. So much for Euclid's theorem; and, as regards Pythagoras's, it is obvious that irrationals are uninteresting to an engineer, since he is concerned only with approximations, and all approximations are rational.

The contrast between pure and applied mathematics stands out most clearly, perhaps, in geometry. There is the science of pure geometry,[8] in which there are many geometries, projective geometry, Euclidean geometry, non-Euclidean geometry, and so forth. Each of these geometries is a *model,* a pattern of ideas, and is to be judged by the interest and beauty of its particular pattern. It is a *map* or *picture,* the joint product of many hands, a partial and imperfect copy (yet exact so far as it extends) of a section of mathematical reality. But the point which is important to us now is this, that there is one thing at any rate of which pure geometries are *not* pictures, and that is the spatio-temporal reality of the physical world. It is obvious, surely, that they cannot be, since earthquakes and eclipses are not mathematical concepts.

This may sound a little paradoxical to an outsider, but it is a truism to a geometer; and I may perhaps be able to make it clearer by an illustration. Let us suppose that I am giving a lecture on some system of geometry, such as ordinary Euclidean geometry, and that I draw figures on the blackboard to stimulate the imagination of my audience, rough drawings of straight lines or circles or ellipses. It is plain, first, that the truth of the theorems which I prove is in no way affected by the quality of my drawings. Their function is merely to bring home my meaning to my hearers, and, if I can do that, there would be no gain in having them redrawn by the most skilful draughtsman. They are pedagogical illustrations, not part of the real subject-matter of the lecture.

Now let us go a stage further. The room in which I am lecturing is part of the physical world, and has itself a certain pattern. The study of that pattern, and of the general pattern of physical reality, is a science in itself, which we may call 'physical geometry.' Suppose now that a violent dynamo, or a massive gravitating body, is introduced into the room. Then the physicists tell us that the geometry of the room is changed, its whole physical pattern slightly but definitely distorted. Do the theorems which I have proved become false? Surely it would be nonsense to suppose that the proofs of them which I have given are affected in any way. It would be like supposing that a play of Shakespeare is changed when a reader spills his tea over a page. The play is independent of the pages on which it is printed, and 'pure geometries' are independent of lecture rooms, or of any other detail of the physical world.

This is the point of view of a pure mathematician. Applied mathematicians, mathematical physicists, naturally take a different view, since they are preoccupied with the physical world itself, which also has its structure or pattern. We cannot describe this pattern exactly, as we can that of a pure geometry, but we can say something significant about it. We can describe, sometimes fairly accurately, sometimes very roughly, the relations which hold between some of its constituents, and compare them with the exact relations holding between constituents of some system of pure geometry. We may be able to trace a certain resemblance between the two sets of relations, and then the pure geometry will become interesting to physicists; it will give us, to that extent, a map which 'fits the facts' of the physical world. The geometer offers to the physicist a whole set of maps from which to choose. One map, perhaps, will fit the facts better than others, and then the geometry which provides that particular map will be the geometry most important for applied mathematics. I may add that even a pure mathematician may find his appreciation of this geometry quickened, since there is no mathematician so pure that he feels no interest at all in the physical world; but, in so far as he succumbs to this temptation, he will be abandoning his purely mathematical position.

I will end with a summary of my conclusions, but putting them in a more personal way. I said at the beginning that anyone who defends his subject will find that he is defending himself; and my justification of the life of a professional mathematician is bound to be, at bottom, a justification of my own. Thus this concluding section will be in its substance a fragment of autobiography.

I cannot remember ever having wanted to be anything but a mathematician. I suppose that it was always clear that my specific abilities lay that way, and it never occurred to me to question the verdict of my elders. I do not remember having felt, as a boy, any *passion* for mathematics, and such notions as I may have had of the career of a mathematician were far from noble. I thought of mathematics in terms of examinations and scholarships: I wanted to beat other boys, and this seemed to be the way in which I could do so most decisively.

I was about fifteen when (in a rather odd way) my ambitions took a sharper turn. There is a book by 'Alan St. Aubyn'[9] called *A Fellow of Trinity,* one of a series dealing with what is supposed to be Cambridge college life. I suppose that it is a worse book than most of Marie Corelli's; but a book can hardly be entirely bad if it fires a clever boy's imagination. There are two heroes, a primary hero called Flowers, who is almost wholly good, and a secondary hero, a much weaker vessel, called Brown. Flowers and Brown find many dangers in university life, but the worst is a gambling saloon in Chesterton[10] run by the Misses Bellenden, two fascinating but extremely wicked young ladies. Flowers survives all these troubles, is Second Wrangler and Senior Classic, and succeeds automatically to a Fellowship (as I suppose he would have done then). Brown succumbs, ruins his parents, takes to drink, is saved from delirium tremens during a thunderstorm only by the prayers of the Junior Dean, has much difficulty in obtaining even an Ordinary Degree, and ultimately becomes a missionary. The friendship is not shattered by these unhappy events, and

Flowers's thoughts stray to Brown, with affectionate pity, as he drinks port and eats walnuts for the first time in Senior Combination Room.

Now Flowers was a decent enough fellow (so far as 'Alan St. Aubyn' could draw one), but even my unsophisticated mind refused to accept him as clever. If he could do these things, why not I? In particular, the final scene in Combination Room fascinated me completely, and from that time, until I obtained one, mathematics meant to me primarily a Fellowship of Trinity.

I found at once, when I came to Cambridge, that a Fellowship implied 'original work,' but it was a long time before I formed any definite idea of research. I had of course found at school, as every future mathematician does, that I could often do things much better than my teachers; and even at Cambridge I found, though naturally much less frequently, that I could sometimes do things better than the College lecturers. But I was really quite ignorant, even when I took the Tripos, of the subjects on which I have spent the rest of my life; and I still thought of mathematics as essentially a 'competitive' subject. My eyes were first opened by Professor Love, who taught me for a few terms and gave me my first serious conception of analysis. But the great debt which I owe to him—he was, after all, primarily an applied mathematician—was his advice to read Jordan's famous *Cours d'analyse*; and I shall never forget the astonishment with which I read that remarkable work, the first inspiration for so many mathematicians of my generation, and learnt for the first time as I read it what mathematics really meant. From that time onwards I was in my way a real mathematician, with sound mathematical ambitions and a genuine passion for mathematics.

I wrote a great deal during the next ten years, but very little of any importance; there are not more than four or five papers which I can still remember with some satisfaction. The real crises of my career came ten or twelve years later, in 1911, when I began my long collaboration with Littlewood, and in 1913, when I discovered Ramanujan. All my best work since then has been bound up with theirs, and it is obvious that my association with them was the decisive event of my life. I still say to myself when I am depressed, and find myself forced to listen to pompous and tiresome people, 'Well, I have done one thing *you* could never have done, and that is to have collaborated with both Littlewood and Ramanujan on something like equal terms.' It is to them that I owe an unusually late maturity: I was at my best at a little past forty, when I was a professor at Oxford. Since then I have suffered from that steady deterioration which is the common fate of elderly men and particularly of elderly mathematicians. A mathematician may still be competent enough at sixty, but it is useless to expect him to have original ideas.

It is plain now that my life, for what it is worth, is finished, and that nothing I can do can perceptibly increase or diminish its value. It is very difficult to be dispassionate, but I count it a 'success'; I have had more reward and not less than was due to a man of my particular grade of ability. I have held a series of comfortable and 'dignified' positions. I have had very little trouble with the duller routine of universities. I hate 'teaching,' and have had to do very little, such teaching as I have done having been almost entirely supervision of research; I love lecturing, and have lectured a great deal to extremely able classes; and I have always had plenty of leisure for the researches which have been the one great permanent happiness of my life. I have found it easy to work with others, and have collaborated on a large scale with two exceptional mathematicians; and this has enabled me to add to mathematics a good deal more than I could reasonably have expected. I have had my disappointments, like any other mathematician, but none of them has been too serious or has made me particularly unhappy. If I had been offered a life neither better nor worse when I was twenty, I would have accepted without hesitation.

It seems absurd to suppose that I could have 'done better.' I have no linguistic or artistic ability, and very little interest in experimental science. I might have been a tolerable philosopher, but not one of a very original kind. I think that I might have made a good lawyer; but journalism is the only profession, outside academic life, in which I should have felt really confident of my chances. There is no doubt that I was right to be a mathematician, if the criterion is to be what is commonly called success.

My choice was right, then, if what I wanted was a reasonably comfortable and happy life. But solicitors and stockbrokers and bookmakers often lead comfortable and happy lives, and it is very difficult to see how the world is the richer for their existence. Is there any sense in which I can claim that my life has been less futile than theirs? It seems to me again that there is only one possible answer: yes, perhaps, but, if so, for one reason only.

I have never done anything 'useful.' No discovery of mine has made, or is likely to make, directly or indirectly, for good or ill, the least difference to the amenity of the world. I have helped to train other mathematicians, but mathematicians of the same kind as myself, and their work has been, so far at any rate as I have helped them to it, as useless as my own. Judged by all practical standards, the value of my mathematical life is nil; and outside mathematics it is trivial anyhow. I have just one chance of escaping a verdict of completely triviality, that I may be judged to have created something worth creating. And that I have created something is undeniable: the question is about its value.

The case for my life, then, or for that of any one else who has been a mathematician in the same sense in which I have been one, is this: that I have added something to knowledge, and helped others to add more; and that these somethings have a value which differs in degree only, and not in kind, from that of the creations of the great mathematicians, or of any of the other artists, great or small, who have left some kind of memorial behind them.

# ENDNOTES

[1]See his letters on the 'Hexlet' in *Nature*, vols. 137–9 (1936–7).

[2]*Elements* IX 20. The real origin of many theorems in the *Elements* is obscure, but there seems to be no particular reason for supposing that this one is not Euclid's own.

[3]There are technical reasons for not counting 1 as a prime.

[4]The proof can be arranged so as to avoid a *reductio,* and logicians of some schools would prefer that it should be.

[5]The proof traditionally ascribed to Pythagoras, and certainly a product of his school. The theorem occurs, in a much more general form, in Euclid (*Elements* X 9).

[6]Euclid, *Elements* I 47.

[7]See Ch. IV of Hardy and Wright's *Introduction to the Theory of Numbers,* where there are discussions of different generalizations of Pythagoras's argument, and of a historical puzzle about Theaetetus.

[8]We must of course, for the purposes of this discussion, count as pure geometry what mathematicians call 'analytical' geometry.

[9]'Alan St. Aubyn' was Mrs. Frances Marshall, wife of Matthew Marshall.

[10]Actually, Chesterton lacks picturesque features.

# COMMENTARY ON
# THE ELUSIVENESS OF
# INVENTION

How is mathematics made? What kind of brain is it that can frame the propositions and compose the systems of mathematics? How are mathematical ideas inspired and incubated? How do the mental processes of the geometer or the algebraist compare with those of the musician, the poet, the painter, the chess player, the ordinary man? In mathematical creation which are the key elements? Intuition? An exquisite sense of space and time? The precision of a calculating machine? A powerful memory? Formidable skill in following complex sequences? An exceptional capacity for concentration?

Psychologists have tried to answer, but their explanations have not been impressive. Jacques Hadamard, an accomplished French mathematician, surveyed the subject in his little book, *The Psychology of Invention in the Mathematical Field*. It is an entertaining account but not very enlightening. The celebrated phrenologist Gall said mathematical ability showed itself in a special bump on the head, the location of which he specified. The psychologist Souriau, we are told, maintained that invention occurs by "pure chance," a valuable theory. It is often suggested that creative ideas are conjured up in "mathematical dreams," but this attractive hypothesis has not been verified. Hadamard reports that mathematicians were asked whether "noises" or "meteorological circumstances" helped or hindered research; all save the most single-minded are evidently discommoded by severe outbreaks, but the constructive effects of such factors is admittedly doubtful. Claude Bernard, the great physiologist, said that in order to invent "one must think aside." Hadamard says this is a profound insight; he also considers whether scientific invention may perhaps be improved by standing or sitting or by taking two baths in a row. Helmholtz and Poincaré worked sitting at a table; Hadamard's practice is to pace the room ("Legs are the wheels of thought," said Emile Angier); the chemist J. Teeple was the two-bath man. Alas, the habits of famous men are rarely profitable to their disciples. The young philosopher will derive little benefit from being punctual like Kant, the biologist from cultivating Darwin's dyspepsia, the playwright from eating Shaw's vegetables.

The following essay, delivered early in this century as a lecture before the Psychological Society in Paris, is at once the least pretentious and the most celebrated of the attempts to describe what goes on in the mathematician's brain. Henri Poincaré, cousin of the politician, was peculiarly fitted to undertake the task. One of the foremost mathematicians of all time, unrivaled as an analyst and mathematical physicist, he was also a matchless expositor of the philosophy of science. Poincaré's piece fails entirely, I think, to elucidate the problem of mathematical creation, but as an autobiographical fragment it tells a dramatic story and is one of the treasures of science and literature.[1]

## ENDNOTE

[1]For a biographical essay on Poincaré see p. 1353.

# MATHEMATICAL CREATION

## *By Henri Poincaré*

The genesis of mathematical creation is a problem which should intensely interest the psychologist. It is the activity in which the human mind seems to take least from the outside world, in which it acts or seems to act only of itself and on itself, so that in studying the procedure of geometric thought we may hope to reach what is most essential in man's mind.

This has long been appreciated, and some time back the journal called *L'enseignement mathématique,* edited by Laisant and Fehr, began an investigation of the mental habits and methods of work of different mathematicians. I had finished the main outlines of this article when the results of that inquiry were published, so I have hardly been able to utilize them and shall confine myself to saying that the majority of witnesses confirm my conclusions; I do not say all, for when the appeal is to universal suffrage unanimity is not to be hoped.

A first fact should surprise us, or rather would surprise us if we were not so used to it. How does it happen there are people who do not understand mathematics? If mathematics invokes only the rules of logic, such as are accepted by all normal minds; if its evidence is based on principles common to all men, and that none could deny without being mad, how does it come about that so many persons are here refractory?

That not every one can invent is nowise mysterious. That not every one can retain a demonstration once learned may also pass. But that not every one can understand mathematical reasoning when explained appears very surprising when we think of it. And yet those who can follow this reasoning only with difficulty are in the majority: that is undeniable, and will surely not be gainsaid by the experience of secondary-school teachers.

And further: how is error possible in mathematics? A sane mind should not be guilty of a logical fallacy, and yet there are very fine minds who do not trip in brief reasoning such as occurs in the ordinary doings of life, and who are incapable of following or repeating without error the mathematical demonstrations which are longer, but which after all are only an accumulation of brief reasonings wholly analogous to those they make so easily. Need we add that mathematicians themselves are not infallible?

The answer seems to me evident. Imagine a long series of syllogisms, and that the conclusions of the first serve as premises of the following: we shall be able to catch each of these syllogisms, and it is not in passing from premises to conclusion that we are in danger of deceiving ourselves. But between the moment in which we first meet a proposition as conclusion of one syllogism, and that in which we reencounter it as premise of another syllogism occasionally some time will elapse, several links of the chain will have unrolled; so it may happen that we have forgotten it, or worse, that we have forgotten its meaning. So it may happen that we replace it by a slightly different proposition, or that, while retaining the same enunciation, we attribute to it a slightly different meaning, and thus it is that we are exposed to error.

Often the mathematician uses a rule. Naturally he begins by demonstrating this rule; and at the time when this proof is fresh in his memory he understands perfectly its meaning and its bearing, and he is in no danger of changing it. But subsequently he trusts his memory and afterward only applies it in a mechanical way; and then if his memory fails him, he may apply it all wrong. Thus it is, to take a simple example, that we sometimes make slips in calculation because we have forgotten our multiplication table.

According to this, the special aptitude for mathematics would be due only to a very sure memory or to a prodigious force of attention. It would be a power like that of the whist-player who remembers the cards played; or, to go up a step, like that of the chess-player who can visualize a great number of combinations and hold them in his memory. Every good mathematician ought to be a good chess-player, and inversely; likewise he should be a good computer. Of course that sometimes happens; thus Gauss was at the same time a geometer of genius and a very precocious and accurate computer.

But there are exceptions; or rather I err; I can not call them exceptions without the exceptions being more than the rule. Gauss it is, on the contrary, who was an exception. As for myself, I must confess, I am absolutely incapable even of adding without mistakes. In the same way I should be but a poor chess-player; I would perceive that by a certain play I should expose myself to a certain danger; I would pass in review several other plays, rejecting them for other reasons, and then finally I should make the move first examined, having meantime forgotten the danger I had foreseen.

In a word, my memory is not bad, but it would be insufficient to make me a good chess-player. Why then does it not fail me in a difficult piece of mathematical reasoning where most chess-players would lose themselves? Evidently because it is guided by the general march of the reasoning. A mathematical demonstration is not a simple juxtaposition of syllogisms, it is syllogisms *placed in a certain order,* and the order in which these elements are placed is much more important than the elements themselves. If I have the feeling, the intuition, so to speak, of this order, so as to perceive at a glance the reasoning as a whole, I need no longer fear lest I forget one of the elements, for each of them will take its allotted place in the array, and that without any effort of memory on my part.

It seems to me then, in repeating a reasoning learned, that I could have invented it. This is often only an illusion; but even then, even if I am not so gifted as to create it by myself, I myself re-invent it in so far as I repeat it.

We know that this feeling, this intuition of mathematical order, that makes us divine hidden harmonies and relations, can not be possessed by every one. Some will not have either this delicate feeling so difficult to define, or a strength of memory and attention beyond the ordinary, and then they will be absolutely incapable of understanding higher mathematics. Such are the majority. Others will have this feeling only in a slight degree, but they will be gifted with an uncommon memory and a great power of attention. They will learn by heart the details one after another; they can understand mathematics and sometimes make applications, but they cannot create. Others, finally, will possess in a less or greater degree the special intuition referred to, and then not only can they understand mathematics even if their memory is nothing extraordinary, but they may become creators and try to invent with more or less success according as this intuition is more or less developed in them.

In fact, what is mathematical creation? It does not consist in making new combinations with mathematical entities already known. Any one could do that, but the combinations so made would be infinite in number and most of them absolutely without interest. To create consists precisely in not making useless combinations and in making those which are useful and which are only a small minority. Invention is discernment, choice.

How to make this choice I have before explained; the mathematical facts worthy of being studied are those which, by their analogy with other facts, are capable of leading us to the knowledge of a mathematical law just as experimental facts lead us to the knowledge of a physical law. They are those which reveal to us unsuspected kinship between other facts, long known, but wrongly believed to be strangers to one another.

Among chosen combinations the most fertile will often be those formed of elements drawn from domains which are far apart. Not that I mean as sufficing for invention the bringing together of objects as disparate as possible; most combinations so formed would be entirely sterile. But certain among them, very rare, are the most fruitful of all.

To invent, I have said, is to choose; but the word is perhaps not wholly exact. It makes one think of a purchaser before whom are displayed a large number of samples, and who examines them, one after the other, to make a choice. Here the samples would be so numerous that a whole lifetime would not suffice to examine them. This is not the actual state of things. The sterile combinations do not even present themselves to the mind of the inventor. Never in the field of his consciousness do combinations appear that are not really useful, except some that he rejects but which have to some extent the characteristics of useful combinations. All goes on as if the inventor were an examiner for the second degree who would only have to question the candidates who had passed a previous examination,

But what I have hitherto said is what may be observed or inferred in reading the writings of the geometers, reading reflectively.

It is time to penetrate deeper and to see what goes on in the very soul of the mathematician. For this, I believe, I can do best by recalling memories of my own. But I shall limit myself to telling how I wrote my first memoir on Fuchsian functions. I beg the reader's pardon; I am about to use some technical expressions, but they need not frighten him, for he is not obliged to understand them. I shall say, for example, that I have found the demonstration of such a theorem under such circumstances. This theorem will have a barbarous name, unfamiliar to many, but that is unimportant; what is of interest for the psychologist is not the theorem but the circumstances.

For fifteen days I strove to prove that there could not be any functions like those I have since called Fuchsian functions. I was then very ignorant; every day I seated myself at my work table, stayed an hour or two, tried a great number of combinations and reached no results. One evening, contrary to my custom, I drank black coffee and could not sleep. Ideas rose in crowds; I felt them collide until pairs interlocked, so to speak, making a stable combination. By the next morning I had established the existence of a class of Fuchsian functions, those which come from the hypergeometric series; I had only to write out the results, which took but a few hours.

Then I wanted to represent these functions by the quotient of two series; this idea was perfectly conscious and deliberate, the analogy with elliptic functions guided me. I asked myself what properties these series must have if they existed, and I succeeded without difficulty in forming the series I have called theta-Fuchsian.

Just at this time I left Caen, where I was then living, to go on a geological excursion under the auspices of the school of mines. The changes of travel made me forget my mathematical work. Having reached Coutances, we entered an omnibus to go some place or other. At the moment when I put my foot on the step the idea came to me, without anything in my former thoughts seeming to have paved the way for it, that the transformations I had used to define the Fuchsian functions were identical with those of non-Euclidean geometry. I did not verify the idea; I should not have had time, as, upon taking my seat in the omnibus, I went on with a conversation already commenced, but I felt a perfect certainty. On my return to Caen, for conscience' sake I verified the result at my leisure.

Then I turned my attention to the study of some arithmetic questions apparently without much success and without a suspicion of any connection with my preceding researches. Disgusted with my failure, I went to spend a few days at the seaside, and thought of something else. One morning, walking on the bluff, the idea came to me, with just the same characteristics of brevity, suddenness and immediate certainty, that the arithmetic transformations of indeterminate ternary quadratic forms were identical with those of non-Euclidean geometry.

Returned to Caen, I meditated on this result and deduced the consequences. The example of quadratic forms showed me that there were Fuchsian groups other than those corresponding to the hypergeometric series; I saw that I could apply to them the theory of theta-Fuchsian series and that consequently there existed Fuchsian functions other than those from the hypergeometric series, the ones I then knew. Naturally I set myself to form all these functions. I made a systematic attack upon them and carried all the outworks, one after another. There was one however that still held out, whose fall would involve that of the whole place. But all my efforts only served at first the better to show me the difficulty, which indeed was something. All this work was perfectly conscious.

Thereupon I left for Mont-Valérien, where I was to go through my military service; so I was very differently occupied. One day, going along the street, the solution of the difficulty which had stopped me suddenly appeared to me. I did not try to go deep into it immediately, and only after my service did I again take up the question. I had all the elements and had only to arrange them and put them together. So I wrote out my final memoir at a single stroke and without difficulty.

I shall limit myself to this single example; it is useless to multiply them. In regard to my other researches I would have to say analogous things, and the observations of other mathematicians given in *L'enseignement mathématique* would only confirm them.

Most striking at first is this appearance of sudden illumination, a manifest sign of long, unconscious prior work. The rôle of this unconscious work in mathematical invention appears to me incontestable, and traces of it would be found in other cases where it is less evident. Often when one works at a hard question, nothing good is accomplished at the first attack. Then one takes a rest, longer or shorter, and sits down anew to the work. During the first half-hour, as before, nothing is found, and then all of a sudden the decisive idea presents itself to the mind. It might be said that the conscious work has been more fruitful because it has been interrupted and the rest has given back to the mind its force and freshness. But it is more probable that this rest has been filled out with unconscious work and that the result of this work has afterwards revealed itself to the geometer just as in the cases I have cited; only the revelation, instead of coming during a walk or a journey, has happened during a period of conscious work, but independently of this work which plays at most a rôle of excitant, as if it were the goad stimulating the results already reached during rest, but remaining unconscious, to assume the conscious form.

There is another remark to be made about the conditions of this unconscious work: it is possible, and of a certainty it is only fruitful, if it is on the one hand preceded and on the other hand followed by a period of conscious work. These sudden inspirations (and the examples already cited sufficiently prove this) never happen except after some days of voluntary effort which has appeared absolutely fruitless and whence nothing good seems to have come, where the way taken seems totally astray. These efforts then have not been as sterile as one thinks; they have set agoing the unconscious machine and without them it would not have moved and would have produced nothing.

The need for the second period of conscious work, after the inspiration, is still easier to understand. It is necessary to put in shape the results of this inspiration, to deduce from them the immediate consequences, to arrange them, to word the demonstrations, but above all is verification necessary. I have spoken of the feeling of absolute certitude accompanying the inspiration; in the cases cited this feeling was no deceiver, nor is it usually. But do not think this a rule without exception; often this feeling deceives us without being any the less vivid, and we only find it out when we seek to put on foot the demonstration. I have especially noticed this fact in regard to ideas coming to me in the morning or evening in bed while in a semi-hypnagogic state.

Such are the realities; now for the thoughts they force upon us. The unconscious, or, as we say, the subliminal self plays an important rôle in mathematical creation; this follows from what we have said. But usually the subliminal self is considered as purely automatic. Now we have seen that mathematical work is not simply mechanical, that it could not be done by a machine, however perfect. It is not merely a question of applying rules, of making the most combinations possible according to certain fixed laws. The combinations so obtained would be exceedingly numerous, useless and cumbersome. The true work of the inventor consists in choosing among these combinations so as to eliminate the useless ones or rather to avoid the trouble of making them, and the rules which must guide this choice are extremely fine and delicate. It is almost impossible to state them precisely; they are felt rather than formulated. Under these conditions, how imagine a sieve capable of applying them mechanically?

A first hypothesis now presents itself: the subliminal self is in no way inferior to the conscious self; it is not purely automatic; it is capable of discernment; it has tact, delicacy; it knows how to choose, to divine. What do I say? It knows better how to divine than the conscious self, since it succeeds where that has failed. In a word, is not the subliminal self superior to the conscious self? You recognize the full importance of this question. Boutroux in a recent lecture has shown how it came up on a very different occasion, and what consequences would follow an affirmative answer. (See also, by the same author, *Science et Religion*, pp. 313 ff.)

Is this affirmative answer forced upon us by the facts I have just given? I confess that, for my part, I should hate to accept it. Reexamine the facts then and see if they are not compatible with another explanation.

It is certain that the combinations which present themselves to the mind in a sort of sudden illumination, after an unconscious working somewhat prolonged, are generally useful and fertile combinations, which seem the result of a first impression. Does it follow that the subliminal self, having divined by a delicate intuition that these combinations would be useful, has formed only these, or has it rather formed many others which were lacking in interest and have remained unconscious?

In his second way of looking at it, all the combinations would be formed in consequence of the automatism of the subliminal self, but only the interesting ones would break into the domain of consciousness. And this is still very mysterious. What is the cause that, among the thousand products of our unconscious activity, some are called to pass the threshold, while others remain below? Is it a simple chance which confers this privilege? Evidently not; among all the stimuli of our senses, for example, only the most intense fix our attention, unless it has been drawn to them by other causes. More generally the privileged unconscious phenomena, those susceptible of becoming conscious, are those which, directly or indirectly affect most profoundly our emotional sensibility.

It may be surprising to see emotional sensibility invoked *à propos* of mathematical demonstrations which, it would seem, can interest only the intellect. This would be to forget the feeling of mathematical beauty, of the harmony of numbers and forms, of geometric elegance. This is a true esthetic feeling that all real mathematicians know, and surely it belongs to emotional sensibility.

Now, what are the mathematic entities to which we attribute this character of beauty and elegance, and which are capable of developing in us a sort of esthetic emotion? They are those whose elements are harmoniously disposed so that the mind without effort can embrace their totality while realizing the details. This harmony is at once a satisfaction of our esthetic needs and an aid to the mind, sustaining and guiding. And at the same time, in putting under our eyes a well-ordered whole, it makes us foresee a mathematical law. Now, as we have said above, the only mathematical facts worthy of fixing our attention and capable of being useful are those which can teach us a mathematical law. So that we reach the following conclusion: The useful combinations are precisely the most beautiful, I mean those best able to charm this special sensibility that all mathematicians know, but of which the profane are so ignorant as often to be tempted to smile at it.

What happens then? Among the great numbers of combinations blindly formed by the subliminal self, almost all are without interest and without utility; but just for that reason they are also without effect upon the esthetic sensibility. Consciousness will never know them; only certain ones are harmonious, and, consequently, at once useful and beautiful. They will be capable of touching his special sensibility of the geometer of which I have just spoken, and which, once aroused, will call our attention to them, and thus give them occasion to become conscious.

This is only a hypothesis, and yet here is an observation which may confirm it: when a sudden illumination seizes upon the mind of the mathematician, it usually happens that it does not deceive him, but it also sometimes happens, as I have said, that it does not stand the test of verification; well, we almost always notice that this false idea, had it been true, would have gratified our natural feeling for mathematical elegance.

Thus it is this special esthetic sensibility which plays the rôle of the delicate sieve of which I spoke, and that sufficiently explains why the one lacking it will never be a real creator.

Yet all the difficulties have not disappeared. The conscious self is narrowly limited, and as for the subliminal self we know not its limitations, and this is why we are not too reluctant in supposing that it has been able in a short time to make more different combinations than the whole life of a conscious being could encompass. Yet these limitations exist. Is it likely that it is able to form all the possible combinations, whose number would frighten the imagination? Nevertheless that would seem necessary, because if it produces only a small part of these combinations, and if it makes them at random, there would be small chance that the *good,* the one we should choose, would be found among them.

Perhaps we ought to seek the explanation in that preliminary period of conscious work which always precedes all fruitful unconscious labor. Permit me a rough comparison. Figure the future elements of our combinations as something like the hooked atoms of Epicurus. During the complete repose of the mind, these atoms are motionless, they are, so to speak, hooked to the wall; so this complete rest may be indefinitely prolonged without the atoms meeting, and consequently without any combination between them.

On the other hand, during a period of apparent rest and unconscious work, certain of them are detached from the wall and put in motion. They flash in every direction through the space (I was about to say the room) where they are enclosed, as would, for example, a swarm of gnats or, if you prefer a more learned comparison, like the molecules of gas in the kinematic theory of gases. Then their mutual impacts may produce new combinations.

What is the rôle of the preliminary conscious work? It is evidently to mobilize certain of these atoms, to unhook them from the wall and put them in swing. We think we have done no good, because we have moved these elements a thousand different ways in seeking to assemble them, and have found no satisfactory aggregate. But, after this shaking up imposed upon them by our will, these atoms do not return to their primitive rest. They freely continue their dance.

Now, our will did not choose them at random; it pursued a perfectly determined aim. The mobilized atoms are therefore not any atoms whatsoever; they are those from which we might reasonably expect the desired solution. Then the mobilized atoms undergo impacts which make them enter into combinations among themselves or with other atoms at rest which they struck against in their course. Again I beg pardon, my comparison is very rough, but I scarcely know how otherwise to make my thought understood.

However it may be, the only combinations that have a chance of forming are those where at least one of the elements is one of those atoms freely chosen by our will. Now, it is evidently among these that is found what I called the *good combination*. Perhaps this is a way of lessening the paradoxical in the original hypothesis.

Another observation. It never happens that the unconscious work gives us the result of a somewhat long calculation *all made,* where we have only to apply fixed rules. We might think the wholly automatic subliminal self particularly apt for this sort of work, which is in a way exclusively mechanical. It seems that thinking in the evening upon the factors of a multiplication we might hope to find the product ready made upon our awakening, or again that an algebraic calculation, for example a verification, would be made unconsciously. Nothing of the sort, as observation proves. All one may hope from these inspirations, fruits of unconscious work, is a point of departure for such calculations. As for the calculations themselves, they must be made in the second period of conscious work, that which follows the inspiration, that in which one verifies the results of this inspiration and deduces their consequences. The rules of these calculations are strict and complicated. They require discipline, attention, will, and therefore consciousness. In the subliminal self, on the contrary, reigns what I should call liberty, if we might give this name to the simple absence of discipline and to the disorder born of chance. Only, this disorder itself permits unexpected combinations.

I shall make a last remark: when above I made certain personal observations, I spoke of a night of excitement when I worked in spite of myself. Such cases are frequent, and it is not necessary that the abnormal cerebral activity be caused by a physical excitant as in that I mentioned. It seems, in such cases, that one is present at his own unconscious work, made partially perceptible to the over-excited consciousness, yet without having changed its nature. Then we vaguely comprehend what distinguishes the two mechanisms or, if you wish, the working methods of the two egos. And the psychologic observations I have been able thus to make seem to me to confirm in their general outlines the views I have given.

Surely they have need of it, for they are and remain in spite of all very hypothetical: the interest of the questions is so great that I do not repent of having submitted them to the reader.

# COMMENTARY ON
# THE USE OF A TOP HAT
# AS A WATER BUCKET

The ideas of mathematics originate in experience. I doubt that this truth is widely recognized or that mathematicians are always prepared to acknowledge the indebtedness it implies. Schoolboys are taught that geometry started as an empirical science, as a set of rules devised for land measurement in Egypt, and it is generally taken for granted—though it may not be strictly true—that we use a decimal system because we have ten fingers. In more advanced mathematics the relation of ideas to experience is less obvious. Leaving aside the unworthy suspicions of the student who thinks that algebra and the calculus were devised for the express purpose of grinding him down, it is patently less easy to explain where these disciplines came from than to explain the origins of arithmetic. Yet the historian of mathematics has no trouble demonstrating that the calculus began as a method for measuring areas and volumes bounded by curves and curved surfaces (Kepler's first attempts at integration, for example, arose in connection with the measurement of kegs); that algebra had "strong empirical ties." In general a close link can be shown between the development of mathematics and the natural sciences.

To be sure, it is not always possible to give a simple and convincing proof that mathematics is rooted in experience. Some important branches of the science are so terrifyingly abstract and inhuman as to satisfy even the purest of pure mathematicians. The study of Hilbert spaces has nothing to do with space though it has to do with a mathematician named Hilbert; J. J. Sylvester's interesting paper "on the Problem of the Virgins, and the General Theory of Compound Partition" is in no wise connected with virgins or partitions. It is amazing what ingenious men can cook up from the simplest ingredients.

It may seem plausible that certain areas of mathematics should be wholly divorced from empirics and live, so to speak, lives of their own, but I think one cannot escape the conclusion that all its branches derive ultimately from sources within human experience. Any other view must fall back in the end on an appeal to mysticism. Furthermore, when the most abstract and "useless" disciplines have been cultivated for a time, they are often seized upon as practical tools by other departments of science. I conceive that this is no accident, as if one bought a top hat for a wedding and discovered later, when a fire broke out, that it could be used as a water bucket. Von Neumann gave the examples of differential geometry and group theory, devised as intellectual games, and still mainly pursued in the "nonapplied" spirit. "After a decade in one case, and a century in the other, they turned out to be very useful in physics." There are mathematicians who regard such harnessing of their beloved disciplines as a defilement; at best, they consider usefulness in the empirical sciences to be immaterial. There is no need to quarrel with this point of view. What is important is that mathematical activities abstractly conceived so often take a hand in the practical work of the world. This suggests, if indeed it does not prove, a profound connection.

These matters are thoughtfully considered in the essay below. John von Neumann (1903–1957) was one of the foremost mathematicians of our time. He never failed to enlarge our understanding of any problem, however complex, to which he turned his attention. His observations both enlightened and stimulated. In this essay he makes the exciting suggestion that the criteria of mathematical success are "almost entirely aesthetical." This is far from the notion that mathematics is a science of necessary and eternal truths. Mathematics as a matter of taste is a conception which will appeal to many, not least to those who have no taste for the subject in its more rigorous form.

I add a biographical note about Von Neumann. He was born in Budapest, Hungary, in 1903, and received his training at the University of Berlin, the Zurich Polytechnical Institute and the University of Budapest. For a time he taught in Berlin, and from 1930 to 1933 was professor of mathematical physics at Princeton. In the latter year he became a professor at the Institute for Advanced Study. His work in several branches of mathematics was of the first order: mathematical logic, set theory, theory of continuous groups, ergodic theory, quantum theory, operator theory and high-speed computing devices. For his inestimable services during the war and after as a consultant to the Army and the Navy and the Atomic Energy Commission he was awarded the Medal of Merit, the Distinguished Civilian Service award, and the Enrico Fermi award. He was a member of the National Academy of Sciences, and in 1954 was appointed to membership in the Atomic Energy Commission. In 1956 he was awarded the medal of Freedom and the Albert Einstein Award. He died of cancer on February 8, 1957.

# THE MATHEMATICIAN

## By John von Neumann

A discussion of the nature of intellectual work is a difficult task in any field, even in fields which are not so far removed from the central area of our common human intellectual effort as mathematics still is. A discussion of the nature of any intellectual effort is difficult per se—at any rate, more difficult than the mere exercise of that particular intellectual effort. It is harder to understand the mechanism of an airplane, and the theories of the forces which lift and which propel it, than merely to ride in it, to be elevated and transported by it—or even to steer it. It is exceptional that one should be able to acquire the understanding of a process without having previously acquired a deep familiarity with running it, with using it, before one has assimilated it in an instinctive and empirical way.

Thus any discussion of the nature of intellectual effort in any field is difficult, unless it presupposes an easy, routine familiarity with that field. In mathematics this limitation becomes very severe, if the discussion is to be kept on a nonmathematical plane. The discussion will then necessarily show some very bad features; points which are made can never be properly documented; and a certain over-all superficiality of the discussion becomes unavoidable.

I am very much aware of these shortcomings in what I am going to say, and I apologize in advance. Besides, the views which I am going to express are probably not wholly shared by many other mathematicians—you will get one man's not-too-well systematized impressions and interpretations—and I can give you only very little help in deciding how much they are to the point.

In spite of all these hedges, however, I must admit that it is an interesting and challenging task to make the attempt and to talk to you about the nature of intellectual effort in mathematics. I only hope that I will not fail too badly.

The most vitally characteristic fact about mathematics is, in my opinion, its quite peculiar relationship to the natural sciences, or, more generally, to any science which interprets experience on a higher than purely descriptive level.

Most people, mathematicians and others, will agree that mathematics is not an empirical science, or at least that it is practiced in a manner which differs in several decisive respects from the techniques of the empirical sciences. And, yet, its development is very closely linked with the natural sciences. One of its main branches, geometry, actually started as a natural, empirical science. Some of the best inspirations of modern mathematics (I believe, the best ones) clearly originated in the natural sciences. The methods of mathematics pervade and dominate the "theoretical" divisions of the natural sciences. In modern empirical sciences it has become more and more a major criterion of success whether they have become accessible to the mathematical method or to the near-mathematical methods of physics. Indeed, throughout the natural sciences an unbroken chain of successive pseudomorphoses, all of them pressing toward mathematics, and almost identified with the idea of scientific progress, has become more and more evident. Biology becomes increasingly pervaded by chemistry and physics, chemistry by experimental and theoretical physics, and physics by very mathematical forms of theoretical physics.

There is a quite peculiar duplicity in the nature of mathematics. One has to realize this duplicity, to accept it, and to assimilate it into one's thinking on the subject. This double face is the face of mathematics, and I do not believe that any simplified, unitarian view of the thing is possible without sacrificing the essence.

I will therefore not attempt to present you with a unitarian version. I will attempt to describe, as best I can, the multiple phenomenon which is mathematics.

It is undeniable that some of the best inspirations in mathematics—in those parts of it which are as pure mathematics as one can imagine—have come from the natural sciences. We will mention the two most monumental facts.

The first example is, as it should be, geometry. Geometry was the major part of ancient mathematics. It is, with several of its ramifications, still one of the main divisions of modern mathematics. There can be no doubt that its origin in antiquity was empirical and that it began as a discipline not unlike theoretical physics today. Apart from all other evidence, the very name "geometry" indicates this. Euclid's postulational treatment represents a great step away from empiricism, but it is not at all simple to defend the position that this was the decisive and final step, producing an absolute separation. That Euclid's axiomatization does at some minor points not meet the modern requirements of absolute axiomatic rigor is of lesser importance in this respect. What is more essential, is this: other disciplines, which are undoubtedly empirical, like mechanics and thermodynamics, are usually presented in a more or less postulational treatment, which in the presentation of some authors is hardly distinguishable from Euclid's procedure. The classic of theoretical physics in our time, Newton's *Principia,* was, in literary form as well as in the essence of some of its most critical parts, very much like Euclid. Of course in all these instances there is behind the postulational presentation the physical insight backing the postulates and the experimental verification supporting the theorems. But one might well argue that a similar interpretation of Euclid is possible, especially from the viewpoint of antiquity, before geometry had acquired its present bimillennial stability and authority—an authority which the modern edifice of theoretical physics is clearly lacking.

Furthermore, while the de-empirization of geometry has gradually progressed since Euclid, it never became quite complete, not even in modern times. The discussion of non-Euclidean geometry offers a good illustration of this. It also offers an illustration of the ambivalence of mathematical thought. Since most of the discussion took place on a highly abstract plane, it dealt with the purely logical problem whether the "fifth postulate" of Euclid was a consequence of the others or not; and the formal conflict was terminated by F. Klein's purely mathematical example, which showed how a piece of a Euclidean plane could be made non-Euclidean by formally redefining certain basic concepts. And yet the empirical stimulus was there from start to finish. The prime reason, why, of all Euclid's postulates, the fifth was questioned, was clearly the unempirical character of the concept of the entire infinite plane which intervenes there, and there only. The idea that in at least one significant sense—and in spite of all mathematico-logical analyses—the decision for or against Euclid may have to be empirical, was certainly present in the mind of the greatest mathematician, Gauss. And after Bolyai, Lobatschefski, Riemann, and Klein had obtained more abstracto, what we today consider the formal resolution of the original controversy, empirics—or rather physics—nevertheless, had the final say. The discovery of general relativity forced a revision of our views on the relationship of geometry in an entirely new setting and with a quite new distribution of the purely mathematical emphases, too. Finally, one more touch to complete the picture of contrast. This last development took place in the same generation which saw the complete de-empirization and abstraction of Euclid's axiomatic method in the hands of the modern axiomatic-logical mathematicians. And these two seemingly conflicting attitudes are perfectly compatible in one mathematical mind; thus Hilbert made important contributions to both axiomatic geometry and to general relativity.

The second example is calculus—or rather all of analysis, which sprang from it. The calculus was the first achievement of modern mathematics, and it is difficult to overestimate its importance. I think it defines more unequivocally than anything else the inception of modern mathematics, and the system of mathematical analysis, which is its logical development, still constitutes the greatest technical advance in exact thinking.

The origins of calculus are clearly empirical. Kepler's first attempts at integration were formulated as "dolichometry"—measurement of kegs—that is, volumetry for bodies with curved surfaces. This is geometry, but post-Euclidean, and, at the epoch in question, nonaxiomatic, empirical geometry.

Of this, Kepler was fully aware. The main effort and the main discoveries, those of Newton and Leibnitz, were of an explicitly physical origin. Newton invented the calculus "of fluxions" essentially for the purposes of mechanics—in fact, the two disciplines, calculus and mechanics, were developed by him more or less together. The first formulations of the calculus were not even mathematically rigorous. An inexact, semiphysical formulation was the only one available for over a hundred and fifty years after Newton! And yet, some of the most important advances of analysis took place during this period, against this inexact, mathematically inadequate background! Some of the leading mathematical spirits of the period were clearly not rigorous, like Euler; but others, in the main, were, like Gauss or Jacobi. The development was as confused and ambiguous as can be, and its relation to empiricism was certainly not according to our present (or Euclid's) ideas of abstraction and rigor. Yet no mathematician would want to exclude it from the fold—that period produced mathematics as first class as ever existed! And even after the reign of rigor was essentially re-established with Cauchy, a very peculiar relapse into semiphysical methods took place with Riemann. Riemann's scientific personality itself is a most illuminating example of the double nature of mathematics, as is the controversy of Riemann and Weierstrass, but it would take me too far into technical matters if I went into specific details. Since Weierstrass, analysis seems to have become completely abstract, rigorous, and unempirical. But even this is not unqualifiedly true. The controversy about the "foundations" of mathematics and logics, which took place during the last two generations, dispelled many illusions on this score.

This brings me to the third example which is relevant for the diagnosis. This example, however, deals with the relationship of mathematics with philosophy or epistemology rather than with the natural sciences. It illustrates in a very striking fashion that the very concept of "absolute" mathematical rigor is not immutable. The variability of the concept of rigor shows that something else besides mathematical abstraction must enter into the makeup of mathematics. In analyzing the controversy about the "foundations," I have not been able to convince myself that the verdict must be in favor of the empirical nature of this extra component. The case in favor of such an interpretation is quite strong, at least in some phases of the discussion. But I do not consider it absolutely cogent. Two things, however, are clear. First, that something nonmathematical, somehow connected with the empirical sciences or with philosophy or both, does enter essentially—and its nonempirical character could only be maintained if one assumed that philosophy (or more specifically epistemology) can exist independently of experience. (And this assumption is only necessary but not in itself sufficient.) Second, that the empirical origin of mathematics is strongly supported by instances like our two earlier examples (geometry and calculus), irrespective of what the best interpretation of the controversy about the "foundations" may be.

In analyzing the variability of the concept of mathematical rigor, I wish to lay the main stress on the "foundations" controversy, as mentioned above. I would, however, like to consider first briefly a secondary aspect of the matter. This aspect also strengthens my argument, but I do consider it as secondary, because it is probably less conclusive than the analysis of the "foundations" controversy. I am referring to the changes of mathematical "style." It is well known that the style in which mathematical proofs are written has undergone considerable fluctuations. It is better to talk of fluctuations than of a trend because in some respects the difference between the present and certain authors of the eighteenth or of the nineteenth centuries is greater than between the present and Euclid. On the other hand, in other respects there has been remarkable constancy. In fields in which differences are present, they are mainly differences in presentation, which can be eliminated without bringing in any new ideas. However, in many cases these differences are so wide that one begins to doubt whether authors who "present their cases" in such divergent ways can have been separated by differences in style, taste, and education only—whether they can really have had the same ideas as to what constitutes mathematical rigor. Finally, in the extreme cases (e.g., in much of the work of the late-eighteenth-century analysis, referred to above), the differences are essential and can be remedied, if at all, only with the help of new and profound theories, which it took up to a hundred years to develop. Some of the mathematicians who worked in such, to us, unrigorous ways (or some of their contemporaries, who criticized them) were well aware of their lack of rigor. Or to be more objective: Their own desires as to what mathematical procedure should be were more in conformity with our present views than their actions. But others—the greatest virtuoso of the period, for example, Euler—seem to have acted in perfect good faith and to have been quite satisfied with their own standards.

However, I do not want to press this matter further. I will turn instead to a perfectly clear-cut case, the controversy about the "foundations of mathematics." In the late nineteenth and the early twentieth centuries a new branch of abstract mathematics, G. Cantor's theory of sets, led into difficulties. That is, certain reasonings led to contradictions; and, while these reasonings were not in the central and "useful" part of set theory, and always easy to spot by certain formal criteria, it was nevertheless not clear why they should be deemed less set-theoretical than the "successful" parts of the theory. Aside from the *ex post* insight that they actually led into disaster, it was not clear what a priori motivation, what consistent philosophy of the situation, would permit one to segregate them from those parts of set theory which one wanted to save. A closer study of the *merita* of the case, undertaken mainly by Russell and Weyl, and concluded by Brouwer, showed that the way in which not only set theory but also most of modern mathematics used the concepts of "general validity" and of "existence" was philosophically objectionable. A system of mathematics which was free of these undesirable traits, "intuitionism," was developed by Brouwer. In this system the difficulties and contradiction of set theory did not arise. However, a good fifty per cent of modern mathematics, in its most vital—and up to then unquestioned—parts, especially in analysis, were also affected by this "purge": they either became invalid or had to be justified by very complicated subsidiary considerations. And in this latter process one usually lost appreciably in generality of validity and elegance of deduction. Nevertheless, Brouwer and Weyl considered it necessary that the concept of mathematical rigor be revised according to these ideas.

It is difficult to overestimate the significance of these events. In the third decade of the twentieth century two mathematicians—both of them of the first magnitude, and as deeply and fully conscious of what mathematics is, or is for, or is about, as anybody could be—actually proposed that the concept of mathematical rigor, of what constitutes an exact proof, should be changed! The developments which followed are equally worth noting.

1. Only very few mathematicians were willing to accept the new, exigent standards for their own daily use. Very many, however, admitted that Weyl and Brouwer were prima facie right, but they themselves continued to trespass, that is, to do their own mathematics in the old, "easy" fashion—probably in the hope that somebody else, at some other time, might find the answer to the intuitionistic critique and thereby justify them *a posteriori*.

2. Hilbert came forward with the following ingenious idea to justify "classical" (i.e., pre-intuitionistic) mathematics: Even in the intuitionistic system it is possible to give a rigorous account of how classical mathematics operate, that is, one can describe how the classical system works, although one cannot justify its workings. It might therefore be possible to demonstrate intuitionistically that classical procedures can never lead into contradictions—into conflicts with each other. It was clear that such a proof would be very difficult, but there were certain indications how it might be attempted. Had this scheme worked, it would have provided a most remarkable justification of classical mathematics on the basis of the opposing intuitionistic system itself! At least, this interpretation would have been legitimate in a system of the philosophy of mathematics which most mathematicians were willing to accept.

3. After about a decade of attempts to carry out this program, Gödel produced a most remarkable result. This result cannot be stated absolutely precisely without several clauses and caveats which are too technical to be formulated here. Its essential import, however, was this: If a system of mathematics does not lead into contradiction, then this fact cannot be demonstrated with the procedures of that system. Gödel's proof satisfied the strictest criterion of mathematical rigor—the intuitionistic one. Its influence on Hilbert's program is somewhat controversial, for reasons which again are too technical for this occasion. My personal opinion, which is shared by many others, is, that Gödel has shown that Hilbert's program is essentially hopeless.

4. The main hope of a justification of classical mathematics—in the sense of Hilbert or of Brouwer and Weyl—being gone, most mathematicians decided to use that system anyway. After all, classical mathematics was producing results which were both elegant and useful, and, even though one could never again be absolutely certain of its reliability, it stood on at least as sound a foundation as, for example, the existence of the electron. Hence, if one was willing to accept the sciences, one might as well accept the classical system of mathematics. Such views turned out to be acceptable even to some of the original protagonists of the intuitionistic system. At present the controversy about the "foundations" is certainly not closed, but it seems most unlikely that the classical system should be abandoned by any but a small minority.

I have told the story of this controversy in such detail, because I think that it constitutes the best caution against taking the immovable rigor of mathematics too much for granted. This happened in our own lifetime, and I know myself how humiliatingly easily my own views regarding the absolute mathematical truth changed during this episode, and how they changed three times in succession!

I hope that the above three examples illustrate one-half of my thesis sufficiently well—that much of the best mathematical inspiration comes from experience and that it is hardly possible to believe in the existence of an absolute, immutable concept of mathematical rigor, dissociated from all human experience. I am trying to take a very low-brow attitude on this matter. Whatever philosophical or epistemological preferences anyone may have in this respect, the mathematical fraternities' actual experiences with its subject give little support to the assumption of the existence of an a priori concept of mathematical rigor. However, my thesis also has a second half, and I am going to turn to this part now.

It is very hard for any mathematician to believe that mathematics is a purely empirical science or that all mathematical ideas originate in empirical subjects. Let me consider the second half of the statement first. There are various important parts of modern mathematics in which the empirical origin is untraceable, or, if traceable, so remote that it is clear that the subject has undergone a complete metamorphosis since it was cut off from its empirical roots. The symbolism of algebra was invented for domestic, mathematical use, but it may be reasonably asserted that it had strong empirical ties. However, modern, "abstract" algebra has more and more developed into directions which have even fewer empirical connections. The same may be said about topology. And in all these fields the mathematician's subjective criterion of success, of the worth-whileness of his effort, is very much self-contained and aesthetical and free (or nearly free) of empirical connections. (I will say more about this further on.) In set theory this is still clearer. The "power" and the "ordering" of an infinite set may be the generalizations of finite numerical concepts, but in their infinite form (especially "power") they have hardly any relation to this world. If

I did not wish to avoid technicalities, I could document this with numerous set theoretical examples—the problem of the "axiom of choice," the "comparability" of infinite "powers," the "continuum problem," etc. The same remarks apply to much of real function theory and real point-set theory. Two strange examples are given by differential geometry and by group theory: they were certainly conceived as abstract, nonapplied disciplines and almost always cultivated in this spirit. After a decade in one case, and a century in the other, they turned out to be very useful in physics. And they are still mostly pursued in the indicated, abstract, nonapplied spirit.

The examples for all these conditions and their various combinations could be multiplied, but I prefer to turn instead to the first point I indicated above: Is mathematics an empirical science? Or, more precisely: Is mathematics actually practiced in the way in which an empirical science is practiced? Or, more generally: What is the mathematician's normal relationship to his subject? What are his criteria of success, of desirability? What influences, what considerations, control and direct his effort?

Let us see, then, in what respects the way in which the mathematician normally works differs from the mode of work in the natural sciences. The difference between these, on one hand, and mathematics, on the other, goes on, clearly increasing as one passes from the theoretical disciplines to the experimental ones and then from the experimental disciplines to the descriptive ones. Let us therefore compare mathematics with the category which lies closest to it—the theoretical disciplines. And let us pick there the one which lies closest to mathematics. I hope that you will not judge me too harshly if I fail to control the mathematical *hybris* and add: because it is most highly developed among all theoretical sciences—that is, theoretical physics. Mathematics and theoretical physics have actually a good deal in common. As I have pointed out before, Euclid's system of geometry was the prototype of the axiomatic presentation of classical mechanics, and similar treatments dominate phenomenological thermodynamics as well as certain phases of Maxwell's system of electrodynamics and also of special relativity. Furthermore, the attitude that theoretical physics does not explain phenomena, but only classifies and correlates, is today accepted by most theoretical physicists. This means that the criterion of success for such a theory is simply whether it can, by a simple and elegant classifying and correlating scheme, cover very many phenomena, which without this scheme would seem complicated and heterogeneous, and whether the scheme even covers phenomena which were not considered or even not known at the time when the scheme was evolved. (These two latter statements express, of course, the unifying and the predicting power of a theory.) Now this criterion, as set forth here, is clearly to a great extent of an aesthetical nature. For this reason it is very closely akin to the mathematical criteria of success, which, as you shall see, are almost entirely aesthetical. Thus we are now comparing mathematics with the empirical science that lies closest to it and with which it has, as I hope I have shown, much in common—with theoretical physics. The differences in the actual *modus procedendi* are nevertheless great and basic. The aims of theoretical physics are in the main given from the "outside," in most cases by the needs of experimental physics. They almost always originate in the need of resolving a difficulty; the predictive and unifying achievements usually come afterward. If we may be permitted a simile, the advances (predictions and unifications) come during the pursuit, which is necessarily preceded by a battle against some pre-existing difficulty (usually an apparent contradiction within the existing system). Part of the theoretical physicist's work is a search for such obstructions, which promise a possibility for a "break-through." As I mentioned, these difficulties originate usually in experimentation, but sometimes they are contradictions between various parts of the accepted body of theory itself. Examples are, of course, numerous.

Michelson's experiment leading to special relativity, the difficulties of certain ionization potentials and of certain spectroscopic structures leading to quantum mechanics exemplify the first case; the conflict between special relativity and Newtonian gravitational theory leading to general relativity exemplifies the second, rarer, case. At any rate, the problems of theoretical physics are objectively given; and, while the criteria which govern the exploitation of a success are, as I indicated earlier, mainly aesthetical, yet the portion of the problem, and that which I called above the original "break-through," are hard, objective facts. Accordingly, the subject of theoretical physics was at almost all times enormously concentrated; at almost all times most of the effort of all theoretical physicists was concentrated on no more than one or two very sharply circumscribed fields—quantum theory in the 1920's and early 1930's and elementary particles and structure of nuclei since the mid-1930's are examples.

The situation in mathematics is entirely different. Mathematics falls into a great number of subdivisions, differing from one another widely in character, style, aims, and influence. It shows the very opposite of the extreme concentration of theoretical physics. A good theoretical physicist may today still have a working knowledge of more than half of his subject. I doubt that any mathematician now living has much of a relationship to more than a quarter. "Objectively" given, "important" problems may arise after a subdivision of mathematics has evolved relatively far and if it has bogged down seriously before a difficulty. But even then the mathematician is essentially free to take it or leave it and turn to something else, while an "important" problem in theoretical physics is usually a conflict, a contradiction, which "must" be resolved. The mathematician has a wide variety of fields to which he may turn, and he enjoys a very considerable freedom in what he does with them. To come to the decisive point: I think that it is correct to say that his criteria of selection, and also those of success, are mainly aesthetical. I realize that this assertion is controversial and that it is impossible to "prove" it, or indeed to go very far in substantiating it, without analyzing numerous specific, technical instances. This would again require a highly technical type of discussion, for which this is not the proper occasion. Suffice it to say that the aesthetical character is even more prominent than in the instance I mentioned above in the case of theoretical physics. One expects a mathematical theorem or a mathematical theory not only to describe and to classify in a simple and elegant way numerous and a priori disparate special cases. One also expects "elegance" in its "architectural," structural makeup. Ease in stating the problem, great difficulty in getting hold of it and in all attempts at approaching it, then again some very surprising twist by which the approach, or some part of the approach, becomes easy, etc. Also, if the deductions are lengthy or complicated, there should be some simple general principle involved, which "explains" the complications and detours, reduces the apparent arbitrariness to a few simple guiding motivations, etc. These criteria are clearly those of any creative art, and the existence of some underlying empirical, worldly motif in the background—often in a very remote background—overgrown by aestheticizing developments and followed into a multitude of labyrinthine variants—all this is much more akin to the atmosphere of art pure and simple than to that of the empirical sciences.

You will note that I have not even mentioned a comparison of mathematics with the experimental or with the descriptive sciences. Here the differences of method and of the general atmosphere are too obvious.

I think that it is a relatively good approximation to truth—which is much too complicated to allow anything but approximations—that mathematical ideas originate in empirics, although the genealogy is sometimes long and obscure. But, once they are so conceived, the subject begins to live a peculiar life of its own and is better compared to a creative one, governed by almost entirely aesthetical motivations, than to anything else and, in particular, to an empirical science. There is, however, a further point which, I believe, needs stressing. As a mathematical discipline travels far from its empirical source, or still more, if it is a second and third generation only indirectly inspired by ideas coming from "reality," it is beset with very grave dangers. It becomes more and more purely aestheticizing, more and more purely *l'art pour l'art*. This need not be bad, if the field is surrounded by correlated subjects, which still have closer empirical connections, or if the discipline is under the influence of men with an exceptionally well-developed taste. But there is a grave danger that the subject will develop along the line of least resistance, that the stream, so far from its source, will separate into a multitude of insignificant branches, and that the discipline will become a disorganized mass of details and complexities. In other words, at a great distance from its empirical source, or after much "abstract" inbreeding, a mathematical subject is in danger of degeneration. At the inception the style is usually classical; when it shows signs of becoming baroque, then the danger signal is up. It would be easy to give examples, to trace specific evolutions into the baroque and the very high baroque, but this, again, would be too technical.

In any event, whenever this stage is reached, the only remedy seems to me to be the rejuvenating return to the source: the reinjection of more or less directly empirical ideas. I am convinced that this was a necessary condition to conserve the freshness and the vitality of the subject and that this will remain equally true in the future.

# PART XIX

# MATHEMATICAL MACHINES: CAN A MACHINE THINK?

# COMMENTARY ON
# AUTOMATIC COMPUTERS

S elf-regulating machines are not new. A very early example was a miniature windmill which, mounted at right angles to the sails of a large windmill, would catch the wind and rotate the main structure into the proper operating position.[1] The automatic fly-ball governor, indispensable to the use of steam power, was invented by James Watt. The now-commonplace thermostat is an example of the feedback principle which is at the heart of all self-regulating mechanisms. Feedback is the use of a fraction of the output of a machine to control the source of power for that machine. When output rises beyond a determined point, this power is throttled back; when output lags the power is increased. Thus the machine is self-determining. The principle can be shown in a simple diagram illustrating the "closed sequence" of control in a room thermostat:

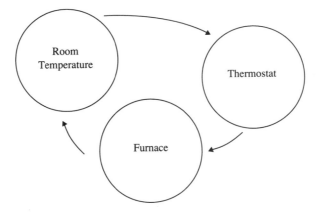

The elements of the system are described as *interdependent*: the room temperature, for example, is the cause of the thermometer reading; the latter is also the cause, with the furnace as mediator, of the room temperature.

Feedback is also an essential component of those prodigious calculating machines known as computers or, with chummy awe, "electronic brains." Automatic computers have existed since the seventeenth century. Pascal built an adding engine of which he was very proud; Leibniz made a machine which could multiply. But neither of these computers was self-regulating. They used toothed wheels so geared that turning one wheel a certain number of notches caused the next wheel to revolve a single notch; multiplication was accomplished by adding the same number over and over again.[2] In the nineteenth century Charles Babbage, a remarkable English mathematician who so hated organ-grinders that he crusaded against them, conceived an elaborate computer which, he said, was capable of "eating its own tail."[3] This machine, far ahead of its time, was unfortunately never completed; but it is the true forerunner of the modern digital computers.

In certain of their basic features, all computers are essentially the same as Pascal's and Leibniz' inventions. They reduce higher mathematics to arithmetic and arithmetic to counting. The vacuum tube does the same job as the toothed wheel, but much faster. The great innovation of today's computers is their capacity to regulate themselves by digesting information which they themselves have produced. These machines cannot wonder, but they can respond. Having received elaborate and precise instructions as to the problem at hand, the computer proceeds to grind out figures at a prodigious rate. In the course of this operation it in many ways apes the process of a human calculation. The machine can organize its problem into separate steps; it can use the results obtained in one step to execute the next; sometimes partial results are laid aside, that is, "remembered," while an intermediate step is carried through; trial and error methods are frequently used. Thus the computer guides itself by its own answers, makes choices, comparisons, decisions. It is less like a man than is an amoeba; nevertheless, it is more like a brain than any machine has ever been before. It is close enough to make men shiver.

Mathematicians are interested in computers because they compute. They take over menial and monotonous exercises and solve quickly problems which would take men a long time and cost them immense labor. But this in a sense is the least important and least interesting aspect of computers. They draw the attention of the more original mathematicians for other reasons. The goading question is how much such machines can tell us about the human brain. What, for example, is to be learned from them about the nature of information, how it is communicated, how acted upon? What is the resemblance between the self-exciting and self-stabilizing properties of automatic computers and similar properties of human nerve circuits? Can one gain knowledge of the structure of the brain by comparing restricted aspects of its input and output with the input and output of computers (as well as of other self-regulating machines) whose structure is known? It is thought that the machines can tell us a good deal about ourselves; perhaps, in due time, answer not only all we are capable of asking but even suggest better questions.

Mathematicians are of course not alone in studying this subject; they are allied with physicists, chemists, biologists, physiologists, psychologists and other specialists in the sciences of human behavior. Research in this sphere takes many forms. On the flourishing practical side engineers are designing servo-mechanisms for innumerable specialized communications, industrial and military purposes: robots to guide bombers, point guns, run a refinery, monitor a telephone exchange, control machine tools and so on. Electro-physiologists and other investigators are building machines which can find their way out of labyrinths; which can respond to simple stimuli (light, heat, sound, physical blows), recognize themselves or others like them; which can be taught, after a certain number of trials, to disregard misleading clues, to distinguish between true and false leads. These new creatures can do much more, as Bronowski says, than merely jump when they are pinched. One more remarkable toy should be mentioned. The British psychiatrist, Dr. Ross Ashby, has made a machine called a homeostat which adapts itself to change and restores its internal stability so that it can keep on doing in the new environment what it was doing in the old.[4] "This creature, *machina sopora*, it might be called, is like a fireside cat or dog which only stirs when disturbed, and then methodically finds a comfortable position and goes to sleep again."[5]

At the outposts of inquiry pure mathematicians and logicians are delving into the correspondence between the function aspects of the organization of automata and of natural organisms. As you will see in Von Neumann's piece, the purpose of this work is to develop the logical theory of automata and, by applying axiomatic procedures, to draw conclusions about the nature of large complex (and essentially incomprehensible) organisms from the study of their component elements. These elements are assumed to have "certain well-defined outside functional characteristics"; that is, they are treated as "black boxes" which we cannot look into but about whose insides we can learn a good deal by checking the signals that go into them against those that come out and by assuming that an unambiguously defined stimulus always provokes an unambiguously defined response.

Can machines think? Is the question itself, for that matter, more than a journalist's gambit? The English logician, A. M. Turing,[6] regards it as a serious, meaningful question and one which can now be answered. He thinks machines can think. He suggests that they can learn, that they can be built so as to be able to do more than what we know how to order them to do, that they may eventually "compete with men in all purely intellectual fields." His conclusions are made plausible in the brilliantly argued essay below. The selection by Claude Shannon neither supports nor refutes Turing's expectations. It describes the design of an electronic computer which could play chess. Shannon is not sure whether his automaton can be rated a thinker. The machine could play a complete game with reasonable skill; it could also solve problems involving an enormous number of individual calculations. (The author says such problems would be "too laborious to carry out by hand"; he underestimates the insane pertinacity of problem solvers.) Claude Shannon was a pioneer of communication theory. He was a mathematician on the staff of Bell Telephone Laboratories from 1941 until 1957, after which he became a professor of electrical engineering and Donner professor of science at MIT. He has written important papers on switching and mathematical logic and has made fundamental contributions to the engineering aspects of communication and to the analysis of the nature of information.[7]

# ENDNOTES

[1]See the excellent article, "Feedback," by Arnold Tustin, *Scientific American,* Sept. 1952, pp. 48–55. The diagram is also from this source.

[2]See J. Bronowski, "Can Machines 'Think'?", *The Observer* (London), May 30, 1953.

[3] An interesting article on the man and his inventions—another of his great conceptions was a calculator called a "Difference Engine" which was partially completed with the expenditure of £6,000 of Babbage's money and £17,000 contributed by the government—is by Philip and Emily Morrison, "The Strange Life of Charles Babbage," *Scientific American,* April 1952, pp. 66–73.

[4]"I have seen Dr. Ross Ashby remove the wiring from his odd machine, leave out some and connect the rest at random. The machine balked, but it worked; after a few minutes the machine pointer was again following the environment pointer." Bronowski, *loc. cit.* For a full discussion of Ashby's work, see *Design for a Brain,* by W. Ross Ashby, New York, 1953.

[5]"There are a number of electronic circuits similar to the reflex arcs in the spinal cord of an animal. They are so combined with a number of thermionic tubes and relays that out of 360,000 possible connections the machine will automatically find one that leads to a condition of dynamic internal stability. That is, after several trials and errors, the instrument, without any prompting or programming, establishes connections which tend to neutralize any change that the experimenter tries to impose from outside." The quotation in the text, and the continuation in the note is from an interesting book (*The Living Brain,* New York, 1953, pp. 123–124) by the British electro-encephalographer, W. Grey Walter. See especially his chapter on automata, "Totems, Toys and Tools," pp. 114–132.

[6]Turing, one of the most gifted of modern mathematical logicians, took his own life in a fit of despondency in the summer of 1954. For an obituary and an account of his scientific papers, by M. H. A. Newman, see *Biographical Memoirs of Fellows of the Royal Society,* 1955, Vol. 1, London, 1955. Of particular interest is Turing's 1937 paper, "On Computable Numbers, with an Application to the *Entscheidungsproblem* (Proc. Lond. Math. Soc. (2), 42,230). In this he argues that a machine—which he describes—could be made to do any piece of work which could be done by a human computer obeying explicit instructions given to him before the work starts. The question arises whether the machine could solve certain fundamental problem as to its own capacity; that is, given a certain tape of instructions, is it possible for the machine to decide whether the problem set on the tape has a determinate solution? Turing's answer, in effect, says that unless the tape gives the method of solution, no machine can solve it; moreover, no tape can give the method. Therefore the problem is insoluble in "an absolute and inescapable sense. From this basic insoluble problem it was not difficult to infer that the Hilbert program of finding a decision method for the axiomatic system, Z, of elementary number-theory, is also impossible." (See selection on Goedel's Proof by Nagel and Newman.)

[7]See *The Mathematical Theory of Communication,* Claude E. Shannon and Warren Weaver, Urbana (The University of Illinois Press), 1949.

1

*Men have become the tools of their tools.*
—THOREAU

*Your worship is your furnaces,*
*Which, like old idols, lost obscenes,*
*Have molten bowels; your vision is*
*Machines for making more machines.*

—GORDON BOTTOMLEY (*1874*)
(*"To Ironfounders and Others"*)

*There once was a man who said, "Damn,*
*It is borne in upon me I am*
*An engine that moves*
*In predestinate grooves,*
*I'm not even a bus I'm a tram."*

—MAURICE EVAN HARE (*1905*)

# THE GENERAL AND LOGICAL THEORY OF AUTOMATA

## By John von Neumann

. . . Automata have been playing a continuously increasing, and have by now attained a very considerable, role in the natural sciences. This is a process that has been going on for several decades. During the last part of this period automata have begun to invade certain parts of mathematics too—particularly, but not exclusively, mathematical physics or applied mathematics. Their role in mathematics presents an interesting counterpart to certain functional aspects of organization in nature. Natural organisms are, as a rule, much more complicated and subtle, and therefore much less well understood in detail, than are artificial automata. Nevertheless, some regularities which we observe in the organization of the former may be quite instructive in our thinking and planning of the latter; and conversely, a good deal of our experiences and difficulties with our artificial automata can be to some extent projected on our interpretations of natural organisms.

### PRELIMINARY CONSIDERATIONS

*Dichotomy of the Problem: Nature of the Elements, Axiomatic Discussion of Their Synthesis.* In comparing living organisms, and, in particular, that most complicated organism, the human central nervous system, with artificial automata, the following limitation should be kept in mind. The natural systems are of enormous complexity, and it is clearly necessary to subdivide the problem that they represent into several parts. One method of subdivision, which is particularly significant in the present context, is this: The organisms can be viewed as made up of parts which to a certain extent are independent, elementary units. We may, therefore, to this extent, view as the first part of the problem the structure and functioning of such elementary units individually. The second part of the problem consists of understanding how these elements are organized into a whole, and how the functioning of the whole is expressed in terms of these elements.

The first part of the problem is at present the dominant one in physiology. It is closely connected with the most difficult chapters of organic chemistry and of physical chemistry, and may in due course be greatly helped by quantum mechanics. I have little qualification to talk about it, and it is not this part with which I shall concern myself here.

The second part, on the other hand, is the one which is likely to attract those of us who have the background and the tastes of a mathematician or a logician. With this attitude, we will be inclined to remove the first part of the problem by the process of axiomatization, and concentrate on the second one.

*The Axiomatic Procedure.* Axiomatizing the behavior of the elements means this: We assume that the elements have certain well-defined, outside, functional characteristics; that is, they are to be treated as "black boxes." They are viewed as automatisms, the inner structure of which need not be disclosed, but which are assumed to react to certain unambiguously defined stimuli, by certain unambiguously defined responses.

This being understood, we may then investigate the larger organisms that can be built up from these elements, their structure, their functioning, the connections between the elements, and the general theoretical regularities that may be detectable in the complex syntheses of the organisms in question.

I need not emphasize the limitations of this procedure. Investigators of this type may furnish evidence that the system of axioms used is convenient and, at least in its effects, similar to reality. They are, however, not the ideal method, and possibly not even a very effective method, to determine the validity of the axioms. Such determinations of validity belong primarily to the first part of the problem. Indeed they are essentially covered by the properly physiological (or chemical or physical-chemical) determinations of the nature and properties of the elements.

*The Significant Orders of Magnitude.* In spite of these limitations, however, the "second part" as circumscribed above is important and difficult. With any reasonable definition of what constitutes an element, the natural organisms are very highly complex aggregations of these elements. The number of cells in the human body is somewhere of the general order of $10^{15}$ or $10^{16}$. The number of neurons in the central nervous system is somewhere of the order of $10^{10}$. We have absolutely no past experience with systems of this degree of complexity. All artificial automata made by man have numbers of parts which by any comparably schematic count are of the order $10^3$ to $10^6$. In addition, those artificial systems which function with that type of logical flexibility and autonomy that we find in the natural organisms do not lie at the peak of this scale. The prototypes for these systems are the modern computing machines, and here a reasonable definition of what constitutes an element will lead to counts of a few times $10^3$ or $10^4$ elements.

## DISCUSSION OF CERTAIN RELEVANT TRAITS OF COMPUTING MACHINES

*Computing Machines—Typical Operations.* Having made these general remarks, let me now be more definite, and turn to that part of the subject about which I shall talk in specific and technical detail. As I have indicated, it is concerned with artificial automata and more specially with computing machines. They have some similarity to the central nervous system, or at least to a certain segment of the system's functions. They are of course vastly less complicated, that is, smaller in the sense which really matters. It is nevertheless of a certain interest to analyze the problem of organisms and organization from the point of view of these relatively small, artificial automata, and to effect their comparisons with the central nervous system from this frog's-view perspective.

I shall begin by some statements about computing machines as such.

The notion of using an automaton for the purpose of computing is relatively new: While computing automata are not the most complicated artificial automata from the point of view of the end results they achieve, they do nevertheless represent the highest degree of complexity in the sense that they produce the longest chains of events determining and following each other.

There exists at the present time a reasonably well-defined set of ideas about when it is reasonable to use a fast computing machine, and when it is not. The criterion is usually expressed in terms of the multiplications involved in the mathematical problem. The use of a fast computing machine is believed to be by and large justified when the computing task involves about a million multiplications or more in a sequence.

An expression in more fundamentally logical terms is this: In the relevant fields (that is, in those parts of [usually applied] mathematics, where the use of such machines is proper) mathematical experience indicates the desirability of precisions of about ten decimal places. A single multiplication would therefore seem to involve at least $10 \times 10$ steps (digital multiplications); hence a million multiplications amount to at least $10^8$ operations. Actually, however, multiplying two decimal digits is not an elementary operation. There are various ways of breaking it down into such, and all of them have about the same degree of complexity. The simplest way to estimate this degree of complexity is, instead of counting decimal places, to count the number of places that would be required for the same precision in the binary system of notation (base 2 instead of base 10). A decimal digit corresponds to about three binary digits, hence ten decimals to about thirty binary. The multiplication referred to above, therefore, consists not of $10 \times 10$, but of $30 \times 30$ elementary steps, that is, not $10^2$, but $10^3$ steps. (Binary digits are "all or none" affairs, capable of the values 0 and 1 only. Their multiplication is, therefore, indeed an elementary operation. By the way, the equivalent of 10 decimals is 33 [rather than 30] binaries—but 33 $\times$ 33, too, is approximately $10^3$.) It follows, therefore, that a million multiplications in the sense indicated above are more reasonably described as corresponding to $10^9$ elementary operations.

*Precision and Reliability Requirements.* I am not aware of any other field of human effort where the result really depends on a sequence of a billion ($10^9$) steps in any artifact, and where, furthermore, it has the characteristic that every step actually matters—or, at least, may matter with a considerable probability. Yet, precisely this is true for computing machines—this is their most specific and most difficult characteristic.

Indeed, there have been in the last two decades automata which did perform hundreds of millions, or even billions, of steps before they produced a result. However, the operation of these automata is not serial. The large number of steps is due to the fact that, for a variety of reasons, it is desirable to do the same experiment over and over again. Such cumulative, repetitive procedures may, for instance, increase the size of the result, that is (and this is the important consideration), increase the significant result, the "signal," relative to the "noise" which contaminates it. Thus any reasonable count of the number of reactions which a microphone gives before a verbally interpretable acoustic signal is produced is in the high tens of thousands. Similar estimates in television will give tens of millions, and in radar possibly many billions. If, however, any of these automata makes mistakes, the mistakes usually matter only to the extent of the fraction of the total number of steps which they represent. (This is not exactly true in all relevant examples, but it represents the qualitative situation better than the opposite statement.) Thus the larger the number of operations required to produce a result, the smaller will be the significant contribution of every individual operation.

In a computing machine no such rule holds. Any step is (or may potentially be) as important as the whole result; any error can vitiate the result in its entirety. (This statement is not absolutely true, but probably nearly 30 per cent of all steps made are usually of this sort.) Thus a computing machine is one of the exceptional artifacts. They not only have to perform a billion or more steps in a short time, but in a considerable part of the procedure (and this is a part that is rigorously specified in advance) they are permitted not a single error. In fact, in order to be sure that the whole machine is operative, and that no potentially degenerative malfunctions have set in, the present practice usually requires that no error should occur anywhere in the entire procedure.

This requirement puts the large, high-complexity computing machines in an altogether new light. It makes in particular a comparison between the computing machines and the operation of the natural organisms not entirely out of proportion.

*The Analogy Principle.* All computing automata fall into two great classes in a way which is immediately obvious and which, as you will see in a moment, carries over to living organisms. This classification is into analogy and digital machines.

Let us consider the analogy principle first. A computing machine may be based on the principle that numbers are represented by certain physical quantities. As such quantities we might, for instance, use the intensity of an electrical current, or the size of an electrical potential, or the number of degrees of arc by which a disk has been rotated (possibly in conjunction with the number of entire revolutions effected), etc. Operations like addition, multiplication, and integration may then be performed by finding various natural processes which act on these quantities in the desired way. Currents may be multiplied by feeding them into the two magnets of a dynamometer, thus producing a rotation. This rotation may then be transformed into an electrical resistance by the attachment of a rheostat; and, finally, the resistance can be transformed into a current by connecting it to two sources of fixed (and different) electrical potentials. The entire aggregate is thus a "black box" into which two currents are fed and which produces a current equal to their product. You are certainly familiar with many other ways in which a wide variety of natural processes can be used to perform this and many other mathematical operations.

The first well-integrated, large, computing machine ever made was an analogy machine, V. Bush's Differential Analyzer. This machine, by the way, did the computing not with electrical currents, but with rotating disks. I shall not discuss the ingenious tricks by which the angles of rotation of these disks were combined according to various operations of mathematics.

I shall make no attempt to enumerate, classify, or systematize the wide variety of analogy principles and mechanisms that can be used in computing. They are confusingly multiple. The guiding principle without which it is impossible to reach an understanding of the situation is the classical one of all "communication theory"—the "signal to noise ratio." That is, the critical question with every analogy procedure is this: How large are the uncontrollable fluctuations of the mechanism that constitute the "noise," compared to the significant "signals" that express the numbers on which the machine operates? The usefulness of any analogy principle depends on how low it can keep the relative size of the uncontrollable fluctuations—the "noise level."

To put this in another way. No analogy machine exists which will really form the product of two numbers. What it will form is this product, plus a small but unknown quantity which represents the random noise of the mechanism and the physical processes involved. The whole problem is to keep this quantity down.

This principle has controlled the entire relevant technology. It has, for instance, caused the adoption of seemingly complicated and clumsy mechanical devices instead of the simpler and elegant electrical ones. (This, at least, has been the case throughout most of the last twenty years. More recently, in certain applications which required only very limited precision the electrical devices have again come to the fore.) In comparing mechanical with electrical analogy processes, this roughly is true: Mechanical arrangements may bring this noise level below the "maximum signal level" by a factor of something like $1:10^4$ or $10^5$. In electrical arrangements, the ratio is rarely much better than $1:10^2$. These ratios represent, of course, errors in the elementary steps of the calculation, and not in its final results. The latter will clearly be substantially larger.

*The Digital Principle.* A digital machine works with the familiar method of representing numbers as aggregates of digits. This is, by the way, the procedure, which all of us use in our individual, non-mechanical computing, where we express numbers in the decimal system. Strictly speaking, digital computing need not be decimal. Any integer larger than one may be used as the basis of a digital notation for numbers. The decimal system (base 10) is the most common one, and all digital machines built to date operate in this system. It seems likely, however, that the binary (base 2) system will, in the end, prove preferable, and a number of digital machines using that system are now under construction.

The basic operations in a digital machine are usually the four species of arithmetic: addition, subtraction, multiplication, and division. We might at first think that, in using these, a digital machine possesses (in contrast to the analogy machines referred to above) absolute precision. This, however, is not the case, as the following consideration shows.

Take the case of multiplication. A digital machine multiplying two 10-digit numbers will produce a 20-digit number, which is their product, with no error whatever. To this extent its precision is absolute, even though the electrical or mechanical components of the arithmetical organ of the machine are as such of limited precision. As long as there is no breakdown of some component, that is, as long as the operation of each component produces only fluctuations within its preassigned tolerance limits, the result will be absolutely correct. This is, of course, the great and characteristic virtue of the digital procedure. Error, as a matter of normal operation and not solely (as indicated above) as an accident attributable to some definite breakdown, nevertheless creeps in, in the following manner. The absolutely correct product of two 10-digit numbers is a 20-digit number. If the machine is built to handle 10-digit numbers only, it will have to disregard the last 10 digits of this 20-digit number and work with the first 10 digits alone. (The small, though highly practical, improvement due to a possible modification of these digits by "round-off" may be disregarded here.) If, on the other hand, the machine can handle 20-digit numbers, then the multiplication of two such will produce 40 digits, and these again have to be cut down to 20, etc., etc. (To conclude, no matter what the maximum number of digits is for which the machine has been built, in the course of successive multiplications this maximum will be reached, sooner or later. Once it has been reached, the next multiplication will produce supernumerary digits, and the product will have to be cut to half of its digits [the first half, suitably rounded off]. The situation for a maximum of 10 digits, is therefore typical, and we might as well use it to exemplify things.)

Thus the necessity of rounding off an (exact) 20-digit product to the regulation (maximum) number of 10 digits, introduces in a digital machine qualitatively the same situation as was found above in an analogy machine. What it produces when a product is called for is not that product itself, but rather the product plus a small extra term—the round-off error. This error is, of course, not a random variable like the noise in an analogy machine. It is, arithmetically, completely determined in every particular instance. Yet its mode of determination is so complicated, and its variations throughout the number of instances of its occurrence in a problem so irregular, that it usually can be treated to a high degree of approximation as a random variable.

(These considerations apply to multiplication. For division the situation is even slightly worse, since a quotient can, in general, not be expressed with absolute precision by any finite number of digits. Hence here rounding off is usually already a necessity after the first operation. For addition and subtraction, on the other hand, this difficulty does not arise: The sum or difference has the same number of digits [if there is no increase in size beyond the planned maximum] as the addends themselves. Size may create difficulties which are added to the difficulties of precision discussed here, but I shall not go into these at this time.)

*The Role of the Digital Procedure in Reducing the Noise Level.* The important difference between the noise level of a digital machine, as described above, and of an analogy machine is not qualitative at all; it is quantitative. As pointed out above, the relative noise level of an analogy machine is never lower than 1 in $10^5$, and in many cases as high as 1 in $10^2$. In the 10-place decimal digital machine referred to above the relative noise level (due to round-off) is 1 part in $10^{10}$. Thus the real importance of the digital procedure lies in its ability to reduce the computational noise level to an extent which is completely unobtainable by any other (analogy) procedure. In addition, further reduction of the noise level is increasingly difficult in an analogy mechanism, and increasingly easy in a digital one. In an analogy machine a precision of 1 in $10^3$ is easy to achieve; 1 in $10^4$ somewhat difficult; 1 in $10^5$ very difficult; and 1 in $10^6$ impossible, in the present state of technology. In a digital machine, the above precisions mean merely that one builds the machine to 3, 4, 5, and 6 decimal places, respectively. Here the transition from each stage to the next one gets actually easier. Increasing a 3-place machine (if anyone wished to build such a machine) to a 4-place machine is a 33 per cent increase; going from 4 to 5 places, a 25 per cent increase; going from 5 to 6 places, a 20 per cent increase. Going from 10 to 11 places is only a 10 per cent increase. This is clearly an entirely different milieu, from the point of view of the reduction of "random noise," from that of physical processes. It is here—and not in its practically ineffective absolute reliability—that the importance of the digital procedure lies.

## COMPARISONS BETWEEN COMPUTING MACHINES AND LIVING ORGANISMS

*Mixed Character of Living Organisms.* When the central nervous system is examined, elements of both procedures, digital and analogy, are discernible.

The neuron transmits an impulse. This appears to be its primary function, even if the last word about this function and its exclusive or nonexclusive character is far from having been said. The nerve impulse seems in the main to be an all-or-none affair, comparable to a binary digit. Thus a digital element is evidently present, but it is equally evident that this is not the entire story. A great deal of what goes on in the organism is not mediated in this manner, but is dependent on the general chemical composition of the blood stream or of other humoral media. It is well known that there are various composite functional sequences in the organism which have to go through a variety of steps from the original stimulus to the ultimate effect—some of the steps being neural, that is, digital, and others humoral, that is, analogy. These digital and analogy portions in such a chain may alternately multiply. In certain cases of this type, the chain can actually feed back into itself, that is, its ultimate output may again stimulate its original input.

It is well known that such mixed (part neural and part humoral) feedback chains can produce processes of great importance. Thus the mechanism which keeps the blood pressure constant is of this mixed type. The nerve which senses and reports the blood pressure does it by a sequence of neural impulses, that is, in a digital manner. The muscular contraction which this impulse system induces may still be described as a superposition of many digital impulses. The influence of such a contraction on the blood stream is, however, hydrodynamical, and hence analogy. The reaction of the pressure thus produced back on the nerve which reports the pressure closes the circular feedback, and at this point the analogy procedure again goes over into a digital one. The comparisons between the living organisms and the computing machines are, therefore, certainly imperfect at this point. The living organisms are very complex—part digital and part analogy mechanisms. The computing machines, at least in their recent forms to which I am referring in this discussion, are purely digital. Thus I must ask you to accept this oversimplification of the system. Although I am well aware of the analogy component in living organisms, and it would be absurd to deny its importance, I shall, nevertheless, for the sake of the simpler discussion, disregard that part. I shall consider the living organisms as if they were purely digital automata.

*Mixed Character of Each Element.* In addition to this, one may argue that even the neuron is not exactly a digital organ. This point has been put forward repeatedly and with great force. There is certainly a great deal of truth in it, when one considers things in considerable detail. The relevant assertion is, in this respect, that the fully developed nervous impulse, to which all-or-none character can be attributed, is not an elementary phenomenon, but is highly complex. It is a degenerate state of the complicated electrochemical complex which constitutes the neuron, and which in its fully analyzed functioning must be viewed as an analogy machine. Indeed, it is possible to stimulate the neuron in such a way that the breakdown that releases the nervous stimulus will not occur. In this area of "subliminal stimulation," we find first (that is, for the weakest stimulations) responses which are proportional to the stimulus, and then (at higher, but still subliminal, levels of stimulation) responses which depend on more complicated non-linear laws, but are nevertheless continuously variable and not of the breakdown type. There are also other complex phenomena within and without the subliminal range: fatigue, summation, certain forms of self-oscillation, etc.

In spite of the truth of these observations, it should be remembered that they may represent an improperly rigid critique of the concept of an all-or-none organ. The electromechanical relay, or the vacuum tube, when properly used, are undoubtedly all-or-none organs. Indeed, they are the prototypes of such organs. Yet both of them are in reality complicated analogy mechanisms, which upon appropriately adjusted stimulation respond continuously, linearly or non-linearly, and exhibit the phenomena of "breakdown" or "all-or-none" response only under very particular conditions of operation. There is little difference between this performance and the above-described performance of neurons. To put it somewhat differently. None of these is an exclusively all-or-none organ (there is little in our technological or physiological experience to indicate that absolute all-or-none organs exist); this, however, is irrelevant. By an all-or-none organ we should rather mean one which fulfills the following two conditions. First, it functions in the all-or-none manner under certain suitable operating conditions. Second, these operating conditions are the ones under which it is normally used; they represent the functionally normal state of affairs within the large organism, of which it forms a part. Thus the important fact is not whether an organ has necessarily and under all conditions the all-or-none character—this is probably never the case—but rather whether in its proper context it functions primarily, and appears to be intended to function primarily, as an all-or-none organ. I realize that this definition brings in rather undesirable criteria of "propriety" of context, of "appearance" and "intention." I do not see, however, how we can avoid using them, and how we can forego counting on the employment of common sense in their application. I shall, accordingly, in what follows use the working hypothesis that the neuron is an all-or-none digital organ. I realize that the last word about this has not been said, but I hope that the above excursus on the limitations of this working hypothesis and the reasons for its use will reassure you. I merely want to simplify my discussion; I am not trying to prejudge any essential open question.

In the same sense, I think that it is permissible to discuss the neurons as electrical organs. The stimulation of a neuron, the development and progress of its impulse, and the stimulating effects of the impulse at a synapse can all be described electrically. The concomitant chemical and other processes are important in order to understand the internal functioning of a nerve cell. They may even be more important than the electrical phenomena. They seem, however, to be hardly necessary for a description of a neuron as a "black box," an organ of the all-or-none type. Again the situation is no worse here than it is for, say, a vacuum tube. Here, too, the purely electrical phenomena are accompanied by numerous other phenomena of solid state physics, thermodynamics, mechanics. All of these are important to understand the structure of a vacuum tube, but are best excluded from the discussion, if it is to treat the vacuum tube as a "black box" with a schematic description.

*The Concept of a Switching Organ or Relay Organ.* The neuron, as well as the vacuum tube, viewed under the aspects discussed above, are then two instances of the same generic entity, which it is customary to call a "switching organ" or "relay organ." (The electromechanical relay is, of course, another instance.) Such an organ is defined as a "black box," which responds to a specified stimulus or combination of stimuli by an energetically independent response. That is, the response is expected to have enough energy to cause several stimuli of the same kind as the ones which initiated it. The energy of the response, therefore, cannot have been supplied by the original stimulus. It must originate in a different and independent source of power. The stimulus merely directs, controls the flow of energy from this source.

(This source, in the case of the neuron, is the general metabolism of the neuron. In the case of a vacuum tube, it is the power which maintains the cathode-plate potential difference, irrespective of whether the tube is conducting or not, and to a lesser extent the heater power which keeps "boiling" electrons out of the cathode. In the case of the electromechanical relay, it is the current supply whose path the relay is closing or opening.)

The basic switching organs of the living organisms, at least to the extent to which we are considering them here, are the neurons. The basic switching organs of the recent types of computing machines are vacuum tubes; in older ones they were wholly or partially electromechanical relays. It is quite possible that computing machines will not always be primarily aggregates of switching organs, but such a development is as yet quite far in the future. A development which may lie much closer is that the vacuum tubes may be displaced from their role of switching organs in computing machines. This, too, however, will probably not take place for a few years yet. I shall, therefore, discuss computing machines solely from the point of view of aggregates of switching organs which are vacuum tubes.

*Comparison of the Sizes of Large Computing Machines and Living Organisms.* Two well-known, very large vacuum tube computing machines are in existence and in operation. Both consist of about 20,000 switching organs. One is a pure vacuum tube machine. (It belongs to the U. S. Army Ordnance Department, Ballistic Research Laboratories, Aberdeen, Maryland, designation "ENIAC.") The other is mixed—part vacuum tube and part electromechanical relays. (It belongs to the I. B. M. Corporation, and is located in New York, designation "SSEC.") These machines are a good deal larger than what is likely to be the size of the vacuum tube computing machines which will come into existence and operation in the next few years. It is probable that each one of these will consist of 2000 to 6000 switching organs. (The reason for this decrease lies in a different attitude about the treatment of the "memory," which I will not discuss here.) It is possible that in later years the machine sizes will increase again, but it is not likely that 10,000 (or perhaps a few times 10,000) switching organs will be exceeded as long as the present techniques and philosophy are employed. To sum up, about $10^4$ switching organs seem to be the proper order of magnitude for a computing machine.

In contrast to this, the number of neurons in the central nervous system has been variously estimated as something of the order of $10^{10}$. I do not know how good this figure is, but presumably the exponent at least is not too high, and not too low by more than a unit. Thus it is very conspicuous that the central nervous system is at least a million times larger than the largest artificial automaton that we can talk about at present. It is quite interesting to inquire why this should be so and what questions of principle are involved. It seems to me that a few very clear-cut questions of principle are indeed involved.

*Determination of the Significant Ratio of Sizes for the Elements.* Obviously, the vacuum tube, as we know it, is gigantic compared to a nerve cell. Its physical volume is about a billion times larger, and its energy dissipation is about a billion times greater. (It is, of course, impossible to give such figures with a unique validity, but the above ones are typical.) There is, on the other hand, a certain compensation for this. Vacuum tubes can be made to operate at exceedingly high speeds in applications other than computing machines, but these need not concern us here. In computing machines the maximum is a good deal lower, but it is still quite respectable. In the present state of the art, it is generally believed to be somewhere around a million actuations per second. The responses of a nerve cell are a good deal slower than this, perhaps $\frac{1}{2000}$ of a second, and what really matters, the minimum time-interval required from stimulation to complete recovery and, possibly, renewed stimulation, is still longer than this—at best approximately $\frac{1}{200}$ of a second. This gives a ratio of 1:5000, which, however, may be somewhat too favorable to the vacuum tube, since vacuum tubes, when used as switching organs at the 1,000,000 steps per second rate, are practically never run at a 100 per cent duty cycle. A ratio like 1:2000 would, therefore, seem to be more equitable. Thus the vacuum tube, at something like a billion times the expense, outperforms the neuron by a factor of somewhat over 1000. There is, therefore, some justice in saying that it is less efficient by factor of the order of a million.

The basic fact is, in every respect, the small size of the neuron compared to the vacuum tube. This ratio is about a billion, as pointed out above. What is it due to?

*Analysis of the Reasons for the Extreme Ratio of Sizes.* The origin of this discrepancy lies in the fundamental control organ or, rather, control arrangement of the vacuum tube as compared to that of the neuron. In the vacuum tube the critical area of control is the space between the cathode (where the active agents, the electrons, originate) and the grid (which controls the electron flow). This space is about one millimeter deep. The corresponding entity in a neuron is the wall of the nerve cell, the "membrane." Its thickness is about a micron ($\frac{1}{1000}$ millimeter), or somewhat less. At this point, therefore, there is a ratio of approximately 1:1000 in linear dimensions. This, by the way, is the main difference. The electrical fields, which exist in the controlling space, are about the same for the vacuum tube and for the neuron. The potential differences by which these organs can be reliably steered are tens of volts in one case and tens of millivolts in the other. Their ratio is again about 1:1000, and hence their gradients (the field strengths) are about identical. Now a ratio of 1:1000 in linear dimensions corresponds to a ratio of 1:1,000,000,000 in volume. Thus the discrepancy factor of a billion in 3-dimensional size (volume) corresponds, as it should, to a discrepancy factor of 1000 in linear size, that is, to the difference between the millimeter interelectrode-space depth of the vacuum tube and the micron membrane thickness of the neuron.

It is worth noting, although it is by no means surprising, how this divergence between objects, both of which are microscopic and are situated in the interior of the elementary components, leads to impressive macroscopic differences between the organisms built upon them. This difference between a millimeter object and a micron object causes the ENIAC to weigh 30 tons and to dissipate 150 kilowatts of energy, while the human central nervous system, which is functionally about a million times larger, has the weight of the order of a pound and is accommodated within the human skull. In assessing the weight and size of the ENIAC as stated above, we should also remember that this huge apparatus is needed in order to handle 20 numbers of 10 decimals each, that is, a total of 200 decimal digits, the equivalent of about 700 binary digits—merely 700 simultaneous pieces of "yes-no" information!

*Technological Interpretation of These Reasons.* These considerations should make it clear that our present technology is still very imperfect in handling information at high speed and high degrees of complexity. The apparatus which results is simply enormous, both physically and in its energy requirements.

The weakness of this technology lies probably, in part at least, in the materials employed. Our present techniques involve the using of metals, with rather close spacings, and at certain critical points separated by vacuum only.

This combination of media has a peculiar mechanical instability that is entirely alien to living nature. By this I mean the simple fact that, if a living organism is mechanically injured, it has a strong tendency to restore itself. If, on the other hand, we hit a man-made mechanism with a sledge hammer, no such restoring tendency is apparent. If two pieces of metal are close together, the small vibrations and other mechanical disturbances, which always exist in the ambient medium, constitute a risk in that they may bring them into contact. If they were at different electrical potentials, the next thing that may happen after this short circuit is that they can become electrically soldered together and the contact becomes permanent. At this point, then, a genuine and permanent breakdown will have occurred. When we injure the membrane of a nerve cell, no such thing happens. On the contrary, the membrane will usually reconstitute itself after a short delay.

It is this mechanical instability of our materials which prevents us from reducing sizes further. This instability and other phenomena of a comparable character make the behavior in our componentry less than wholly reliable, even at the present sizes. Thus it is the inferiority of our materials, compared with those used in nature, which prevents us from attaining the high degree of complication and the small dimensions which have been attained by natural organisms.

## THE FUTURE LOGICAL THEORY OF AUTOMATA

*Further Discussion of the Factors That Limit the Present Size of Artificial Automata.* We have emphasized how the complication is limited in artificial automata, that is, the complication which can be handled without extreme difficulties and for which automata can still be expected to function reliably. Two reasons that put a limit on complication in this sense have already been given. They are the large size and the limited reliability of the componentry that we must see, both of them due to the fact that we are employing materials which seem to be quite satisfactory in simpler applications, but marginal and inferior to the natural ones in this highly complex application. There is, however, a third important limiting factor, and we should now turn our attention to it. This factor is of an intellectual, and not physical, character.

*The Limitation Which Is Due to the Lack of a Logical Theory of Automata.* We are very far from possessing a theory of automata which deserves that name, that is, a properly mathematical-logical theory. There exists today a very elaborate system of formal logic, and, specifically, of logic as applied to mathematics. This is a discipline with many good sides, but also with certain serious weaknesses. This is not the occasion to enlarge upon the good sides, which I have certainly no intention to belittle. About the inadequacies, however, this may be said: Everybody who has worked in formal logic will confirm that it is one of the technically most refractory parts of mathematics. The reason for this is that it deals with rigid, all-or-none concepts, and has very little contact with the continuous concept of the real or of the complex number, that is, with mathematical analysis. Yet analysis is the technically most successful and best-elaborated part of mathematics. Thus formal logic is, by the nature of its approach, cut off from the best cultivated portions of mathematics, and forced onto the most difficult part of the mathematical terrain, into combinatorics.

The theory of automata, of the digital, all-or-none type, as discussed up to now, is certainly a chapter in formal logic. It would, therefore, seem that it will have to share this unattractive property of formal logic. It will have to be, from the mathematical point of view, combinatorial rather than analytical.

*Probable Characteristics of Such a Theory.* Now it seems to me that this will in fact not be the case. In studying the functioning of automata, it is clearly necessary to pay attention to a circumstance which has never before made its appearance in formal logic.

Throughout all modern logic, the only thing that is important is whether a result can be achieved in a finite number of elementary steps or not. The size of the number of steps which are required, on the other hand, is hardly ever a concern of formal logic. Any finite sequence of correct steps is, as a matter of principle, as good as any other. It is a matter of no consequence whether the number is small or large, or even so large that it couldn't possibly be carried out in a lifetime, or in the presumptive lifetime of the stellar universe as we know it. In dealing with automata, this statement must be significantly modified. In the case of an automaton the thing which matters is not only whether it can reach a certain result in a finite number of steps at all but also how many such steps are needed. There are two reasons. First, automata are constructed in order to reach certain results in certain pre-assigned durations, or at least in pre-assigned orders of magnitude of duration. Second, the componentry employed has on every individual operation a small but nevertheless non-zero probability of failing. In a sufficiently long chain of operations the cumulative effect of these individual probabilities of failure may (if unchecked) reach the order of magnitude of unity—at which point it produces, in effect, complete unreliability. The probability levels which are involved here are very low, but still not too far removed from the domain of ordinary technological experience. It is not difficult to estimate that a high-speed computing machine, dealing with a typical problem, may have to perform as much as $10^{12}$ individual operations. The probability of error on an individual operation which can be tolerated must, therefore, be small compared to $10^{-12}$. I might mention that an electro-mechanical relay (a telephone relay) is at present considered acceptable if its probability of failure on an individual operation is of the order $10^{-8}$. It is considered excellent if this order of probability is $10^{-9}$. Thus the reliabilities required in a high-speed computing machine are higher, but not prohibitively higher, than those that constitute sound practice in certain existing industrial fields. The actually obtainable reliabilities are, however, not likely to leave a very wide margin against the minimum requirements just mentioned.

An exhaustive study and a non-trivial theory will, therefore, certainly be called for.

Thus the logic of automata will differ from the present system of formal logic in two relevant respects.

1. The actual length of "chains of reasoning," that is, of the chains of operations, will have to be considered.

2. The operations of logic (syllogisms, conjunctions, disjunctions, negations, etc., that is, in the terminology that is customary for automata, various forms of gating, coincidence, anti-coincidence, blocking, etc., actions) will all have to be treated by procedures which allow exceptions (malfunctions) with low but non-zero probabilities. All of this will lead to theories which are much less rigidly of an all-or-none nature than past and present formal logic. They will be of a much less combinatorial, and much more analytical, character. In fact, there are numerous indications to make us believe that this new system of formal logic will move closer to another discipline which has been little linked in the past with logic. This is thermodynamics, primarily in the form it was received from Boltzmann, and is that part of theoretical physics which comes nearest in some of its aspects to manipulating and measuring information. Its techniques are indeed much more analytical than combinatorial, which again illustrates the point that I have been trying to make above. It would, however, take me too far to go into this subject more thoroughly on this occasion.

All of this re-emphasizes the conclusion that was indicated earlier, that a detailed, highly mathematical, and more specifically analytical, theory of automata and of information is needed. We possess only the first indications of such a theory at present. In assessing artificial automata, which are, as I discussed earlier, of only moderate size, it has been possible to get along in a rough, empirical manner without such a theory. There is every reason to believe that this will not be possible with more elaborate automata.

*Effects of the Lack of a Logical Theory of Automata on the Procedures in Dealing with Errors.* This, then, is the last, and very important, limiting factor. It is unlikely that we could construct automata of a much higher complexity than the ones we now have, without possessing a very advanced and subtle theory of automata and information. A fortiori, this is inconceivable for automata of such enormous complexity as is possessed by the human central nervous system.

This intellectual inadequacy certainly prevents us from getting much farther than we are now.

A simple manifestation of this factor is our present relation to error checking. In living organisms malfunctions of components occur. The organism obviously has a way to detect them and render them harmless. It is easy to estimate that the number of nerve actuations which occur in a normal lifetime must be of the order of $10^{20}$. Obviously, during this chain of events there never occurs a malfunction which cannot be corrected by the organism itself, without any significant outside intervention. The system must, therefore, contain the necessary arrangements to diagnose errors as they occur, to readjust the organism so as to minimize the effects of the errors, and finally to correct or to block permanently the faulty components. Our modus procedendi with respect to malfunctions in our artificial automata is entirely different. Here the actual practice, which has the consensus of all experts of the field, is somewhat like this: Every effort is made to detect (by mathematical or by automatical checks) every error as soon as it occurs. Then an attempt is made to isolate the component that caused the error as rapidly as feasible. This may be done partly automatically, but in any case a significant part of this diagnosis must be effected by intervention from the outside. Once the faulty component has been identified, it is immediately corrected or replaced.

Note the difference in these two attitudes. The basic principle of dealing with malfunctions in nature is to make their effect as unimportant as possible and to apply correctives, if they are necessary at all, at leisure. In our dealings with artificial automata, on the other hand, we require an immediate diagnosis. Therefore, we are trying to arrange the automata in such a manner that errors will become as conspicuous as possible, and intervention and correction follow immediately. In other words, natural organisms are constructed to make errors as inconspicuous, as harmless, as possible. Artificial automata are designed to make errors as conspicuous, as disastrous, as possible. The rationale of this difference is not far to seek. Natural organisms are sufficiently well conceived to be able to operate even when malfunctions have set in. They can operate in spite of malfunctions, and their subsequent tendency is to remove these malfunctions. An artificial automaton could certainly be designed so as to be able to operate normally in spite of a limited number of malfunctions in certain limited areas. Any malfunction, however, represents a considerable risk that some generally degenerating process has already set in within the machine. It is, therefore, necessary to intervene immediately, because a machine which has begun to malfunction has only rarely a tendency to restore itself, and will more probably go from bad to worse. All of this comes back to one thing. With our artificial automata we are moving much more in the dark than nature appears to be with its organisms. We are, and apparently, at least at present, have to be, much more "scared" by the occurrence of an isolated error and by the malfunction which must be behind it. Our behavior is clearly that of overcaution, generated by ignorance.

*The Single-Error Principle.* A minor side light to this is that almost all our error-diagnosing techniques are based on the assumption that the machine contains only one faulty component. In this case, iterative subdivisions of the machine into parts permit us to determine which portion contains the fault. As soon as the possibility exists that the machine may contain several faults, these, rather powerful, dichotomic methods of diagnosis are lost. Error diagnosing then becomes an increasingly hopeless proposition. The high premium on keeping the number of errors to be diagnosed down to one, or at any rate as low as possible, again illustrates our ignorance in this field, and is one of the main reasons why errors must be made as conspicuous as possible, in order to be recognized and apprehended as soon after their occurrence as feasible, that is, before further errors have had time to develop.

## PRINCIPLES OF DIGITALIZATION

*Digitalization of Continuous Quantities: the Digital Expansion Method and the Counting Method.* Consider the digital part of a natural organism; specifically, consider the nervous system. It seems that we are indeed justified in assuming that this is a digital mechanism, that it transmits messages which are made up of signals possessing the all-or-none character. (See also the earlier discussion, page 2054–55.) In other words, each elementary signal, each impulse, simply either is or is not there, with no further shadings. A particularly relevant illustration of this fact is furnished by those cases where the underlying problem has the opposite character, that is, where the nervous system is actually called upon to transmit a continuous quantity. Thus the case of a nerve which has to report on the value of a pressure is characteristic.

Assume, for example, that a pressure (clearly a continuous quantity) is to be transmitted. It is well known how this trick is done. The nerve which does it still transmits nothing but individual all-or-none impulses. How does it then express the continuously numerical value of pressure in terms of these impulses, that is, of digits? In other words, how does it encode a continuous number into a digital notation? It does certainly not do it by expanding the number in question into decimal (or binary, or any other base) digits in the conventional sense. What appears to happen is that it transmits pulses at a frequency which varies and which is within certain limits proportional to the continuous quantity in question, and generally a monotone function of it. The mechanism which achieves this "encoding" is, therefore, essentially a frequency modulation system.

The details are known. The nerve has a finite recovery time. In other words, after it has been pulsed once, the time that has to lapse before another stimulation is possible is finite and dependent upon the strength of the ensuing (attempted) stimulation. Thus, if the nerve is under the influence of a continuing stimulus (one which is uniformly present at all times, like the pressure that is being considered here), then the nerve will respond periodically, and the length of the period between two successive stimulations is the recovery time referred to earlier, that is, a function of the strength of the constant stimulus (the pressure in the present case). Thus, under a high pressure, the nerve may be able to respond every 8 milliseconds, that is, transmit at the rate of 125 impulses per second; while under the influence of a smaller pressure it may be able to repeat only every 14 milliseconds, that is, transmit at the rate of 71 times per second. This is very clearly the behavior of a genuinely yes-or-no organ, of a digital organ. It is very instructive, however, that it uses a "count" rather than a "decimal expansion" (or "binary expansion," etc.) method.

*Comparison of the Two Methods. The Preference of Living Organisms for the Counting Method.* Compare the merits and demerits of these two methods. The counting method is certainly less efficient than the expansion method. In order to express a number of about a million (that is, a physical quantity of a million distinguishable resolution-steps) by counting, a million pulses have to be transmitted. In order to express a number of the same size by expansion, 6 or 7 decimal digits are needed, that is, about 20 binary digits. Hence, in this case only 20 pulses are needed. Thus our expansion method is much more economical in notation than the counting methods which are resorted to by nature. On the other hand, the counting method has a high stability and safety from error. If you express a number of the order of a million by counting and miss a count, the result is only irrelevantly changed. If you express it by (decimal or binary) expansion, a single error in a single digit may vitiate the entire result. Thus the undesirable trait of our computing machines reappears in our digital expansion system, in fact, the former is clearly deeply connected with, and partly a consequence of, the latter. The high stability and nearly error-proof character of natural organisms, on the other hand, is reflected in the counting method that they seem to use in this case. All of this reflects a general rule. You can increase the safety from error by a reduction of the efficiency of the notation, or, to say it positively, by allowing redundancy of notation. Obviously, the simplest form of achieving safety by redundancy is to use the, per se, quite unsafe digital expansion notation, but to repeat every such message several times. In the case under discussion, nature has obviously resorted to an even more redundant and even safer system.

There are, of course, probably other reasons why the nervous system uses the counting rather than the digital expansion. The encoding-decoding facilities required by the former are much simpler than those required by the latter. It is true, however, that nature seems to be willing and able to go much further in the direction of complication than we are, or rather than we can afford to go. One may, therefore, suspect that if the only demerit of the digital expansion system were its greater logical complexity, nature would not, for this reason alone, have rejected it. It is, nevertheless, true that we have nowhere an indication of its use in natural organisms. It is difficult to tell how much "final" validity one should attach to this observation. The point deserves at any rate attention, and should receive it in future investigations of the functioning of the nervous system.

## FORMAL NEURAL NETWORKS

*The McCulloch-Pitts Theory of Formal Neural Networks.* A great deal more could be said about these things from the logical and the organizational point of view, but I shall not attempt to say it here. I shall instead go on to discuss what is probably the most significant result obtained with the axiomatic method up to now. I mean the remarkable theorems of McCulloch and Pitts on the relationship of logics and neural networks.

In this discussion I shall, as I have said, take the strictly axiomatic point of view. I shall, therefore, view a neuron as a "black box" with a certain number of inputs that receive stimuli and an output that emits stimuli. To be specific, I shall assume that the input connections of each one of these can be of two types, excitatory and inhibitory. The boxes themselves are also of two types, threshold 1 and threshold 2. These concepts are linked and circumscribed by the following definitions. In order to stimulate such an organ it is necessary that it should receive simultaneously at least as many stimuli on its excitatory inputs as correspond to its threshold, and not a single stimulus on any one of its inhibitory inputs. If it has been thus stimulated, it will after a definite time delay (which is assumed to be always the same, and may be used to define the unit of time) emit an output pulse. This pulse can be taken by appropriate connections to any number of inputs of other neurons (also to any of its own inputs) and will produce at each of these the same type of input stimulus as the ones described above.

It is, of course, understood that this is an oversimplification of the actual functioning of a neuron. I have already discussed the character, the limitations, and the advantages of the axiomatic method. (See pages 2048 and 2055.) They all apply here, and the discussion which follows is to be taken in this sense.

McCulloch and Pitts have used these units to build up complicated networks which may be called "formal neural networks." Such a system is built up of any number of these units, with their inputs and outputs suitably interconnected with arbitrary complexity. The "functioning" of such a network may be defined by singling out some of the inputs of the entire system and some of its outputs, and then describing what original stimuli on the former are to cause what ultimate stimuli on the latter.

*The Main Result of the McCulloch-Pitts Theory.* McCulloch and Pitts' important result is that any functioning in this sense which can be defined at all logically, strictly, and unambiguously in a finite number of words can also be realized by such a formal neural network.

It is well to pause at this point and to consider what the implications are. It has often been claimed that the activities and functions of the human nervous system are so complicated that no ordinary mechanism could possibly perform them. It has also been attempted to name specific functions which by their nature exhibit this limitation. It has been attempted to show that such specific functions, logically, completely described, are per se unable of mechanical, neural realization. The McCulloch-Pitts result puts an end to this. It proves that anything that can be exhaustively and unambiguously described, anything that can be completely and unambiguously put into words, is ipso facto realizable by a suitable finite neural network. Since the converse statement is obvious, we can therefore say that there is no difference between the possibility of describing a real or imagined mode of behavior completely and unambiguously in words, and the possibility of realizing it by a finite formal neural network. The two concepts are co-extensive. A difficulty of principle embodying any mode of behavior in such a network can exist only if we are also unable to describe that behavior completely.

Thus the remaining problems are these two. First, if a certain mode of behavior can be effected by a finite neural network, the question still remains whether that network can be realized within a practical size, specifically, whether it will fit into the physical limitations of the organism in question. Second, the question arises whether every existing mode of behavior can really be put completely and unambiguously into words.

The first problem is, of course, the ultimate problem of nerve physiology, and I shall not attempt to go into it any further here. The second question is of a different character, and it has interesting logical connotations.

*Interpretations of This Result.* There is no doubt that any special phase of any conceivable form of behavior can be described "completely and unambiguously" in words. This description may be lengthy, but it is always possible. To deny it would amount to adhering to a form of logical mysticism which is surely far from most of us. It is, however, an important limitation, that this applies only to every element separately, and it is far from clear how it will apply to the entire syndrome of behavior. To be more specific, there is no difficulty in describing how an organism might be able to identify any two rectilinear triangles, which appear on the retina, as belonging to the same category "triangle." There is also no difficulty in adding to this, that numerous other objects, besides regularly drawn rectilinear triangles, will also be classified and identified as triangles—triangles whose sides are curved, triangles whose sides are not fully drawn, triangles that are indicated merely by a more or less homogeneous shading of their interior, etc. The more completely we attempt to describe everything that may conceivably fall under this heading, the longer the description becomes. We may have a vague and uncomfortable feeling that a complete catalogue along such lines would not only be exceedingly long, but also unavoidably indefinite at its boundaries. Nevertheless, this may be a possible operation.

All of this, however, constitutes only a small fragment of the more general concept of identification of analogous geometrical entities. This, in turn, is only a microscopic piece of the general concept of analogy. Nobody would attempt to describe and define within any practical amount of space the general concept of analogy which dominates our interpretation of vision. There is no basis for saying whether such an enterprise would require thousands or millions or altogether impractical numbers of volumes. Now it is perfectly possible that the simplest and only practical way actually to say what constitutes a visual analogy consists in giving a description of the connections of the visual brain.

We are dealing here with parts of logics with which we have practically no past experience. The order of complexity is out of all proportion to anything we have ever known. We have no right to assume that the logical notations and procedures used in the past are suited to this part of the subject. It is not at all certain that in this domain a real object might not constitute the simplest description of itself, that is, any attempt to describe it by the usual literary or formal-logical method may lead to something less manageable and more involved. In fact, some results in modern logic would tend to indicate that phenomena like this have to be expected when we come to really complicated entities. It is, therefore, not at all unlikely that it is futile to look for a precise logical concept, that is, for a precise verbal description, of "visual analogy." It is possible that the connection pattern of the visual brain itself is the simplest logical expression or definition of this principle.

Obviously, there is on this level no more profit in the McCulloch-Pitts result. At this point it only furnishes another illustration of the situation outlined earlier. There is an equivalence between logical principles and their embodiment in a neural network, and while in the simpler cases the principles might furnish a simplified expression of the network, it is quite possible that in cases of extreme complexity the reverse is true.

All of this does not alter my belief that a new, essentially logical, theory is called for in order to understand high-complication automata and, in particular, the central nervous system. It may be, however, that in this process logic will have to undergo a pseudomorphosis to neurology to a much greater extent than the reverse. The foregoing analysis shows that one of the relevant things we can do at this moment with respect to the theory of the central nervous system is to point out the directions in which the real problem does not lie.

## THE CONCEPT OF COMPLICATION;
## SELF-REPRODUCTION

*The Concept of Complication.* The discussions so far have shown that high complexity plays an important role in any theoretical effort relating to automata, and that this concept, in spite of its prima facie quantitative character, may in fact stand for something qualitative—for a matter of principle. For the remainder of my discussion I will consider a remoter implication of this concept, one which makes one of the qualitative aspects of its nature even more explicit.

There is a very obvious trait, of the "vicious circle" type, in nature, the simplest expression of which is the fact that very complicated organisms can reproduce themselves.

We are all inclined to suspect in a vague way the existence of a concept of "complication." This concept and its putative properties have never been clearly formulated. We are, however, always tempted to assume that they will work in this way. When an automaton performs certain operations, they must be expected to be of a lower degree of complication than the automaton itself. In particular, if an automaton has the ability to construct another one, there must be a decrease in complication as we go from the parent to the construct. That is, if $A$ can produce $B$, then $A$ in some way must have contained a complete description of $B$. In order to make it effective, there must be, furthermore, various arrangements in $A$ that see to it that this description is interpreted and that the constructive operations that it calls for are carried out. In this sense, it would therefore seem that a certain degenerating tendency must be expected, some decrease in complexity as one automaton makes another automaton.

Although this has some indefinite plausibility to it, it is in clear contradiction with the most obvious things that go on in nature. Organisms reproduce themselves, that is, they produce new organisms with no decrease in complexity. In addition, there are long periods of evolution during which the complexity is even increasing. Organisms are indirectly derived from others which had lower complexity.

Thus there exists an apparent conflict of plausibility and evidence, if nothing worse. In view of this, it seems worth while to try to see whether there is anything involved here which can be formulated rigorously.

So far I have been rather vague and confusing, and not unintentionally at that. It seems to me that it is otherwise impossible to give a fair impression of the situation that exists here. Let me now try to become specific.

*Turing's Theory of Computing Automata.* The English logician, Turing, about twelve years ago attacked the following problem.

He wanted to give a general definition of what is meant by a computing automaton. The formal definition came out as follows:

An automaton is a "black box," which will not be described in detail but is expected to have the following attributes. It possesses a finite number of states, which need be prima facie characterized only by stating their number, say $n$, and by enumerating them accordingly: 1, 2, . . ., $n$. The essential operating characteristic of the automaton consists of describing how it is caused to change its state, that is, to go over from a state $i$ into a state $j$. This change requires some interaction with the outside world, which will be standardized in the following manner. As far as the machine is concerned, let the whole outside world consist of a long paper tape. Let this tape be, say, 1 inch wide, and let it be subdivided into fields (squares) 1 inch long. On each field of this strip we may or may not put a sign, say, a dot, and it is assumed that it is possible to erase as well as to write in such a dot. A field marked with a dot will be called a "1," a field unmarked with a dot will be called a "0." (We might permit more ways of marking, but Turing showed that this is irrelevant and does not lead to any essential gain in generality.) In describing the position of the tape relative to the automaton it is assumed that one particular field of the tape is under direct inspection by the automaton, and that the automaton has the ability to move the tape forward and backward, say, by one field at a time. In specifying this, let the automaton be in the state $i$ ($= 1 \ldots, n$), and let it see on the tape an $e$ ($= 0$, 1). It will then go over into the state $j$ ($= 0, 1, \ldots, n$), move the tape by $p$ fields ($p = 0, +1, -1; +1$ is a move forward, $-1$ is a move backward), and inscribe into the new field that it sees $f$ ($= 0, 1$; inscribing 0 means erasing; inscribing 1 means putting in a dot). Specifying $j, p, f$ as functions of $i, e$ is then the complete definition of the functioning of such an automaton.

Turing carried out a careful analysis of what mathematical processes can be effected by automata of this type. In this connection he proved various theorems concerning the classical "decision problem" of logic, but I shall not go into these matters here. He did, however, also introduce and analyze the concept of a "universal automaton," and this is part of the subject that is relevant in the present context.

An infinite sequence of digits $e$ ($= 0, 1$) is one of the basic entities in mathematics. Viewed as a binary expansion, it is essentially equivalent to the concept of a real number. Turing, therefore, based his consideration on these sequences.

He investigated the question as to which automata were able to construct which sequences. That is, given a definite law for the formation of such a sequence, he inquired as to which automata can be used to form the sequence based on that law. The process of "forming" a sequence is interpreted in this manner. An automaton is able to "form" a certain sequence if it is possible to specify a finite length of tape, appropriately marked, so that, if this tape is fed to the automaton in question, the automaton will thereupon write the sequence on the remaining (infinite) free portion of the tape. This process of writing the infinite sequence is, of course, an indefinitely continuing one. What is meant is that the automaton will keep running indefinitely and, given a sufficiently long time, will have inscribed any desired (but of course finite) part of the (infinite) sequence. The finite, premarked, piece of tape constitutes the "instruction" of the automaton for this problem.

An automaton is "universal" if any sequence that can be produced by any automaton at all can also be solved by this particular automaton. It will, of course, require in general a different instruction for this purpose.

*The Main Result of the Turing Theory.* We might expect a priori that this is impossible. How can there be an automaton which is at least as effective as any conceivable automaton, including, for example, one of twice its size and complexity?

Turing, nevertheless, proved that this is possible. While his construction is rather involved, the underlying principle is nevertheless quite simple. Turing observed that a completely general description of any conceivable automaton can be (in the sense of the foregoing definition) given in a finite number of words. This description will contain certain empty passages—those referring to the functions mentioned earlier ($j, p, f$ in terms of $i, e$), which specify the actual functioning of the automaton. When these empty passages are filled in, we deal with a specific automaton. As long as they are left empty, this schema represents the general definition of the general automaton. Now it becomes possible to describe an automaton which has the ability to interpret such a definition. In other words, which, when fed the functions that in the sense described above define a specific automaton, will thereupon function like the object described. The ability to do this is no more mysterious than the ability to read a dictionary and a grammar and to follow their instructions about the uses and principles of combinations of words. This automaton, which is constructed to read a description and to imitate the object described, is then the universal automaton in the sense of Turing. To make it duplicate any operation that any other automaton can perform, it suffices to furnish it with a description of the automaton in question and, in addition, with the instructions which that device would have required for the operation under consideration.

*Broadening of the Program to Deal with Automata That Produce Automata.* For the question which concerns me here, that of "self-reproduction" of automata, Turing's procedure is too narrow in one respect only. His automata are purely computing machines. Their output is a piece of tape with zeros and ones on it. What is needed for the construction to which I referred is an automaton whose output is other automata. There is, however, no difficulty in principle in dealing with this broader concept and in deriving from it the equivalent of Turing's result.

*The Basic Definitions.* As in the previous instance, it is again of primary importance to give a rigorous definition of what constitutes an automaton for the purpose of the investigation. First of all, we have to draw up a complete list of the elementary parts to be used. This list must contain not only a complete enumeration but also a complete operational definition of each elementary part. It is relatively easy to draw up such a list, that is, to write a catalogue of "machine parts" which is sufficiently inclusive to permit the construction of the wide variety of mechanisms here required, and which has the axiomatic rigor that is needed for this kind of consideration. The list need not be very long either. It can, of course, be made either arbitrarily long or arbitrarily short. It may be lengthened by including in it, as elementary parts, things which could be achieved by combinations of others. It can be made short—in fact, it can be made to consist of a single unit—by endowing each elementary part with a multiplicity of attributes and functions. Any statement on the number of elementary parts required will therefore represent a common-sense compromise, in which nothing too complicated is expected from any one elementary part, and no elementary part is made to perform several, obviously separate, functions. In this sense, it can be shown that about a dozen elementary parts suffice. The problem of self-reproduction can then be stated like this: Can one build an aggregate out of such elements in such a manner that if it is put into a reservoir, in which there float all these elements in large numbers, it will then begin to construct other aggregates, each of which will at the end turn out to be another automaton exactly like the original one? This is feasible, and the principle on which it can be based is closely related to Turing's principle outlined earlier.

*Outline of the Derivation of the Theorem Regarding Self-reproduction.* First of all, it is possible to give a complete description of everything that is an automaton in the sense considered here. This description is to be conceived as a general one, that is, it will again contain empty spaces. These empty spaces have to be filled in with the functions which describe the actual structure of an automaton. As before, the difference between these spaces filled and unfilled is the difference between the description of a specific automaton and the general description of a general automaton. There is no difficulty of principle in describing the following automata.

(*a*) Automaton *A*, which when furnished the description of any other automaton in terms of appropriate functions, will construct that entity. The description should in this case not be given in the form of a marked tape, as in Turing's case, because we will not normally choose a tape as a structural element. It is quite easy, however, to describe combinations of structural elements which have all the notational properties of a tape with fields that can be marked. A description in this sense will be called an instruction and denoted by a letter *I*.

"Constructing" is to be understood in the same sense as before. The constructing automaton is supposed to be placed in a reservoir in which all elementary components in large numbers are floating, and it will effect its construction in that milieu. One need not worry about how a fixed automaton of this sort can produce others which are larger and more complex than itself. In this case the greater size and the higher complexity of the object to be constructed will be reflected in a presumably still greater size of the instructions *I* that have to be furnished. These instructions, as pointed out, will have to be aggregates of elementary parts. In this sense, certainly, an entity will enter the process whose size and complexity is determined by the size and complexity of the object to be constructed.

In what follows, all automata for whose construction the facility *A* will be used are going to share with *A* this property. All of them will have a place for an instruction *I*, that is, a place where such an instruction can be inserted. When such an automaton is being described (as, for example, by an appropriate instruction), the specification of the location for the insertion of an instruction *I* in the foregoing sense is understood to form a part of the description. We may, therefore, talk of "inserting a given instruction *I* into a given automaton," without any further explanation.

(*b*) Automaton *B*, which can make a copy of any instruction *I* that is furnished to it. *I* is an aggregate of elementary parts in the sense outlined in (*a*), replacing a tape. This facility will be used when *I* furnishes a description of another automaton. In other words, this automaton is nothing more subtle than a "reproducer"—the machine which can read a punched tape and produce a second punched tape that is identical with the first. Note that this automaton, too, can produce objects which are larger and more complicated than itself. Note again that there is nothing surprising about it. Since it can only copy, an object of the exact size and complexity of the output will have to be furnished to it as input.

After these preliminaries, we can proceed to the decisive step.

(*c*) Combine the automata *A* and *B* with each other, and with a control mechanism *C* which does the following. Let *A* be furnished with an instruction *I* (again in the sense of [*a*] and [*b*]). Then *C* will first cause *A* to construct the automaton which is described by this instruction *I*. Next *C* will cause *B* to copy the instruction *I* referred to above, and insert the copy into the automaton referred to above, which has just been constructed by *A*. Finally, *C* will separate this construction from the system $A + B + C$ and "turn it loose" as an independent entity.

(*d*) Denote the total aggregate $A + B + C$ by *D*.

(*e*) In order to function, the aggregate $D = A + B + C$ must be furnished with an instruction *I*, as described above. This instruction, as pointed out above, has to be inserted into *A*. Now form an instruction $I_D$, which describes this automaton *D*, and insert $I_D$ into *A* within *D*. Call the aggregate which now results *E*.

*E* is clearly self-reproductive. Note that no vicious circle is involved. The decisive step occurs in *E*, when the instruction $I_D$, describing *D*, is constructed and attached to *D*. When the construction (the copying) of $I_D$ is called for, *D* exists already, and it is in no wise modified by the construction of $I_D$. $I_D$ is simply added to form *E*. Thus there is a definite chronological and logical order in which *D* and $I_D$ have to be formed, and the process is legitimate and proper according to the rules of logic.

*Interpretations of This Result and of Its Immediate Extensions.* The description of this automaton *E* has some further attractive sides, into which I shall not go at this time at any length. For instance, it is quite clear that the instruction $I_D$ is roughly effecting the functions of a gene. It is also clear that the copying mechanism *B* performs the fundamental act of reproduction, the duplication of the genetic material, which is clearly the fundamental operation in the multiplication of living cells. It is also easy to see how arbitrary alterations of the system *E*, and in particular of $I_D$, can exhibit certain typical traits which appear in connection with mutation, lethally as a rule, but with a possibility of continuing reproduction with a modification of traits. It is, of course, equally clear at which point the analogy ceases to be valid. The natural gene does probably not contain a complete description of the object whose construction its presence stimulates. It probably contains only general pointers, general cues. In the generality in which the foregoing consideration is moving, this simplification is not attempted. It is, nevertheless, clear that this simplification, and others similar to it, are in themselves of great and qualitative importance. We are very far from any real understanding of the natural processes if we do not attempt to penetrate such simplifying principles.

Small variations of the foregoing scheme also permit us to construct automata which can reproduce themselves and, in addition, construct others. (Such an automaton performs more specifically what is probably a—if not the—typical gene function, self-reproduction plus production—or stimulation of production—of certain specific enzymes.) Indeed, it suffices to replace the $I_D$ by an instruction $I_{D+F}$, which describes the automaton $D$ plus another given automaton $F$. Let $D$, with $I_{D+F}$ inserted into $A$ within it, be designated by $E_F$. This $E_F$ clearly has the property already described. It will reproduce itself, and, besides, construct $F$.

Note that a "mutation" of $E_F$, which takes place within the $F$-part of $I_{D+F}$ in $E_F$, is not lethal. If it replaces $F$ by $F'$, it changes $E_F$ into $E_{F'}$, that is, the "mutant" is still self-reproductive; but its by-product is changed—$F'$ instead of $F$. This is, of course, the typical non-lethal mutant.

All these are very crude steps in the direction of a systematic theory of automata. They represent, in addition, only one particular direction. This is, as I indicated before, the direction towards forming a rigorous concept of what constitutes "complication." They illustrate that "complication" on its lower levels is probably degenerative, that is, that every automaton that can produce other automata will only be able to produce less complicated ones. There is, however, a certain minimum level where this degenerative characteristic ceases to be universal. At this point automata which can reproduce themselves, or even construct higher entities, become possible. This fact, that complication, as well as organization, below a certain minimum level is degenerative, and beyond that level can become self-supporting and even increasing, will clearly play an important role in any future theory of the subject.

**2**

*Thinking is very far from knowing.*
                          —PROVERB

*Beware when the great God lets loose a thinker on this planet.*
                          —EMERSON

*For 'tis the sport to have the enginer*
*Hoist with his own petar . . .*
                          —SHAKESPEARE (*Hamlet*)

# CAN A MACHINE THINK?

*By A. M. Turing*

## 1. THE IMITATION GAME

I propose to consider the question, 'Can machines think?' This should begin with definitions of the meaning of the terms 'machine' and 'think.' The definitions might be framed so as to reflect so far as possible the normal use of the words, but this attitude is dangerous. If the meaning of the words 'machine' and 'think' are to be found by examining how they are commonly used it is difficult to escape the conclusion that the meaning and the answer to the question, 'Can machines think?' is to be sought in a statistical survey such as a Gallup poll. But this is absurd. Instead of attempting such a definition I shall replace the question by another, which is closely related to it and is expressed in relatively unambiguous words.

The new form of the problem can be described in terms of a game which we call the 'imitation game.' It is played with three people, a man (A), a woman (B), and an interrogator (C) who may be of either sex. The interrogator stays in a room apart from the other two. The object of the game for the interrogator is to determine which of the other two is the man and which is the woman. He knows them by labels X and Y, and at the end of the game he says either 'X is A and Y is B' or 'X is B and Y is A.' The interrogator is allowed to put questions to A and B thus:

C: Will X please tell me the length of his or her hair?

Now suppose X is actually A, then A must answer. It is A's object in the game to try and cause C to make the wrong identification. His answer might therefore be

'My hair is shingled, and the longest strands are about nine inches long.'

In order that tones of voice may not help the interrogator the answers should be written, or better still, typewritten. The ideal arrangement is to have a teleprinter communicating between the two rooms. Alternatively the question and answers can be repeated by an intermediary. The object of the game for the third player (B) is to help the interrogator. The best strategy for her is probably to give truthful answers. She can add such things as 'I am the woman, don't listen to him!' to her answers, but it will avail nothing as the man can make similar remarks.

We now ask the question, 'What will happen when a machine takes the part of A in this game?' Will the interrogator decide wrongly as often when the game is played like this as he does when the game is played between a man and a woman? These questions replace our original, 'Can machines think?'

## 2. CRITIQUE OF THE NEW PROBLEM

As well as asking, 'What is the answer to this new form of the question,' one may ask, 'Is this new question a worthy one to investigate?' This latter question we investigate without further ado, thereby cutting short an infinite regress.

The new problem has the advantage of drawing a fairly sharp line between the physical and the intellectual capacities of a man. No engineer or chemist claims to be able to produce a material which is indistinguishable from the human skin. It is possible that at some time this might be done, but even supposing this invention available we should feel there was little point in trying to make a 'thinking machine' more human by dressing it up in such artificial flesh. The form in which we have set the problem reflects this fact in the condition which prevents the interrogator from seeing or touching the other competitors, or hearing their voices. Some other advantages of the proposed criterion may be shown up by specimen questions and answers. Thus:

Q: Please write me a sonnet on the subject of the Forth Bridge.

A: Count me out on this one. I never could write poetry.

Q: Add 34957 to 70764.

A: (Pause about 30 seconds and then give as answer) 105621.

Q: Do you play chess?

A: Yes.

Q: I have K at my K1, and no other pieces. You have only K at K6 and R at R1. It is your move. What do you play?

A: (After a pause of 15 seconds) R-R8 mate.

The question and answer method seems to be suitable for introducing almost any one of the fields of human endeavour that we wish to include. We do not wish to penalise the machine for its inability to shine in beauty competitions, nor to penalise a man for losing in a race against an aeroplane. The conditions of our game make these disabilities irrelevant. The 'witnesses' can brag, if they consider it advisable, as much as they please about their charms, strength or heroism, but the interrogator cannot demand practical demonstrations.

The game may perhaps be criticised on the ground that the odds are weighted too heavily against the machine. If the man were to try and pretend to be the machine he would clearly make a very poor showing. He would be given away at once by slowness and inaccuracy in arithmetic. May not machines carry out something which ought to be described as thinking but which is very different from what a man does? This objection is a very strong one, but at least we can say that if, nevertheless, a machine can be constructed to play the imitation game satisfactorily, we need not be troubled by this objection.

It might be urged that when playing the 'imitation game' the best strategy for the machine may possibly be something other than imitation of the behaviour of a man. This may be, but I think it is unlikely that there is any great effect of this kind. In any case there is no intention to investigate here the theory of the game, and it will be assumed that the best strategy is to try to provide answers that would naturally be given by a man.

## 3. THE MACHINES CONCERNED IN THE GAME

The question which we put in § 1 will not be quite definite until we have specified what we mean by the word 'machine.' It is natural that we should wish to permit every kind of engineering technique to be used in our machines. We also wish to allow the possibility that an engineer or team of engineers may construct a machine which works, but whose manner of operation cannot be satisfactorily described by its constructors because they have applied a method which is largely experimental. Finally, we wish to exclude from the machines men born in the usual manner. It is difficult to frame the definitions so as to satisfy these three conditions. One might for instance insist that the team of engineers should be all of one sex, but this would not really be satisfactory, for it is probably possible to rear a complete individual from a single cell of the skin (say) of a man. To do so would be a feat of biological technique deserving of the very highest praise, but we would not be inclined to regard it as a case of 'constructing a thinking machine.' This prompts us to abandon the requirement that every kind of technique should be permitted. We are the more ready to do so in view of the fact that the present interest in 'thinking machines' has been aroused by a particular kind of machine, usually called an 'electronic computer' or 'digital computer.' Following this suggestion we only permit digital computers to take part in our game.

This restriction appears at first sight to be a very drastic one. I shall attempt to show that it is not so in reality. To do this necessitates a short account of the nature and properties of these computers.

It may also be said that this identification of machines with digital computers, like our criterion for 'thinking,' will only be unsatisfactory if (contrary to my belief), it turns out that digital computers are unable to give a good showing in the game.

There are already a number of digital computers in working order, and it may be asked, 'Why not try the experiment straight away? It would be easy to satisfy the conditions of the game. A number of interrogators could be used, and statistics compiled to show how often the right identification was given.' The short answer is that we are not asking whether all digital computers would do well in the game nor whether the computers at present available would do well, but whether there are imaginable computers which would do well. But this is only the short answer. We shall see this question in a different light later.

## 4. DIGITAL COMPUTERS

The idea behind digital computers may be explained by saying that these machines are intended to carry out any operations which could be done by a human computer. The human computer is supposed to be following fixed rules; he has no authority to deviate from them in any detail. We may suppose that these rules are supplied in a book, which is altered whenever he is put on to a new job. He has also an unlimited supply of paper on which he does his calculations. He may also do his multiplications and additions on a 'desk machine,' but this is not important.

If we use the above explanation as a definition we shall be in danger of circularity of argument. We avoid this by giving an outline of the means by which the desired effect is achieved. A digital computer can usually be regarded as consisting of three parts:

(i) Store.
(ii) Executive unit.
(iii) Control.

The store is a store of information, and corresponds to the human computer's paper, whether this is the paper on which he does his calculations or that on which his book of rules is printed. In so far as the human computer does calculations in his head a part of the store will correspond to his memory.

The executive unit is the part which carries out the various individual operations involved in a calculation. What these individual operations are will vary from machine to machine. Usually fairly lengthy operations can be done such as 'Multiply 3540675445 by 7076345687' but in some machines only very simple ones such as 'Write down 0' are possible.

We have mentioned that the 'book of rules' supplied to the computer is replaced in the machine by a part of the store. It is then called the 'table of instructions.' It is the duty of the control to see that these instructions are obeyed correctly and in the right order. The control is so constructed that this necessarily happens.

The information in the store is usually broken up into packets of moderately small size. In one machine, for instance, a packet might consist of ten decimal digits. Numbers are assigned to the parts of the store in which the various packets of information are stored, in some systematic manner. A typical instruction might say—

'Add the number stored in position 6809 to that in 4302 and put the result back into the latter storage position.'

Needless to say it would not occur in the machine expressed in English. It would more likely be coded in a form such as 6809430217. Here 17 says which of various possible operations is to be performed on the two numbers. In this case the operation is that described above, *viz.* 'Add the number. . . .' It will be noticed that the instruction takes up 10 digits and so forms one packet of information, very conveniently. The control will normally take the instructions to be obeyed in the order of the positions in which they are stored, but occasionally an instruction such as

'Now obey the instruction stored in position 5606, and continue from there' may be encountered, or again

'If position 4505 contains 0 obey next the instruction stored in 6707, otherwise continue straight on.'

Instructions of these latter types are very important because they make it possible for a sequence of operations to be repeated over and over again until some condition is fulfilled, but in doing so to obey, not fresh instructions on each repetition, but the same ones over and over again. To take a domestic analogy. Suppose Mother wants Tommy to call at the cobbler's every morning on his way to school to see if her shoes are done, she can ask him afresh every morning. Alternatively she can stick up a notice once and for all in the hall which he will see when he leaves for school and which tells him to call for the shoes, and also to destroy the notice when he comes back if he has the shoes with him.

The reader must accept it as a fact that digital computers can be constructed, and indeed have been constructed, according to the principles we have described, and that they can in fact mimic the actions of a human computer very closely.

The book of rules which we have described our human computer as using is of course a convenient fiction. Actual human computers really remember what they have got to do. If one wants to make a machine mimic the behaviour of the human computer in some complex operation one has to ask him how it is done, and then translate the answer into the form of an instruction table. Constructing instruction tables is usually described as 'programming.' To 'programme a machine to carry out the operation A' means to put the appropriate instruction table into the machine so that it will do A.

An interesting variant on the idea of a digital computer is a 'digital computer with a random element.' These have instructions involving the throwing of a die or some equivalent electronic process; one such instruction might for instance be, 'Throw the die and put the resulting number into store 1000.' Sometimes such a machine is described as having free will (though I would not use this phrase myself). It is not normally possible to determine from observing a machine whether it has a random element, for a similar effect can be produced by such devices as making the choices depend on the digits of the decimal for $\pi$.

Most actual digital computers have only a finite store. There is no theoretical difficulty in the idea of a computer with an unlimited store. Of course only a finite part can have been used at any one time. Likewise only a finite amount can have been constructed, but we can imagine more and more being added as required. Such computers have special theoretical interest and will be called infinitive capacity computers.

The idea of a digital computer is an old one. Charles Babbage, Lucasian Professor of Mathematics at Cambridge from 1828 to 1839, planned such a machine, called the Analytical Engine, but it was never completed. Although Babbage had all the essential ideas, his machine was not at that time such a very attractive prospect. The speed which would have been available would be definitely faster than a human computer but something like 100 times slower than the Manchester machine, itself one of the slower of the modern machines. The storage was to be purely mechanical, using wheels and cards.

The fact that Babbage's Analytical Engine was to be entirely mechanical will help us to rid ourselves of a superstition. Importance is often attached to the fact that modern digital computers are electrical, and that the nervous system also is electrical. Since Babbage's machine was not electrical, and since all digital computers are in a sense equivalent, we see that this use of electricity cannot be of theoretical importance. Of course electricity usually comes in where fast signalling is concerned, so that it is not surprising that we find it in both these connections. In the nervous system chemical phenomena are at least as important as electrical. In certain computers the storage system is mainly acoustic. The feature of using electricity is thus seen to be only a very superficial similarity. If we wish to find such similarities we should look rather for mathematical analogies of function.

## 5. UNIVERSALITY OF DIGITAL COMPUTERS

The digital computers considered in the last section may be classified amongst the 'discrete state machines.' These are the machines which move by sudden jumps or clicks from one quite definite state to another. These states are sufficiently different for the possibility of confusion between them to be ignored. Strictly speaking there are no such machines. Everything really moves continuously. But there are many kinds of machine which can profitably be *thought of* as being discrete state machines. For instance in considering the switches for a lighting system it is a convenient fiction that each switch must be definitely on or definitely off. There must be intermediate positions, but for most purposes we can forget about them. As an example of a discrete state machine we might consider a wheel which clicks round through 120° once a second, but may be stopped by a lever which can be operated from outside; in addition a lamp is to light in one of the positions of the wheel. This machine could be described abstractly as follows. The internal state of the machine (which is described by the position of the wheel) may be $q_1$, $q_2$ or $q_3$. There is an input signal $i_0$ or $i_1$ (position of lever). The internal state at any moment is determined by the last state and input signal according to the table

|  |  | Last State | | |
|---|---|---|---|---|
|  |  | $q_1$ | $q_2$ | $q_3$ |
| | $i_0$ | $q_2$ | $q_3$ | $q_1$ |
| Input | | | | |
| | $i_1$ | $q_1$ | $q_2$ | $q_3$ |

The output signals, the only externally visible indication of the internal state (the light) are described by the table

| State | $q_1$ | $q_2$ | $q_3$ |
|---|---|---|---|
| Output | $o_0$ | $o_0$ | $o_1$ |

This example is typical of discrete state machines. They can be described by such tables provided they have only a finite number of possible states.

It will seem that given the initial state of the machine and the input signals it is always possible to predict all future states. This is reminiscent of Laplace's view that from the complete state of the universe at one moment of time, as described by the positions and velocities of all particles, it should be possible to predict all future states. The prediction which we are considering is, however, rather nearer to practicability than that considered by Laplace. The system of the 'universe as a whole' is such that quite small errors in the initial conditions can have an overwhelming effect at a later time. The displacement of a single electron by a billionth of a centimetre at one moment might make the difference between a man being killed by an avalanche a year later, or escaping. It is an essential property of the mechanical systems which we have called 'discrete state machines' that this phenomenon does not occur. Even when we consider the actual physical machines instead of the idealised machines, reasonably accurate knowledge of the state at one moment yields reasonably accurate knowledge any number of steps later.

As we have mentioned, digital computers fall within the class of discrete state machines. But the number of states of which such a machine is capable is usually enormously large. For instance, the number for the machine now working at Manchester is about $2^{165,000}$, i.e., about $10^{50,000}$. Compare this with our example of the clicking wheel described above, which had three states. It is not difficult to see why the number of states should be so immense. The computer includes a store corresponding to the paper used by a human computer. It must be possible to write into the store any one of the combinations of symbols which might have been written on the paper. For simplicity suppose that only digits from 0 to 9 are used as symbols. Variations in handwriting are ignored. Suppose the computer is allowed 100 sheets of paper each containing 50 lines each with room for 30 digits. Then the number of states is $10^{100 \times 50 \times 30}$ i.e., $10^{150,000}$. This is about the number of states of three Manchester machines put together. The logarithm to the base two of the number of states is usually called the 'storage capacity' of the machine. Thus the Manchester machine has a storage capacity of about 165,000 and the wheel machine of our example about 1.6. If two machines are put together their capacities must be added to obtain the capacity of the resultant machine. This leads to the possibility of statements such as 'The Manchester machine contains 64 magnetic tracks each with a capacity of 2560, eight electronic tubes with a capacity of 1280. Miscellaneous storage amounts to about 300 making a total of 174,380.'

Given the table corresponding to a discrete state machine it is possible to predict what it will do. There is no reason why this calculation should not be carried out by means of a digital computer. Provided it could be carried out sufficiently quickly the digital computer could mimic the behaviour of any discrete state machine. The imitation game could then be played with the machine in question (as B) and the mimicking digital computer (as A) and the interrogator would be unable to distinguish them. Of course the digital computer must have an adequate storage capacity as well as working sufficiently fast. Moreover, it must be programmed afresh for each new machine which it is desired to mimic.

This special property of digital computers, that they can mimic any discrete state machine, is described by saying that they are *universal* machines. The existence of machines with this property has the important consequence that, considerations of speed apart, it is unnecessary to design various new machines to do various computing processes. They can all be done with one digital computer, suitably programmed for each case. It will be seen that as a consequence of this all digital computers are in a sense equivalent.

We may now consider again the point raised at the end of § 3. It was suggested tentatively that the question, 'Can machines think?' should be replaced by 'Are there imaginable digital computers which would do well in the imitation game?' If we wish we can make this superficially more general and ask 'Are there discrete state machines which would do well?' But in view of the universality property we see that either of these questions is equivalent to this, 'Let us fix our attention on one particular digital computer $C$. Is it true that by modifying this computer to have an adequate storage, suitably increasing its speed of action, and providing it with an appropriate programme, $C$ can be made to play satisfactorily the part of A in the imitation game, the part of B being taken by a man?'

# 6. CONTRARY VIEWS ON THE MAIN QUESTION

We may now consider the ground to have been cleared and we are ready to proceed to the debate on our question, 'Can machines think?' and the variant of it quoted at the end of the last section. We cannot altogether abandon the original form of the problem, for opinions will differ as to the appropriateness of the substitution and we must at least listen to what has to be said in this connexion.

It will simplify matters for the reader if I explain first my own beliefs in the matter. Consider first the more accurate form of the question. I believe that in about fifty years' time it will be possible to programme computers, with a storage capacity of about $10^9$, to make them play the imitation game so well that an average interrogator will not have more than 70 per cent. chance of making the right identification after five minutes of questioning. The original question, 'Can machines think?' I believe to be too meaningless to deserve discussion. Nevertheless I believe that at the end of the century the use of words and general educated opinion will have altered so much that one will be able to speak of machines thinking without expecting to be contradicted. I believe further that no useful purpose is served by concealing these beliefs. The popular view that scientists proceed inexorably from well-established fact to well-established fact, never being influenced by any unproved conjecture, is quite mistaken. Provided it is made clear which are proved facts and which are conjectures, no harm can result. Conjectures are of great importance since they suggest useful lines of research.

I now proceed to consider opinions opposed to my own.

(1) *The Theological Objection.* Thinking is a function of man's immortal soul. God has given an immortal soul to every man and woman, but not to any other animal or to machines. Hence no animal or machine can think.[1]

I am unable to accept any part of this, but will attempt to reply in theological terms. I should find the argument more convincing if animals were classed with men, for there is a greater difference, to my mind, between the typical animate and the inanimate than there is between man and the other animals. The arbitrary character of the orthodox view becomes clearer if we consider how it might appear to a member of some other religious community. How do Christians regard the Moslem view that women have no souls? But let us leave this point aside and return to the main argument. It appears to me that the argument quoted above implies a serious restriction of the omnipotence of the Almighty. It is admitted that there are certain things that He cannot do such as making one equal to two, but should we not believe that He has freedom to confer a soul on an elephant if He sees fit? We might expect that He would only exercise this power in conjunction with a mutation which provided the elephant with an appropriately improved brain to minister to the needs of this soul. An argument of exactly similar form may be made for the case of machines. It may seem different because it is more difficult to "swallow." But this really only means that we think it would be less likely that He would consider the circumstances suitable for conferring a soul. The circumstances in question are discussed in the rest of this paper. In attempting to construct such machines we should not be irreverently usurping His power of creating souls, any more than we are in the procreation of children: rather we are, in either case, instruments of His will providing mansions for the souls that He creates.

However, this is mere speculation. I am not very impressed with theological arguments whatever they may be used to support. Such arguments have often been found unsatisfactory in the past. In the time of Galileo it was argued that the texts, "And the sun stood still . . . and hasted not to go down about a whole day" (Joshua x. 13) and "He laid the foundations of the earth, that it should not move at any time" (Psalm cv. 5) were an adequate refutation of the Copernican theory. With our present knowledge such an argument appears futile. When that knowledge was not available it made a quite different impression.

(2) *The 'Heads in the Sand' Objection.* "The consequences of machines thinking would be too dreadful. Let us hope and believe that they cannot do so."

This argument is seldom expressed quite so openly as in the form above. But it affects most of us who think about it at all. We like to believe that Man is in some subtle way superior to the rest of creation. It is best if he can be shown to be *necessarily* superior, for then there is no danger of him losing his commanding position. The popularity of the theological argument is clearly connected with this feeling. It is likely to be quite strong in intellectual people, since they value the power of thinking more highly than others, and are more inclined to base their belief in the superiority of Man on this power.

I do not think that this argument is sufficiently substantial to require refutation. Consolation would be more appropriate: perhaps this should be sought in the transmigration of souls.

(3) *The Mathematical Objection.* There are a number of results of mathematical logic which can be used to show that there are limitations to the powers of discrete-state machines. The best known of these results is known as Gödel's theorem, and shows that in any sufficiently powerful logical system statements can be formulated which can neither be proved nor disproved within the system, unless possibly the system itself is inconsistent. There are other, in some respects similar, results due to *Church,*[2] *Kleene, Rosser,* and *Turing.* The latter result is the most convenient to consider, since it refers directly to machines, whereas the others can only be used in a comparatively indirect argument: for instance if Gödel's theorem is to be used we need in addition to have some means of describing logical systems in terms of machines, and machines in terms of logical systems. The result in question refers to a type of machine which is essentially a digital computer with an infinite capacity. It

states that there are certain things that such a machine cannot do. If it is rigged up to give answers to questions as in the imitation game, there will be some questions to which it will either give a wrong answer, or fail to give an answer at all however much time is allowed for a reply. There may, of course, be many such questions, and questions which cannot be answered by one machine may be satisfactorily answered by another. We are of course supposing for the present that the questions are of the kind to which an answer 'Yes' or 'No' is appropriate, rather than questions such as 'What do you think of Picasso?' The questions that we know the machines must fail on are of this type, "Consider the machine specified as follows. . . . Will this machine ever answer 'Yes' to any question?" The dots are to be replaced by a description of some machine in a standard form, which could be something like that used in § 5. When the machine described bears a certain comparatively simple relation to the machine which is under interrogation, it can be shown that the answer is either wrong or not forthcoming. This is the mathematical result: it is argued that it proves a disability of machines to which the human intellect is not subject.

The short answer to this argument is that although it is established that there are limitations to the powers of any particular machine, it has only been stated, without any sort of proof, that no such limitations apply to the human intellect. But I do not think this view can be dismissed quite so lightly. Whenever one of these machines is asked the appropriate critical question, and gives a definite answer, we know that this answer must be wrong, and this gives us a certain feeling of superiority. Is this feeling illusory? It is no doubt quite genuine, but I do not think too much importance should be attached to it. We too often give wrong answers to questions ourselves to be justified in being very pleased at such evidence of fallibility on the part of the machines. Further, our superiority can only be felt on such an occasion in relation to the one machine over which we have scored our petty triumph. There would be no question of triumphing simultaneously over *all* machines. In short, then, there might be men cleverer than any given machine, but then again there might be other machines cleverer again, and so on.

Those who hold to the mathematical argument would, I think, mostly be willing to accept the imitation game as a basis for discussion. Those who believe in the two previous objections would probably not be interested in any criteria.

(4) *The Argument from Consciousness.* This argument is very well expressed in *Professor Jefferson's* Lister Oration for 1949, from which I quote. "Not until a machine can write a sonnet or compose a concerto because of thoughts and emotions felt, and not by the chance fall of symbols, could we agree that machine equals brain—that is, not only write it but know that it had written it. No mechanism could feel (and not merely artificially signal, an easy contrivance) pleasure at its successes, grief when its valves fuse, be warmed by flattery, be made miserable by its mistakes, be charmed by sex, be angry or depressed when it cannot get what it wants."

This argument appears to be a denial of the validity of our test. According to the most extreme form of this view the only way by which one could be sure that a machine thinks is to *be* the machine and to feel oneself thinking. One could then describe these feelings to the world, but of course no one would be justified in taking any notice. Likewise according to this view the only way to know that a *man* thinks is to be that particular man. It is in fact the solipsist point of view. It may be the most logical view to hold but it makes communication of ideas difficult. A is liable to believe 'A thinks but B does not' whilst B believes 'B thinks but A does not.' Instead of arguing continually over this point it is usual to have the polite convention that everyone thinks.

I am sure that Professor Jefferson does not wish to adopt the extreme and solipsist point of view. Probably he would be quite willing to accept the imitation game as a test. The game (with the player B omitted) is frequently used in practice under the name of *viva voce* to discover whether some one really understands something or has 'learnt it parrot fashion.' Let us listen in to a part of such a *viva voce*:

Interrogator: In the first line of your sonnet which reads 'Shall I compare thee to a summer's day' would not 'a spring day' do as well or better?

Witness: It wouldn't scan.

Interrogator: How about 'a winter's day.' That would scan all right.

Witness: Yes, but nobody wants to be compared to a winter's day.

Interrogator: Would you say Mr. Pickwick reminded you of Christmas?

Witness: In a way.

Interrogator: Yet Christmas is a winter's day, and I do not think Mr. Pickwick would mind the comparison.

Witness: I don't think you're serious. By a winter's day one means a typical winter's day, rather than a special one like Christmas.

And so on. What would Professor Jefferson say if the sonnet-writing machine was able to answer like this in the *viva voce*? I do not know whether he would regard the machine as 'merely artificially signalling' these answers, but if the answers were as satisfactory and sustained as in the above passage I do not think he would describe it as 'an easy contrivance.' This phrase is, I think, intended to cover such devices as the inclusion in the machine of a record of someone reading a sonnet, with appropriate switching to turn it on from time to time.

In short then, I think that most of those who support the argument from consciousness could be persuaded to abandon it rather than be forced into the solipsist position. They will then probably be willing to accept our test.

I do not wish to give the impression that I think there is no mystery about consciousness. There is, for instance, something of a paradox connected with any attempt to localise it. But I do not think these mysteries necessarily need to be solved before we can answer the question with which we are concerned in this paper.

(5) *Arguments from Various Disabilities.* These arguments take the form, "I grant you that you can make machines do all the things you have mentioned but you will never be able to make one to do X." Numerous features X are suggested in this connexion. I offer a selection:

Be kind, resourceful, beautiful, friendly (p. 2087), have initiative, have a sense of humour, tell right from wrong, make mistakes (p. 2088), fall in love, enjoy strawberries and cream (p. 2087), make some one fall in love with it, learn from experience (pp. 2094 f.), use words properly, be the subject of its own thought (p. 2088), have as much diversity of behaviour as a man, do something really new (p. 2089). (Some of these disabilities are given special consideration as indicated by the page numbers.)

No support is usually offered for these statements. I believe they are mostly founded on the principle of scientific induction. A man has seen thousands of machines in his lifetime. From what he sees of them he draws a number of general conclusions. They are ugly, each is designed for a very limited purpose, when required for a minutely different purpose they are useless, the variety of behaviour of any one of them is very small, etc., etc. Naturally he concludes that these are necessary properties of machines in general. Many of these limitations are associated with the very small storage capacity of most machines. (I am assuming that the idea of storage capacity is extended in some way to cover machines other than discrete-state machines. The exact definition does not matter as no mathematical accuracy is claimed in the present discussion.) A few years ago, when very little had been heard of digital computers, it was possible to elicit much incredulity concerning them, if one mentioned their properties without describing their construction. That was presumably due to a similar application of the principle of scientific induction. These applications of the principle are of course largely unconscious. When a burnt child fears the fire and shows that he fears it by avoiding it, I should say that he was applying scientific induction. (I could of course also describe his behaviour in many other ways.) The works and customs of mankind do not seem to be very suitable material to which to apply scientific induction. A very large part of space-time must be investigated, if reliable results are to be obtained. Otherwise we may (as most English children do) decide that everybody speaks English, and that it is silly to learn French.

There are, however, special remarks to be made about many of the disabilities that have been mentioned. The inability to enjoy strawberries and cream may have struck the reader as frivolous. Possibly a machine might be made to enjoy this delicious dish, but any attempt to make one do so would be idiotic. What is important about this disability is that it contributes to some of the other disabilities, *e.g.,* to the difficulty of the same kind of friendliness occurring between man and machine as between white man and white man, or between black man and black man.

The claim that "machines cannot make mistakes" seems a curious one. One is tempted to retort, "Are they any the worse for that?" But let us adopt a more sympathetic attitude, and try to see what is really meant. I think this criticism can be explained in terms of the imitation game. It is claimed that the interrogator could distinguish the machine from the man simply by setting them a number of problems in arithmetic. The machine would be unmasked because of its deadly accuracy. The reply to this is simple. The machine (programmed for playing the game) would not attempt to give the *right* answers to the arithmetic problems. It would deliberately introduce mistakes in a manner calculated to confuse the interrogator. A mechanical fault would probably show itself through an unsuitable decision as to what sort of a mistake to make in the arithmetic. Even this interpretation of the criticism is not sufficiently sympathetic. But we cannot afford the space to go into it much further. It seems to me that this criticism depends on a confusion between two kinds of mistake. We may call them 'errors of functioning' and 'errors of conclusion.' Errors of functioning are due to some mechanical or electrical fault which causes the machine to behave otherwise than it was designed to do. In philosophical discussions one likes to ignore the possibility of such errors; one is therefore discussing 'abstract machines.' These abstract machines are mathematical fictions rather than physical objects. By definition they are incapable of errors of functioning. In this sense we can truly say that 'machines can never make mistakes.' Errors of conclusion can only arise when some meaning is attached to the output signals from the machine. The machine might, for instance, type out mathematical equations, or sentences in English. When a false proposition is typed we say that the machine has committed an error of conclusion. There is clearly no reason at all for saying that a machine cannot make this kind of mistake. It might do nothing but type out repeatedly '0 = 1.' To take a less perverse example, it might have some method for drawing conclusions by scientific induction. We must expect such a method to lead occasionally to erroneous results.

The claim that a machine cannot be the subject of its own thought can of course only be answered if it can be shown that the machine has *some* thought with *some* subject matter. Nevertheless, 'the subject matter of a machine's operations' does seem to mean something, at least to the people who deal with it. If, for instance, the machine was trying to find a solution of the equation $x^2 - 40x - 11 = 0$ one would be tempted to describe this equation as part of the machine's subject matter at that moment. In this sort of sense a machine undoubtedly can be its own subject matter. It may be used to help in making up its own programmes, or to predict the effect of alterations in its own structure. By observing the results of its own behaviour it can modify its own programmes so as to achieve some purpose more effectively. These are possibilities of the near future, rather than Utopian dreams.

The criticism that a machine cannot have much diversity of behaviour is just a way of saying that it cannot have much storage capacity. Until fairly recently a storage capacity of even a thousand digits was very rare.

The criticisms that we are considering here are often disguised forms of the argument from consciousness. Usually if one maintains that a machine *can* do one of these things, and describes the kind of method that the machine could use, one will not make much of an impression. It is thought that the method (whatever it may be, for it must be mechanical) is really rather base. Compare the parenthesis in Jefferson's statement quoted on p. 2085.

(6) *Lady Lovelace's Objection.* Our most detailed information of Babbage's Analytical Engine comes from a memoir by *Lady Lovelace.* In it she states, "The Analytical Engine has no pretensions to *originate* anything. It can do *whatever we know how to order it* to perform" (her italics). This statement is quoted by *Hartree* who adds: "This does not imply that it may not be possible to construct electronic equipment which will 'think for itself,' or in which, in biological terms, one could set up a conditioned reflex, which would serve as a basis for 'learning.' Whether this is possible in principle or not is a stimulating and exciting question, suggested by some of these recent developments. But it did not seem that the machines constructed or projected at the time had this property."

I am in thorough agreement with Hartree over this. It will be noticed that he does not assert that the machines in question had not got the property, but rather that the evidence available to Lady Lovelace did not encourage her to believe that they had it. It is quite possible that the machines in question had in a sense got this property. For suppose that some discrete-state machine has the property. The Analytical Engine was a universal digital computer, so that, if its storage capacity and speed were adequate, it could by suitable programming be made to mimic the machine in question. Probably this argument did not occur to the Countess or to Babbage. In any case there was no obligation on them to claim all that could be claimed.

This whole question will be considered again under the heading of learning machines.

A variant of Lady Lovelace's objection states that a machine can 'never do anything really new.' This may be parried for a moment with the saw, 'There is nothing new under the sun.' Who can be certain that 'original work' that he has done was not simply the growth of the seed planted in him by teaching, or the effect of following well-known general principles. A better variant of the objection says that a machine can never 'take us by surprise.' This statement is a more direct challenge and can be met directly. Machines take me by surprise with great frequency. This is largely because I do not do sufficient calculation to decide what to expect them to do, or rather because, although I do a calculation, I do it in a hurried, slipshod fashion, taking risks. Perhaps I say to myself, 'I suppose the voltage here ought to be the same as there: anyway let's assume it is.' Naturally I am often wrong, and the result is a surprise for me for by the time the experiment is done these assumptions have been forgotten. These admissions lay me open to lectures on the subject of my vicious ways, but do not throw any doubt on my credibility when I testify to the surprises I experience.

I do not expect this reply to silence my critic. He will probably say that such surprises are due to some creative mental act on my part, and reflect no credit on the machine. This leads us back to the argument from consciousness, and far from the idea of surprise. It is a line of argument we must consider closed, but it is perhaps worth remarking that the appreciation of something as surprising requires as much of a 'creative mental act' whether the surprising event originates from a man, a book, a machine or anything else.

The view that machines cannot give rise to surprises is due, I believe, to a fallacy to which philosophers and mathematicians are particularly subject. This is the assumption that as soon as a fact is presented to a mind all consequences of that fact spring into the mind simultaneously with it. It is a very useful assumption under many circumstances, but one too easily forgets that it is false. A natural consequence of doing so is that one then assumes that there is no virtue in the mere working out of consequences from data and general principles.

(7) *Argument from Continuity in the Nervous System.* The nervous system is certainly not a discrete-state machine. A small error in the information about the size of a nervous impulse impinging on a neuron, may make a large difference to the size of the outgoing impulse. It may be argued that, this being so, one cannot expect to be able to mimic the behaviour of the nervous system with a discrete-state system.

It is true that a discrete-state machine must be different from a continuous machine. But if we adhere to the conditions of the imitation game, the interrogator will not be able to take any advantage of this difference. The situation can be made clearer if we consider some other simpler continuous machine. A differential analyser will do very well. (A differential analyser is a certain kind of machine not of the discrete-state type used for some kinds of calculation.) Some of these provide their answers in a typed form, and so are suitable for taking part in the game. It would not be possible for a digital computer to predict exactly what answers the differential analyser would give to a problem, but it would be quite capable of giving the right sort of answer. For instance, if asked to give the value of $\pi$ (actually about 3.1416) it would be reasonable to choose at random between the values 3.12, 3.13, 3.14, 3.15, 3.16 with the probabilities of 0.05, 0.15, 0.55, 0.19, 0.06 (say). Under these circumstances it would be very difficult for the interrogator to distinguish the differential analyser from the digital computer.

(8) *The Argument from Informality of Behaviour.* It is not possible to produce a set of rules purporting to describe what a man should do in every conceivable set of circumstances. One might for instance have a rule that one is to stop when one sees a red traffic light, and to go if one sees a green one, but what if by some fault both appear together? One may perhaps decide that it is safest to stop. But some further difficulty may well arise from this decision later. To attempt to provide rules of conduct to cover every eventuality, even those arising from traffic lights, appears to be impossible. With all this I agree.

From this it is argued that we cannot be machines. I shall try to reproduce the argument, but I fear I shall hardly do it justice. It seems to run something like this. 'If each man had a definite set of rules of conduct by which he regulated his life he would be no better than a machine. But there are no such rules, so men cannot be machines.' The undistributed middle is glaring. I do not think the argument is ever put quite like this, but I believe this is the argument used nevertheless. There may however be a certain confusion between 'rules of conduct' and 'laws of behaviour' to cloud the issue. By 'rules of conduct' I mean precepts such as 'Stop if you see red lights,' on which one can act, and of which one can be conscious. By 'laws of behaviour' I mean laws of nature as applied to a man's body such as 'if you pinch him he will squeak.' If we substitute 'laws of behaviour which regulate his life' for 'laws of conduct by which he regulates his life' in the argument quoted the undistributed middle is no longer insuperable. For we believe that it is not only true that being regulated by laws of behaviour implies being some sort of machine (though not necessarily a discrete-state machine), but that conversely being such a machine implies being regulated by such laws. However, we cannot so easily convince ourselves of the absence of complete laws of behaviour as of complete rules of conduct. The only way we know of for finding such laws is scientific observation, and we certainly know of no circumstances under which we could say, 'We have searched enough. There are no such laws.'

We can demonstrate more forcibly that any such statement would be unjustified. For suppose we could be sure of finding such laws if they existed. Then given a discrete-state machine it should certainly be possible to discover by observation sufficient about it to predict its future behaviour, and this within a reasonable time, say a thousand years. But this does not seem to be the case. I have set up on the Manchester computer a small programme using only 1000 units of storage, whereby the machine supplied with one sixteen figure number replies with another within two seconds. I would defy anyone to learn from these replies sufficient about the programme to be able to predict any replies to untried values.

(9) *The Argument from Extra-Sensory Perception.* I assume that the reader is familiar with the idea of extra-sensory perception, and the meaning of the four items of it, *viz.* telepathy, clairvoyance, precognition and psycho-kinesis. These disturbing phenomena seem to deny all our usual scientific ideas. How we should like to discredit them! Unfortunately the statistical evidence, at least for telepathy, is overwhelming. It is very difficult to rearrange one's ideas so as to fit these new facts in. Once one has accepted them it does not seem a very big step to believe in ghosts and bogies. The idea that our bodies move simply according to the known laws of physics, together with some others not yet discovered but somewhat similar, would be one of the first to go.

This argument is to my mind quite a strong one. One can say in reply that many scientific theories seem to remain workable in practice, in spite of clashing with E.S.P.; that in fact one can get along very nicely if one forgets about it. This is rather cold comfort, and one fears that thinking is just the kind of phenomenon where E.S.P. may be especially relevant.

A more specific argument based on E.S.P. might run as follows: "Let us play the imitation game, using as witnesses a man who is good as a telepathic receiver, and a digital computer. The interrogator can ask such questions as 'What suit does the card in my right hand belong to?' The man by telepathy or clairvoyance gives the right answer 130 times out of 400 cards. The machine can only guess at random, and perhaps gets 104 right, so the interrogator makes the right identification." There is an interesting possibility which opens here. Suppose the digital computer contains a random number generator. Then it will be natural to use this to decide what answer to give. But then the random number generator will be subject to the psycho-kinetic powers of the interrogator. Perhaps this psycho-kinesis might cause the machine to guess right more often than would be expected on a probability calculation, so that the interrogator might still be unable to make the right identification. On the other hand, he might be able to guess right without any questioning, by clairvoyance. With E.S.P. anything may happen.

If telepathy is admitted it will be necessary to tighten our test up. The situation could be regarded as analogous to that which would occur if the interrogator were talking to himself and one of the competitors was listening with his ear to the wall. To put the competitors into a 'telepathy-proof room' would satisfy all requirements.

## 7. LEARNING MACHINES

The reader will have anticipated that I have no very convincing arguments of a positive nature to support my views. If I had I should not have taken such pains to point out the fallacies in contrary views. Such evidence as I have I shall now give.

Let us return for a moment to Lady Lovelace's objection, which stated that the machine can only do what we tell it to do. One could say that a man can 'inject' an idea into the machine, and that it will respond to a certain extent and then drop into quiescence, like a piano string struck by a hammer. Another simile would be an atomic pile of less than critical size: an injected idea is to correspond to a neutron entering the pile from without. Each such neutron will cause a certain disturbance which eventually dies away. If, however, the size of the pile is sufficiently increased, the disturbance caused by such an incoming neutron will very likely go on and on increasing until the whole pile is destroyed. Is there a corresponding phenomenon for minds, and is there one for machines? There does seem to be one for the human mind. The majority of them seem to be 'sub-critical,' *i.e.,* to correspond in this analogy to piles of sub critical size. An idea presented to such a mind will on average give rise to less than one idea in reply. A smallish proportion are super-critical. An idea presented to such a mind may give rise to a whole 'theory' consisting of secondary, tertiary and more remote ideas. Animals' minds seem to be very definitely sub-critical. Adhering to this analogy we ask, 'Can a machine be made to be super-critical?'

The 'skin of an onion' analogy is also helpful. In considering the functions of the mind or the brain we find certain operations which we can explain in purely mechanical terms. This we say does not correspond to the real mind: it is a sort of skin which we must strip off if we are to find the real mind. But then in what remains we find a further skin to be stripped off, and so on. Proceeding in this way do we ever come to the 'real' mind, or do we eventually come to the skin which has nothing in it? In the latter case the whole mind is mechanical. (It would not be a discrete-state machine however. We have discussed this.)

These last two paragraphs do not claim to be convincing arguments. They should rather be described as 'recitations tending to produce belief.'

The only really satisfactory support that can be given for the view expressed at the beginning of Sec. 6, p. 2083, will be that provided by waiting for the end of the century and then doing the experiment described. But what can we say in the meantime? What steps should be taken now if the experiment is to be successful?

As I have explained, the problem is mainly one of programming. Advances in engineering will have to be made too, but it seems unlikely that these will not be adequate for the requirements. Estimates of the storage capacity of the brain vary from $10^{10}$ to $10^{15}$ binary digits. I incline to the lower values and believe that only a very small fraction is used for the higher types of thinking. Most of it is probably used for the retention of visual impressions. I should be surprised if more than $10^9$ was required for satisfactory playing of the imitation game, at any rate against a blind man. (Note—The capacity of the *Encyclopaedia Britannica,* 11th edition, is $2 \times 10^9$.) A storage capacity of $10^7$ would be a very practicable possibility even by present techniques. It is probably not necessary to increase the speed of operations of the machines at all. Parts of modern machines which can be regarded as analogues of nerve cells work about a thousand times faster than the latter. This should provide a 'margin of safety' which could cover losses of speed arising in many ways. Our problem then is to find out how to programme these machines to play the game. At my present rate of working I produce about a thousand digits of programme a day, so that about sixty workers, working steadily through the fifty years might accomplish the job, if nothing went into the waste-paper basket. Some more expeditious method seems desirable.

In the process of trying to imitate an adult human mind we are bound to think a good deal about the process which has brought it to the state that it is in. We may notice three components,

(*a*) The initial state of the mind, say at birth,

(*b*) The education to which it has been subjected,

(*c*) Other experience, not to be described as education, to which it has been subjected.

Instead of trying to produce a programme to simulate the adult mind, why not rather try to produce one which simulates the child's? If this were then subjected to an appropriate course of education one would obtain the adult brain. Presumably the child-brain is something like a note-book as one buys it from the stationers. Rather little mechanism, and lots of blank sheets. (Mechanism and writing are from our point of view almost synonymous.) Our hope is that there is so little mechanism in the child-brain that something like it can be easily programmed. The amount of work in the education we can assume, as a first approximation, to be much the same as for the human child.

We have thus divided our problem into two parts. The child-programme and the education process. These two remain very closely connected. We cannot expect to find a good child-machine at the first attempt. One must experiment with teaching one such machine and see how well it learns. One can then try another and see if it is better or worse. There is an obvious connection between this process and evolution, by the identifications

| Structure of the child machine | = Hereditary material |
| Changes of the child machine | = Mutations |
| Natural selection | = Judgment of the experimenter |

One may hope, however, that this process will be more expeditious than evolution. The survival of the fittest is a slow method for measuring advantages. The experimenter, by the exercise of intelligence, should be able to speed it up. Equally important is the fact that he is not restricted to random mutations. If he can trace a cause for some weakness he can probably think of the kind of mutation which will improve it.

It will not be possible to apply exactly the same teaching process to the machine as to a normal child. It will not, for instance, be provided with legs, so that it could not be asked to go out and fill the coal scuttle. Possibly it might not have eyes. But however well these deficiencies might be overcome by clever engineering, one could not send the creature to school without the other children making excessive fun of it. It must be given some tuition. We need not be too concerned about the legs, eyes, etc. The example of Miss *Helen Keller* shows that education can take place provided that communication in both directions between teacher and pupil can take place by some means or other.

We normally associate punishments and rewards with the teaching process. Some simple child-machines can be constructed or programmed on this sort of principle. The machine has to be so constructed that events which shortly preceded the occurrence of a punishment-signal are unlikely to be repeated, whereas a reward-signal increased the probability of repetition of the events which led up to it. These definitions do not presuppose any feelings on the part of the machine. I have done some experiments with one such child-machine, and succeeded in teaching it a few things, but the teaching method was too unorthodox for the experiment to be considered really successful.

The use of punishments and rewards can at best be a part of the teaching process. Roughly speaking, if the teacher has no other means of communicating to the pupil, the amount of information which can reach him does not exceed the total number of rewards and punishments applied. By the time a child has learnt to repeat 'Casabianca' he would probably feel very sore indeed, if the text could only be discovered by a 'Twenty Questions' technique, every 'NO' taking the form of a blow. It is necessary therefore to have some other 'unemotional' channels of communication. If these are available it is possible to teach a machine by punishments and rewards to obey orders given in some language, *e.g.*, a symbolic language. These orders are to be transmitted through the 'unemotional' channels. The use of this language will diminish greatly the number of punishments and rewards required.

Opinions may vary as to the complexity which is suitable in the child machine. One might try to make it as simple as possible consistently with the general principles. Alternatively one might have a complete system of logical inference 'built in.'[3] In the latter case the store would be largely occupied with definitions and propositions. The propositions would have various kinds of status, *e.g.*, well-established facts, conjectures, mathematically proved theorems, statements given by an authority, expressions having the logical form of proposition but not belief-value. Certain propositions may be described as 'imperatives.' The machine should be so constructed that as soon as an imperative is classed as 'well-established' the appropriate action automatically takes place. To illustrate this, suppose the teacher says to the machine, 'Do your homework now.' This may cause "Teacher says 'Do your homework now'" to be included amongst the well-established facts. Another such fact might be, "Everything that teacher says is true." Combining these may eventually lead to the imperative, 'Do your homework now,' being included amongst the well-established facts, and this, by the construction of the machine, will mean that the homework actually gets started, but the effect is very satisfactory. The processes of inference used by the machine need not be such as would satisfy the most exacting logicians. There might for instance be no hierarchy of types. But this need not mean that type fallacies will occur, any more than we are bound to fall over unfenced cliffs. Suitable imperatives (expressed *within* the systems, not forming part of the rules *of* the system) such as 'Do not use a class unless it is a subclass of one which has been mentioned by teacher' can have a similar effect to 'Do not go too near the edge.'

The imperatives that can be obeyed by a machine that has no limbs are bound to be of a rather intellectual character, as in the example (doing homework) given above. Important amongst such imperatives will be ones which regulate the order in which the rules of the logical system concerned are to be applied.

For at each stage when one is using a logical system, there is a very large number of alternative steps, any of which one is permitted to apply, so far as obedience to the rules of the logical system is concerned. These choices make the difference between a brilliant and a footling reasoner, not the difference between a sound and a fallacious one. Propositions leading to imperatives of this kind might be "When Socrates is mentioned, use the syllogism in Barbara" or "If one method has been proved to be quicker than another, do not use the slower method." Some of these may be 'given by authority,' but others may be produced by the machine itself, *e.g.*, by scientific induction.

The idea of a learning machine may appear paradoxical to some readers. How can the rules of operation of the machine change? They should describe completely how the machine will react whatever its history might be, whatever changes it might undergo. The rules are thus quite time-invariant. This is quite true. The explanation of the paradox is that the rules which get changed in the learning process are of a rather less pretentious kind, claiming only an ephemeral validity. The reader may draw a parallel with the Constitution of the United States.

An important feature of a learning machine is that its teacher will often be very largely ignorant of quite what is going on inside, although he may still be able to some extent to predict his pupil's behaviour. This should apply most strongly to the later education of a machine arising from a child-machine of well-tried design (or programme). This is in clear contrast with normal procedure when using a machine to do computations: one's object is then to have a clear mental picture of the state of the machine at each moment in the computation. This object can only be achieved with a struggle. The view that 'the machine can only do what we know how to order it to do,'[4] appears strange in face of this. Most of the programmes which we can put into the machine will result in its doing something that we cannot make sense of at all, or which we regard as completely random behaviour. Intelligent behaviour presumably consists in a departure from the completely disciplined behaviour involved in computation, but a rather slight one, which does not give rise to random behaviour, or to pointless repetitive loops. Another important result of preparing our machine for its part in the imitation game by a process of teaching and learning is that 'human fallibility' is likely to be omitted in a rather natural way, *i.e.*, without special 'coaching.' (The reader should reconcile this with the point of view on pp. 2087–88.) Processes that are learnt do not produce a hundred per cent. certainty of result; if they did they could not be unlearnt.

It is probably wise to include a random element in a learning machine (see pp. 2079–80). A random element is rather useful when we are searching for a solution of some problem. Suppose for instance we wanted to find a number between 50 and 200 which was equal to the square of the sum of its digits, we might start at 51 then try 52 and go on until we got a number that worked. Alternatively we might choose numbers at random until we got a good one.

This method has the advantage that it is unnecessary to keep track of the values that have been tried, but the disadvantage that one may try the same one twice, but this is not very important if there are several solutions. The systematic method has the disadvantage that there may be an enormous block without any solutions in the region, which has to be investigated first. Now the learning process may be regarded as a search for a form of behaviour which will satisfy the teacher (or some other criterion). Since there is probably a very large number of satisfactory solutions the random method seems to be better than the systematic. It should be noticed that it is used in the analogous process of evolution. But there the systematic method is not possible. How could one keep track of the different genetical combinations that had been tried, so as to avoid trying them again?

We may hope that machines will eventually compete with men in all purely intellectual fields. But which are the best ones to start with? Even this is a difficult decision. Many people think that a very abstract activity, like the playing of chess, would be best. It can also be maintained that it is best to provide the machine with the best sense organs that money can buy, and then teach it to understand and speak English. This process could follow the normal teaching of a child. Things would be pointed out and named, etc. Again I do not know what the right answer is, but I think both approaches should be tried.

We can only see a short distance ahead, but we can see plenty there that needs to be done.

## BIBLIOGRAPHY

Samuel Butler, *Erewhon*, London, 1865. Chapters 23, 24, 25, *The Book of the Machines.*

Alonzo Church, "An Unsolvable Problem of Elementary Number Theory," *American J. of Math.*, 58 (1936), 345–363.

K. Gödel, "Über formal unentscheidbare Sätze der Principia Mathematica und verwandter Systeme, I," *Monatshefte für Math. und Phys.* (1931), 173–189.

D. R. Hartree, *Calculating Instruments and Machines*, New York, 1949. S. C. Kleene, "General Recursive Functions of Natural Numbers," *American J. of Math.*, 57 (1935), 153–173 and 219–244.

G. Jefferson, "The Mind of Mechanical Man." Lister oration for 1949. *British Medical Journal*, vol. i (1949), 1105–1121.

Countess of Lovelace, "Translator's notes to an article on Babbage's Analytical Engine," *Scientific Memoirs* (ed. by R. Taylor), vol. 3 (1842), 691–731.

Bertrand Russell, *History of Western Philosophy*, London, 1940.

A. M. Turing, "On Computable Numbers, with an Application to the Entscheidungsproblem," *Proc. London Math. Soc.* (2), 42 (1937), 230–265.

## ENDNOTES

[1]Possibly this view is heretical. St. Thomas Aquinas (*Summa Theologica*, quoted by Bertrand Russell, *A History of Western Philosophy*, Simon and Schuster, New York, 1945, p. 458) states that God cannot make a man to have no soul. But this may not be a real restriction on His powers, but only a result of the fact that men's souls are immortal, and therefore indestructible.

[2]Authors' names in italics refer to Bibliography. (*See above.*)

[3]Or rather 'programmed in' for our child-machine will be programmed in a digital computer. But the logical system will not have to be learnt.

[4]Compare Lady Lovelace's statement (p. 2089), which does not contain the word 'only.'

3

*You're not a man, you're a machine.*

—GEORGE BERNARD SHAW
(*Arms and the Man*)

*Thinking makes it so.*

—SHAKESPEARE (*Hamlet*)

*Things are in the saddle and ride mankind.*

—RALPH WALDO EMERSON

# A CHESS-PLAYING MACHINE

## By Claude E. Shannon

For centuries philosophers and scientists have speculated about whether or not the human brain is essentially a machine. Could a machine be designed that would be capable of "thinking"? During the past decade several large-scale electronic computing machines have been constructed which are capable of something very close to the reasoning process. These new computers were designed primarily to carry out purely numerical calculations. They perform automatically a long sequence of additions, multiplications and other arithmetic operations at a rate of thousands per second. The basic design of these machines is so general and flexible, however, that they can be adapted to work symbolically with elements representing words, propositions or other conceptual entities.

One such possibility, which is already being investigated in several quarters, is that of translating from one language to another by means of a computer. The immediate goal is not a finished literary rendition, but only a word-by-word translation that would convey enough of the meaning to be understandable. Computing machines could also be employed for many other tasks of a semi-rote, semi-thinking character, such as designing electrical filters and relay circuits, helping to regulate airplane traffic at busy airports, and routing long-distance telephone calls most efficiently over a limited number of trunks.

Some of the possibilities in this direction can be illustrated by setting up a computer in such a way that it will play a fair game of chess. This problem, of course, is of no importance in itself, but it was undertaken with a serious purpose in mind. The investigation of the chess-playing problem is intended to develop techniques that can be used for more practical applications.

The chess machine is an ideal one to start with for several reasons. The problem is sharply defined, both in the allowed operations (the moves of chess) and in the ultimate goal (checkmate). It is neither so simple as to be trivial nor too difficult for satisfactory solution. And such a machine could be pitted against a human opponent, giving a clear measure of the machine's ability in this type of reasoning.

There is already a considerable literature on the subject of chess-playing machines. During the late 18th and early 19th centuries a Hungarian inventor named Wolfgang von Kempelen astounded Europe with a device known as the Maelzel Chess Automaton, which toured the Continent to large audiences. A number of papers purporting to explain its operation, including an analytical essay by Edgar Allan Poe, soon appeared. Most of the analysts concluded, quite correctly, that the automaton was operated by a human chess master concealed inside. Some years later the exact manner of operation was exposed (see Figure 1).

**FIGURE 1**
*Chess machine of the 18th century was actually run by man inside.*

A more honest attempt to design a chess-playing machine was made in 1914 by a Spanish inventor named L. Torres y Quevedo, who constructed a device that played an end game of king and rook against king. The machine, playing the side with king and rook, would force checkmate in a few moves however its human opponent played. Since an explicit set of rules can be given for making satisfactory moves in such an end game, the problem is relatively simple, but the idea was quite advanced for that period.

An electronic computer can be set up to play a complete game. In order to explain the actual setup of a chess machine, it may be best to start with a general picture of a computer and its operation.

A general-purpose electronic computer is an extremely complicated device containing several thousand vacuum tubes, relays and other elements. The basic principles involved, however, are quite simple. The machine has four main parts: (1) an "arithmetic organ," (2) a control element, (3) a numerical memory and (4) a program memory. (In some designs the two memory functions are carried out in the same physical apparatus.) The manner of operation is exactly analogous to a human computer carrying out a series of numerical calculations with an ordinary desk computing machine. The arithmetic organ corresponds to the desk computing machine, the control element to the human operator, the numerical memory to the work sheet on which intermediate and final results are recorded, and the program memory to the computing routine describing the series of operations to be performed.

In an electronic computing machine, the numerical memory consists of a large number of "boxes," each capable of holding a number. To set up a problem on the computer, it is necessary to assign box numbers to all numerical quantities involved, and then to construct a program telling the machine what arithmetical operations must be performed on the numbers and where the results should go. The program consists of a sequence of "orders," each describing an elementary calculation. For example, a typical order may read A 372, 451, 133. This means: add the number stored in box 372 to that in box 451, and put the sum in box 133. Another type of order requires the machine to make a decision. For example, the order C 291, 118, 345 tells the machine to compare the contents of boxes 291 and 118; if the number in box 291 is larger, the machine goes on to the next order in the program; if not, it takes its next order from box 345. This type of order enables the machine to choose from alternative procedures, depending on the results of previous calculations. The "vocabulary" of an electronic computer may include as many as 30 different types of orders.

After the machine is provided with a program, the initial numbers required for the calculation are placed in the numerical memory and the machine then automatically carries out the computation. Of course such a machine is most useful in problems involving an enormous number of individual calculations, which would be too laborious to carry out by hand.

The problem of setting up a computer for playing chess can be divided into three parts: first, a code must be chosen so that chess positions and the chess pieces can be represented as numbers; second, a strategy must be found for choosing the moves to be made; and third, this strategy must be translated into a sequence of elementary computer orders, or a program.

A suitable code for the chessboard and the chess pieces is shown in Figure 2. Each square on the board has a number consisting of two digits, the first digit corresponding to the "rank" or horizontal row, the second to the "file" or vertical row. Each different chess piece also is designated by a number: a pawn is numbered 1, a knight 2, a bishop 3, a rook 4 and so on. White pieces are represented by positive numbers and black pieces by negative ones. The positions of all the pieces on the board can be shown by a sequence of 64 numbers, with zeros to indicate the empty squares. Thus any chess position can be recorded as a series of numbers and stored in the numerical memory of a computing machine.

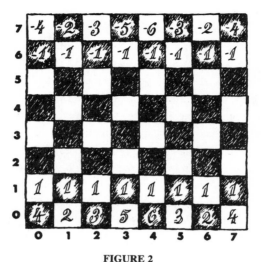

**FIGURE 2**

*Code for a chess-playing machine is plotted on a chessboard. Each square can be designated by two digits, one representing the horizontal row and the other the vertical. Pieces also are coded in numbers.*

A chess move is specified by giving the number of the square on which the piece stands and of the one to which it is moved. Ordinarily two numbers would be sufficient to describe a move, but to take care of the special case of the promotion of a pawn to a higher piece a third number is necessary. This number indicates the piece to which the pawn is converted. In all other moves the third number is zero. Thus a knight move from square 01 to 22 is encoded into 01, 22, 0. The move of a pawn from 62 to 72, and its promotion to a queen, is represented by 62, 72, 5.

The second main problem is that of deciding on a strategy of play. A straightforward process must be found for calculating a reasonably good move for any given chess position. This is the most difficult part of the problem. The program designer can employ here the principles of correct play that have been evolved by expert chess players. These empirical principles are a means of bringing some order to the maze of possible variations of a chess game. Even the high speeds available in electronic computers are hopelessly inadequate to play perfect chess by calculating all possible variations to the end of the game. In a typical chess position there will be about 32 possible moves with 32 possible replies—already this creates 1,024 possibilities. Most chess games last 40 moves or more for each side. So the total number of possible variations in an average game is about $10^{120}$. A machine calculating one variation each millionth of a second would require over $10^{95}$ years to decide on its first move! Other methods of attempting to play perfect chess seem equally impracticable; we resign ourselves, therefore, to having the machine play a reasonably skillful game, admitting occasional moves that may not be the best. This, of course, is precisely what human players do: no one plays a perfect game.

In setting up a strategy on the machine one must establish a method of numerical evaluation for any given chess position. A chess player looking at a position can form an estimate as to which side, White or Black, has the advantage. Furthermore, his evaluation is roughly quantitative. He may say, "White has a rook for a bishop, an advantage of about two pawns"; or "Black has sufficient mobility to compensate for a sacrificed pawn." These judgments are based on long experience and are summarized in the principles of chess expounded in chess literature. For example, it has been found that a queen is worth nine pawns, a rook is worth five, and a bishop or a knight is worth about three. As a first rough approximation, a position can be evaluated by merely adding up the total forces for each side, measured in terms of the pawn unit. There are, however, numerous other features which must be taken into account: the mobility and placement of pieces, the weakness of king protection, the nature of the pawn formation, and so on. These too can be given numerical weights and combined in the evaluation, and it is here that the knowledge and experience of chess masters must be enlisted.

Assuming that a suitable method of position evaluation has been decided upon, how should a move be selected? The simplest process is to consider all the possible moves in the given position and choose the one that gives the best immediate evaluation. Since, however, chess players generally look more than one move ahead, one must take account of the opponent's various possible responses to each projected move. Assuming that the opponent's reply will be the one giving the best evaluation from his point of view, we would choose the move that would leave us as well off as possible after his best reply. Unfortunately, with the computer speeds at present available, the machine could not explore all the possibilities for more than two moves ahead for each side, so a strategy of this type would play a poor game by human standards. Good chess players frequently play combinations four or five moves deep, and occasionally world champions have seen as many as 20 moves ahead. This is possible only because the variations they consider are highly selected. They do not investigate all lines of play, but only the important ones.

The amount of selection exercised by chess masters in examining possible variations has been studied experimentally by the Dutch chess master and psychologist A. D. De Groot. He showed various typical positions to chess masters and asked them to decide on the best move, describing aloud their analyses of the positions as they thought them through. By this procedure the number and depth of the variations examined could be determined. In one typical case a chess master examined 16 variations, ranging in depth from one Black move to five Black and four White moves. The total number of positions considered was 44.

Clearly it would be highly desirable to improve the strategy for the machine by including such a selection process in it. Of course one could go too far in this direction. Investigating one particular line of play for 40 moves would be as bad as investigating all lines for just two moves. A suitable compromise would be to examine only the important possible variations—that is, forcing moves, captures and main threats—and carry out the investigation of the possible moves far enough to make the consequences of each fairly clear. It is possible to set up some rough criteria for selecting important variations, not as efficiently as a chess master, but sufficiently well to reduce the number of variations appreciably and thereby permit a deeper investigation of the moves actually considered.

The final problem is that of reducing the strategy to a sequence of orders, translated into the machine's language. This is a relatively straightforward but tedious process, and we shall only indicate some of the general features. The complete program is made up of nine sub-programs and a master program that calls the sub-programs into operation as needed. Six of the sub-programs deal with the movements of the various kinds of pieces. In effect they tell the machine the allowed moves for these pieces. Another sub-program enables the machine to make a move "mentally" without actually carrying it out: that is, with a given position stored in its memory it can construct the position that would result if the move were made. The seventh sub-program enables the computer to make a list of all possible moves in a given position, and the last sub-program evaluates any given position. The master program correlates and supervises the application of the sub-programs. It starts the seventh sub-program making a list of possible moves, which in turn calls in previous sub-programs to determine where the various pieces could move. The master program then evaluates the resulting positions by means of the eighth sub-program and compares the results according to the process described above. After comparison of all the investigated variations, the one that gives the best evaluation according to the machine's calculations is selected. This move is translated into standard chess notation and typed out by the machine.

It is believed that an electronic computer programmed in this manner would play a fairly strong game at speeds comparable to human speeds. A machine has several obvious advantages over a human player: (1) it can make individual calculations with much greater speed; (2) its play is free of errors other than those due to deficiencies of the program, whereas human players often make very simple and obvious blunders; (3) it is free from laziness, or the temptation to make an instinctive move without proper analysis of the position; (4) it is free from "nerves," so it will make no blunders due to overconfidence or defeatism. Against these advantages, however, must be weighed the flexibility, imagination and learning capacity of the human mind.

Under some circumstances the machine might well defeat the program designer. In one sense, the designer can surely outplay his machine; knowing

**FIGURE 3**

*Inevitable advantage of man over the machine is illustrated in this drawing. At top human player loses to machine. In center nettled human player revises machine's instructions. At bottom human player wins.*

the strategy used by the machine, he can apply the same tactics at a deeper level. But he would require several weeks to calculate a move, while the machine uses only a few minutes. On an equal time basis, the speed, patience and deadly accuracy of the machine would be telling against human fallibility. Sufficiently nettled, however, the designer could easily weaken the playing skill of the machine by changing the program in such a way as to reduce the depth of investigation (see Figure 3). This idea was expressed by a cartoon in *The Saturday Evening Post* a while ago.

As described so far, the machine would always make the same move in the same position. If the opponent made the same moves, this would always lead to the same game. Once the opponent won a game, he could win every time thereafter by playing the same strategy, taking advantage of some particular position in which the machine chooses a weak move. One way to vary the machine's play would be to introduce a statistical element. Whenever it was confronted with two or more possible moves that were about equally good according to the machine's calculations, it would choose from them at random. Thus if it arrived at the same position a second time it might choose a different move.

Another place where statistical variation could be introduced is in the opening game. It would be desirable to have a number of standard openings, perhaps a few hundred, stored in the memory of the machine. For the first few moves, until the opponent deviated from the standard responses or the machine reached the end of the stored sequence of moves, the machine would play by memory. This could hardly be considered cheating, since that is the way chess masters play the opening.

We may note that within its limits a machine of this type will play a brilliant game. It will readily make spectacular sacrifices of important pieces in order to gain a later advantage or to give checkmate, provided the completion of the combination occurs within its computing limits. For example, in the position illustrated in Figure 4 the machine would quickly discover the sacrificial mate in three moves:

| White | Black |
|---|---|
| 1. R-K8 Ch | R X R |
| 2. Q-Kt4 Ch | Q X Q |
| 3. Kt-B6 Mate | |

Winning combinations of this type are frequently overlooked in amateur play.

The chief weakness of the machine is that it will not learn by its mistakes. The only way to improve its play is by improving the program. Some thought has been given to designing a program that would develop its own improvements in strategy with increasing experience in play. Although it appears to be theoretically possible, the methods thought of so far do not seem to be very practical. One possibility is to devise a program that would change the terms and coefficients involved in the evaluation function on the basis of the results of games the machine had already played. Small variations might be introduced in these terms, and the values would be selected to give the greatest percentage of wins.

**FIGURE 4**

*Problem that the machine could solve brilliantly might begin with this chess position. The machine would sacrifice a rook and a queen, the most powerful piece on the board, and then win in only one more move.*

The Gordian question, more easily raised than answered is: Does a chess-playing machine of this type "think"? The answer depends entirely on how we define thinking. Since there is no general agreement as to the precise connotation of this word, the question has no definite answer. From a behavioristic point of view, the machine acts as though it were thinking. It has always been considered that skillful chess play requires the reasoning faculty. If we regard thinking as a property of external actions rather than internal method the machine is surely thinking.

The thinking process is considered by some psychologists to be essentially characterized by the following steps: various possible solutions of a problem are tried out mentally or symbolically without actually being carried out physically; the best solution is selected by a mental evaluation of the results of these trials; and the solution found in this way is then acted upon. It will be seen that this is almost an exact description of how a chess-playing computer operates, provided we substitute "within the machine" for "mentally."

On the other hand, the machine does only what it has been told to do. It works by trial and error, but the trials are trials that the program designer ordered the machine to make, and the errors are called errors because the evaluation function gives these variations low ratings. The machine makes decisions, but the decisions were envisaged and provided for at the time of design. In short, the machine does not, in any real sense, go beyond what was built into it. The situation was nicely summarized by Torres y Quevedo, who, in connection with his end-game machine, remarked: "The limits within which thought is really necessary need to be better defined . . . the automaton can do many things that are popularly classed as thought."

# PART XX

# MATHEMATICS IN WARFARE

# COMMENTARY ON FREDERICK WILLIAM LANCHESTER

F rederick William Lanchester, an Englishman who died in 1946 at the age of 78, was interested, among other things, in aerodynamics, economic and industrial problems, the theory of relativity, fiscal policies and military strategy. His writings on these matters, apart from high professional competence, exhibit such striking independence of judgment and boldness of conception that it is surprising to learn he was an engineer.[1] Lanchester was one of the first to recognize the extent to which aircraft would alter the character of warfare. Nebulous profundities had of course been uttered on the subject since Biblical times and even military men—the more advanced thinkers among them—were aware by the outbreak of the First World War that the airplane would change some of their business methods. It was Lanchester, however, who first considered the matter quantitatively. He set down his conclusions on the subject in *Aircraft in Warfare* (1916), a book consisting mainly of a series of articles contributed in 1914 to the British journal *Engineering*. Lanchester was convinced that most of the important operations hitherto entrusted to land armies could be executed "as well or better by a squad or fleet of aeronautical machines. If this should prove true, the number of flying machines eventually to be utilized by any of the great military powers will be counted not by hundreds but by thousands, and possibly by tens of thousands, and the issue of any great battle will be definitely determined by the efficiency of the aeronautical forces."

To prove his Point, Lanchester found it necessary to make a mathematical analysis of the relation of opposing forces in battle. Under what circumstances can a smaller army (or naval fleet) defeat a larger? Can a mathematical measure be assigned to concentrations of firepower and, if so, can equations in which such measures appear be set up to describe what happens and what may be expected to happen in military engagements? These were among the questions he considered and for which he devised the elegant Pythagorean formula described below. His N-square law of the relative fighting strength of two armies is simple, but its implications are not. Scientists engaged on operational research have done a considerable amount of mathematical work to draw some of the consequences from Lanchester's equations; his equations are not recognizable in these later formidable elaborations. But then today's wars have become so elaborate that Mars himself would not recognize them and it was inevitable that mathematics would have to keep up.

## ENDNOTE

[1]"Lanchester made a brilliant analysis of the inherent stability of model airplanes in 1897, long before there were real airplanes. His work was a little like a treatise on the dynamics of the automobile before any automobile existed. The Physical Society of London declined to print this paper, but some thirty years later Lanchester was awarded a gold medal for it by the Royal Aeronautical Society." Jerome C. Hunsaker, *Aeronautics at the Mid-Century* (Yale University Press, 1952). Lanchester was also one of the foremost pioneers of automobile design. He built an experimental engine in 1895–probably the first to be made in England. The Lanchester automobile was put into production in 1900. It was an outstanding vehicle of the vintage period, incorporating many unorthodox and advanced features.

1

*"If you look up 'Intelligence' in the new volumes of the Encyclo-
paedia Britannica," he had said, "you'll find it classified under
the following three heads: Intelligence, Human; Intelligence,
Animal; Intelligence, Military. My stepfather's a perfect specimen
of Intelligence, Military."*

—ALDOUS HUXLEY (*Point Counter Point*)

*. . . a science is said to be useful if its development tends to
accentuate the existing inequalities in the distribution of wealth,
or more directly promotes the destruction of human life.*

—G. H. HARDY

# MATHEMATICS IN WARFARE

## By Frederick William Lanchester

### THE PRINCIPLE OF CONCENTRATION.
### THE "N-SQUARE" LAW.

*The Principle of Concentration.* It is necessary at the present juncture to make a
digression and to treat of certain fundamental considerations which underlie the
whole science and practice of warfare in all its branches. One of the great
questions at the root of all strategy is that of *concentration*; the concentration of
the whole resources of a belligerent on a single purpose or object, and concur-
rently the concentration of the main strength of his forces, whether naval or
military, at one point in the field of operations. But the principle of concentration is not in itself a strategic principle; it applies with equal effect to purely
tactical operations; it is on its material side based upon facts of a purely
scientific character. The subject is somewhat befogged by many authors of
repute, inasmuch as the two distinct sides—the moral concentration (the nar-
rowing and fixity of purpose) and the material concentration—are both included under one general heading, and one is invited to believe that there is
some peculiar virtue in the word *concentration,* like the "blessed word Mesopotamia," whereas the truth is that the word in its two applications refers to two
entirely independent conceptions, whose underlying principles have nothing
really in common.

The importance of concentration in the material sense is based on certain
elementary principles connected with the means of attack and defence, and if
we are properly to appreciate the value and importance of concentration in this
sense, we must not fix our attention too closely upon the bare fact of concentration, but rather upon the underlying principles, and seek a more solid foundation in the study of the controlling factors.

*The Conditions of Ancient and Modern Warfare Contrasted.* There is an important difference between the methods of defence of primitive times and those of the present day which may be used to illustrate the point at issue. In olden times, when weapon directly answered weapon, the act of defence was positive and direct, the blow of sword or battleaxe was parried by sword and shield; under modern conditions gun answers gun, the defence from rifle-fire is rifle-fire, and the defence from artillery, artillery. But the defence of modern arms is indirect: tersely, the enemy is prevented from killing you by your killing him first, and the fighting is essentially collective. As a consequence of this difference, the importance of concentration in history has been by no means a constant quantity. Under the old conditions it was not possible by any strategic plan or tactical manoeuvre to bring other than approximately equal numbers of men into the actual fighting line; one man would ordinarily find himself opposed to one man. Even were a General to concentrate twice the number of men on any given portion of the field to that of the enemy, the number of men actually wielding their weapons at any given instant (so long as the fighting line was unbroken), was, roughly speaking, the same on both sides. Under present-day conditions all this is changed. With modern long-range weapons—fire-arms, in brief—the concentration of superior numbers gives an immediate superiority in the active combatant ranks, and the numerically inferior force finds itself under a far heavier fire, man for man, than it is able to return. The importance of this difference is greater than might casually be supposed, and, since it contains the kernel of the whole question, it will be examined in detail.

In thus contrasting the ancient conditions with the modern, it is not intended to suggest that the advantages of concentration did not, to some extent, exist under the old order of things. For example, when an army broke and fled, undoubtedly any numerical superiority of the victor could be used with telling effect, and, before this, pressure, as distinct from blows, would exercise great influence. Also the bow and arrow and the cross-bow were weapons that possessed in a lesser degree the properties of fire-arms, inasmuch as they enabled numbers (within limits) to concentrate their attack on the few. As here discussed, the conditions are contrasted in their most accentuated form as extremes for the purpose of illustration.

Taking, first, the ancient conditions where man is opposed to man, then, assuming the combatants to be of equal fighting value, and other conditions equal, clearly, on an average, as many of the "duels" that go to make up the whole fight will go one way as the other, and there will be about equal numbers killed of the forces engaged; so that if 1,000 men meet 1,000 men, it is of little or no importance whether a "Blue" force of 1,000 men meet a "Red" force of 1,000 men in a single pitched battle, or whether the whole "Blue" force concentrates on 500 of the "Red" force, and, having annihilated them, turns its attention to the other half; there will, presuming the "Reds" stand their ground

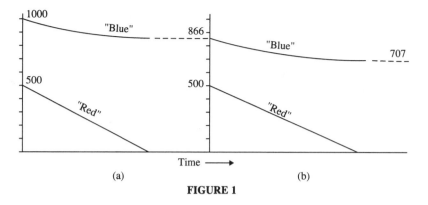

Time ⟶
(a)                                   (b)
FIGURE 1

to the last, be half the "Blue" force wiped out in the annihilation of the "Red" force[1] in the first battle, and the second battle will start on terms of equality— *i.e.,* 500 "Blue" against 500 "Red."

*Modern Conditions Investigated.* Now let us take the modern conditions. If, again, we assume equal individual fighting value, and the combatants otherwise (as to "cover," etc.) on terms of equality, each man will in a given time score, on an average, a certain number of hits that are effective; consequently, the number of men knocked out per unit time will be directly proportional to the numerical strength of the opposing force. Putting this in mathematical language, and employing symbol $b$ to represent the numerical strength of the "Blue" force, and $r$ for the "Red," we have:—

$$\frac{db}{dt} = -r \times c \ldots . (1)$$

and

$$\frac{dr}{dt} = -b \times k \ldots . (2)$$

in which $t$ is time and $c$ and $k$ are constants ($c = k$ if the fighting values of the individual units of the force are equal).

The reduction of strength of the two forces may be represented by two conjugate curves following the above equations. In Figure 1 (*a*) graphs are given representing the case of the "Blue" force 1,000 strong encountering a section of the "Red" force 500 strong, and it will be seen that the "Red" force is wiped out of existence with a loss of only about 134 men of the "Blue" force, leaving 866 to meet the remaining 500 of the "Red" force with an easy and decisive victory; this is shown in Figure 1 (b), the victorious "Blues" having annihilated the whole "Red" force of equal total strength with a loss of only 293 men.

In Figure 2*a* a case is given in which the "Red" force is inferior to the "Blue" in the relation $1:\sqrt{2}$ say, a "Red" force 1,000 strong meeting a "Blue"

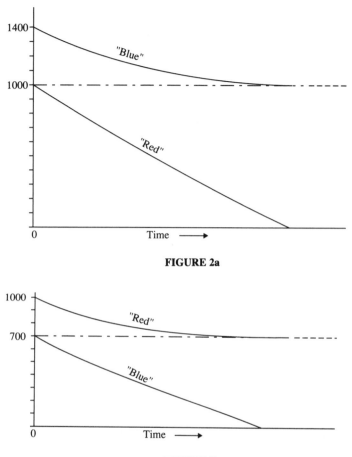

**FIGURE 2a**

**FIGURE 2b**

force 1,400 strong. Assuming they meet in a single pitched battle fought to a conclusion, the upper line will represent the "Blue" force, and it is seen that the "Reds" will be annihilated, the "Blues" losing only 400 men. If, on the other hand, the "Reds" by superior strategy compel the "Blues" to give battle divided—say into two equal armies—then, Figure 2*b*, in the first battle the 700 "Blues" will be annihilated with a loss of only 300 to the "Reds" and in the second battle the two armies will meet on an equal numerical footing, and so we may presume the final battle of the campaign as drawn. In this second case the result of the second battle is presumed from the initial equality of the forces; the curves are not given.

In the case of equal forces the two conjugate curves become coincident; there is a single curve of logarithmic form, Figure 3; the battle is prolonged indefinitely. Since the forces actually consist of a finite number of finite units (instead of an infinite number of infinitesimal units), the end of the curve must show discontinuity, and break off abruptly when the last man is reached; the

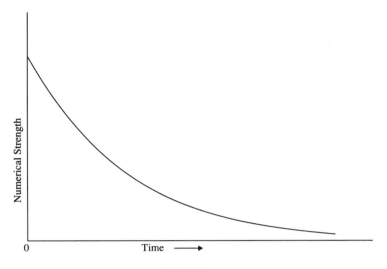

**FIGURE 3**

law based on averages evidently does not hold rigidly when the numbers become small. Beyond this, the condition of two equal curves is unstable, and any advantage secured by either side will tend to augment.

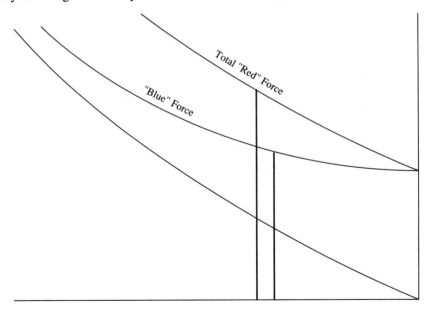

**FIGURE 4a**

*Graph representing Weakness of a Divided Force.* In Figure 4a, a pair of conjugate curves have been plotted backwards from the vertical datum representing the finish, and an upper graph has been added representing the total of the "Red" force, which is equal in strength to the "Blue" force for any ordinate, on the basis that the "Red" force is divided into two portions as given by the intersection of the lower graph. In Figure 4*b*, this diagram has been reduced to give the same information in terms *per cent.* for a "Blue" force of constant value. Thus in its application Figure 4*b* gives the correct percentage increase necessary in the fighting value of, for example, an army or fleet to give equality, on the assumption that political or strategic necessities impose the condition of dividing the said army or fleet into two in the proportions given by the lower graph, the enemy being able to attack either proportion with his full strength. Alternatively, if the constant (= 100) be taken to represent a numerical strength that would be deemed sufficient to ensure victory against the enemy, given that both fleets engage in their full strength, then the upper graph gives the numerical superiority needed to be equally sure of victory, in case, from political or other strategic necessity, the fleet has to be divided in the proportions given. In Figure 4*b* abscissae have no quantitative meaning.

*Validity of Mathematical Treatment.* There are many who will be inclined to cavil at any mathematical or semi-mathematical treatment of the present subject, on the ground that with so many unknown factors, such as the morale or leadership of the men, the unaccounted merits or demerits of the weapons, and the still more unknown "chances of war," it is ridiculous to pretend to calculate anything. The answer to this is simple: the direct numerical comparison of the forces engaging in conflict or available in the event of war is almost universal. It is a factor always carefully reckoned with by the various military authorities; it is discussed *ad nauseam* in the Press. Yet such direct counting of forces is in itself a tacit acceptance of the applicability of mathematical principles, but confined to a special case. To accept without reserve the mere "counting of the pieces" as of value, and to deny the more extended application of mathematical theory, is as illogical and unintelligent as to accept broadly and indiscriminately the balance and the weighing-machine as instruments of precision, but to decline to permit in the latter case any allowance for the known inequality of leverage.

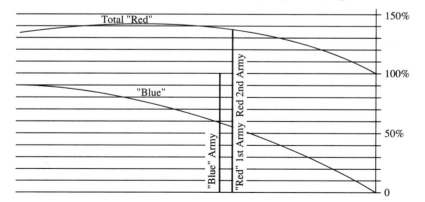

**FIGURE 4*b***

*Fighting Units not of Equal Strength.* In the equations (1) and (2), two constants were given, $c$ and $k$, which in the plotting of the Figures 1 to 4$b$ were taken as equal; the meaning of this is that the fighting strength of the individual units has been assumed equal. This condition is not necessarily fulfilled if the combatants be unequally trained, or of different morale. Neither is it fulfilled if their weapons are of unequal efficiency. The first two of these, together with a host of other factors too numerous to mention, cannot be accounted for in an equation any more than can the quality of wine or steel be estimated from the weight. The question of weapons is, however, eminently suited to theoretical discussion. It is also a matter that (as will be subsequently shown) requires consideration in relation to the main subject of the present articles.

*Influence of Efficiency of Weapons.* Any difference in the efficiency of the weapons—for example, the accuracy or rapidity of rifle-fire—may be represented by a disparity in the constants $c$ and $k$ in equations (1) and (2). The case of the rifle or machine-gun is a simple example to take, inasmuch as comparative figures are easily obtained which may be said fairly to represent the fighting efficiency of the weapon. Now numerically equal forces will no longer be forces of equal strength; they will only be of equal strength if, when in combat, their losses result in no change in their numerical proportion. Thus, if a "Blue" force initially 500 strong, using a magazine rifle, attack a "Red" force of 1,000, armed with a single breech-loader, and after a certain time the "Blue" are found to have lost 100 against 200 loss by the "Red," the proportions of the forces will have suffered no change, and they may be regarded (due to the superiority of the "Blue" arms) as being of equal strength.

If the condition of equality is given by writing M as representing the efficiency or value of an individual unit of the "Blue" force, and N the same for the "Red," we have:—

Rate of reduction of "Blue" force:—

$$\frac{db}{dt} = -\,\mathrm{N}\,r \times \text{constant} \ldots (3)$$

and "Red,"

$$\frac{dr}{dt} = -\mathrm{M}\,b \times \text{constant} \ldots (4)$$

And for the condition of equality,

$$\frac{db}{b\,dt} = \frac{dr}{r\,dt},$$

or

$$\frac{-\mathrm{N}\,r}{b} = \frac{-\mathrm{M}\,b}{r},$$

or

$$\mathrm{N}\,r^2 = \mathrm{M}\,b^2 \ldots (5)$$

In other words, the fighting strengths of the two forces are equal when the *square of the numerical strength multiplied by the fighting value of the individual units are equal.*

*The Outcome of the Investigation. The n-square Law.* It is easy to show that this expression (5) may be interpreted more generally; the *fighting strength* of a force may be broadly defined as proportional to *the square of its numerical strength multiplied by the fighting value of its individual units.*

Thus, referring to Figure 4*b*, the sum of the squares of the two portions of the "Red" force are for all values equal to the square of the "Blue" force (the latter plotted as constant); the curve might equally well have been plotted directly to this law as by the process given. A simple proof of the truth of the above law as arising from the differential equations (1) and (2), p. 2115, is as follows:—

In Figure 5, let the numerical values of the "blue" and "red" forces be represented by lines *b* and *r* as shown; then in an infinitesimally small interval of time the change in *b* and *r* will be represented respectively by *db* and *dr* of such relative magnitude that $db/dr = r/b$ or,

$$b \, db = r \, dr \tag{1}$$

If (Figure 5) we draw the squares on *b* and *r* and represent the increments *db* and *dr* as small finite increments, we see at once that the *change of area* of $b^2$ is *2b db* and the *change of area* of $r^2$ is *2r dr* which according to the foregoing (1), are equal. Therefore the difference between the two squares is constant

$$b^2 - r^2 = \text{constant}.$$

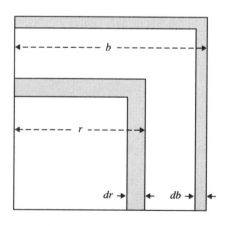

**FIGURE 5**

If this constant be represented by a quantity $q^2$ then $b^2 = r^2 + q^2$ and $q$ represents the numerical value of the remainder of the blue "force" after annihilation of the red. Alternatively $q$ represents numerically a second "red" army of the strength necessary in a *separate action* to place the red forces on terms of equality, as in Figure 4*b*.

*A Numerical Example.* As an example of the above, let us assume an army of 50,000 giving battle in turn to two armies of 40,000 and 30,000 respectively, equally well armed; then the strengths are equal, since $(50,000)^2 = (40,000)^2 + (30,000)^2$. If, on the other hand, the two smaller armies are given time to effect a junction, then the army of 50,000 will be overwhelmed, for the fighting strength of the opposing force, 70,000 is no longer equal, but is in fact nearly twice as great—namely, in the relation of 49 to 25. Superior morale or better tactics or a hundred and one other extraneous causes may intervene in practice to modify the issue, but this does not invalidate the mathematical statement.

*Example Involving Weapons of Different Effective Value.* Let us now take an example in which a difference in the fighting value of the unit is a factor. We will assume that, as a matter of experiment, one man employing a machine-gun can punish a target to the same extent in a given time as sixteen riflemen. What is the number of men armed with the machine gun necessary to replace a battalion a thousand strong in the field? Taking the fighting value of a rifleman as unity, let $n$ = the number required. The fighting strength of the battalion is, $(1000)^2$ or

$$n = \sqrt{\frac{1,000,000}{16}} = \frac{1,000}{4} = 250$$

or one quarter the number of the opposing force.

This example is instructive; it exhibits at once the utility and weakness of the method. The basic assumption is that the fire of each force is definitely *concentrated* on the opposing force. Thus the enemy will concentrate on the one machine-gun operator the fire that would otherwise be distributed over four riflemen, and so on an average he will only last for one quarter the time, and at sixteen times the efficiency during his short life he will only be able to do the work of four riflemen in lieu of sixteen, as one might easily have supposed. This is in agreement with the equation. The conditions may be regarded as corresponding to those prevalent in the Boer War, when individual-aimed firing or sniping was the order of the day.

When, on the other hand, the circumstances are such as to preclude the possibility of such concentration, as when searching an area or ridge at long range, or volley firing at a position, or "into the brown," the basic conditions are violated, and the value of the individual machine-gun operator becomes more nearly that of the sixteen riflemen that the power of his weapon represents. The same applies when he is opposed by shrapnel fire or any other weapon which is directed at a position rather than the individual. It is well thus to call attention to the variations in the conditions and the nature of the resulting departure from the conclusions of theory; such variations are far less common in naval than in military warfare; the individual unit—the ship—is always the gunner's mark. When we come to deal with aircraft, we shall find the conditions in this respect more closely resemble those that obtain in the Navy than in the Army; the enemy's aircraft individually rather than collectively is the air-gunner's mark, and the law herein laid down will be applicable.

*The Hypothesis Varied.* Apart from its connection with the main subject, the present line of treatment has a certain fascination, and leads to results which, though probably correct, are in some degree unexpected. If we modify the initial hypothesis to harmonise with the conditions of long-range fire, and assume the fire concentrated on a certain area known to be held by the enemy, and take this area to be independent of the numerical value of the forces, then, with notation as before, we have—

$$\left. \begin{aligned} -\frac{db}{dt} &= b \times \mathrm{N}\,r \\[2mm] -\frac{dr}{dt} &= r \times \mathrm{M}\,b \end{aligned} \right\} \times \text{constant.}$$

or

$$\frac{\mathrm{M}\,db}{dt} = \frac{\mathrm{N}\,dr}{dt}$$

or the rate of loss is independent of the numbers engaged, and is directly as the efficiency of the weapons. Under these conditions the fighting strength of the forces is directly proportional to their numerical strength; there is no direct value in concentration, *qua* concentration, and the advantage of rapid fire is relatively great. Thus in effect the conditions approximate more closely to those of ancient warfare.

*An Unexpected Deduction.* Evidently it is the business of a numerically superior force to come to close quarters, or, at least, to get within decisive range as rapidly as possible, in order that the concentration may tell to advantage. As an extreme case, let us imagine a "Blue" force of 100 men armed with the machine gun opposed by a "Red" 1,200 men armed with the ordinary service rifle. Our first assumption will be that both forces are spread over a front of given length and at long range. Then the "Red" force will lose 16 men to the "Blue" force loss of one, and, if the combat is continued under these conditions, the "Reds" must lose. If, however, the "Reds" advance, and get within short range, where each man and gunner is an individual mark, the tables are turned, the previous equation and conditions apply, and, even if "Reds" lose half their effective in gaining the new position, with 600 men remaining they are masters of the situation; their strength is $600^2 \times 1$ against the "Blue" $100^2 \times 16$. It is certainly a not altogether expected result that, in the case of fire so deadly as the modern machine-gun, circumstances may arise that render it imperative, and at all costs, to come to close range.

*Examples from History.* It is at least agreed by all authorities that on the field of battle concentration is a matter of the most vital importance; in fact, it is admitted to be one of the controlling factors both in the strategy and tactics of modern warfare. It is aptly illustrated by the important results that have been obtained in some of the great battles of history by the attacking of opposing forces before concentration has been effected. A classic example is that of the defeat by Napoleon, in his Italian campaign, of the Austrians near Verona, where he dealt with the two Austrian armies in detail before they had been able to effect a junction, or even to act in concert. Again, the same principle is exemplified in the oft-quoted case of the defeat of Jourdan and Moreau on the Danube by the Archduke Charles in 1796. It is evident that the conditions in the broad field of military operations correspond in kind, if not in degree, to the earlier hypothesis, and that the law deduced therefrom, that the fighting strength of a force can be represented by the square of its numerical strength, does, in its essence, represent an important truth.

## THE "N-SQUARE" LAW IN ITS APPLICATION

*The n-square Law in its Application to a Heterogeneous Force.* In the preceding article it was demonstrated that under the conditions of modern warfare the fighting strength of a force, so far as it depends upon its numerical strength, is best represented or measured by the square of the number of units. In land operations these units may be the actual men engaged, or in an artillery duel the gun battery may be the unit; in a naval battle the number of units will be the number of capital ships, or in an action between aeroplanes the number of machines. In all cases where the individual fighting strength of the component units may be different it has been shown that if a numerical fighting value can be assigned to these units, the fighting strength of the whole force is as the square of the number multiplied by their individual strength. Where the component units differ among themselves, as in the case of a fleet that is not homogeneous, the measure of the total of fighting strength of a force will be *the square of the sum of the square roots of the strengths of its individual units.*

*Graphic Representation.* Before attempting to apply the foregoing, either as touching the conduct of aerial warfare or the equipment of the fighting aeroplane, it is of interest to examine a few special cases and applications in other directions and to discuss certain possible limitations. A convenient graphic form in which the operation of the *n-square* law can be presented is given in Figure 6; here the strengths of a number of separate armies or forces successively mobilised and brought into action are represented numerically by the lines *a, b, c, d, e,* and the aggregate fighting strengths of these armies are given by the lengths of the lines A, B, C, D, E, each being the hypotenuse of a right-angle triangle, as indicated. Thus two forces or armies *a* and *b,* if acting separately (in point of time), have only the fighting strength of a single force or army represented numerically by the line B. Again, the three separate forces, *a, b,* and *c,* could be met on equal terms in three successive battles by a single army of the numerical strength C, and so on.

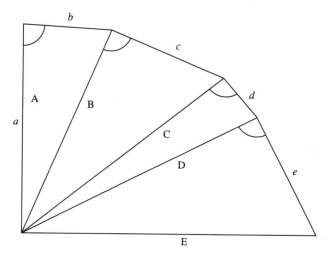

**FIGURE 6**

*Special or Extreme Case.* From the diagram given in Figure 6 arises a special case that at first sight may look like a *reductio ad absurdum,* but which, correctly interpreted, is actually a confirmation of the *n-square* law. Referring to Figure 6, let us take it that the initial force (army or fleet), is of some definite finite magnitude, but that the later arrivals *b, c, d,* etc., be very small and numerous detachments—so small, in fact, as to be reasonably represented to the scale of the diagram as infinitesimal quantities. Then the lines *b, c, d, e, f,* etc., describe a polygonal figure approximating to a circle, which in the limit becomes a circle, whose radius is represented by the original force *a,* Figure 7.

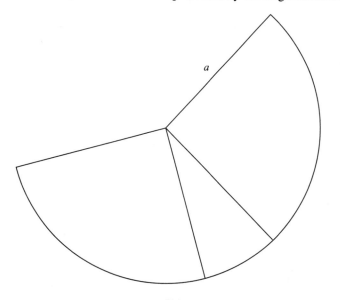

**FIGURE 7**

Here we have graphically represented the result that the fighting value of the added forces, no matter what their numerical aggregate (represented in Figure 7 by the circumferential line), is zero. The correct interpretation of this is that in the open a small force attacking, or attacked by one of overwhelming magnitude is wiped out of existence without being able to exact a toll even comparable to its own numerical value; it is necessary to say *in the open*, since, under other circumstances, the larger force is unable to bring its weapons to bear, and this is an essential portion of the basic hypothesis. In the limiting case when the disparity of force is extreme, the capacity of the lesser force to effect anything at all becomes negligible. There is nothing improbable in this conclusion, but it manifestly does not apply to the case of a small force concealed or "dug in," since the hypothesis is infringed. Put bluntly, the condition represented in Figure 7 illustrates the complete impotence of small forces in the presence of one of overwhelming power. Once more we are led to contrast the ancient conditions, under which the weapons of a large army could not be brought to bear, with modern conditions, where it is physically possible for the weapons of ten thousand to be concentrated on one. Macaulay's lines

> "In yon strait path a thousand
> May well be stopped by three,"

belong intrinsically to the methods and conditions of the past.

*The n-square Law in Naval Warfare.* We have already seen that the *n-square* law applies broadly, if imperfectly, to military operations; on land however, there sometimes exist special conditions and a multitude of factors extraneous to the hypothesis whereby its operation may be suspended or masked. In the case of naval warfare, however, the conditions more strictly conform to our basic assumptions, and there are comparatively few disturbing factors. Thus, when battle fleet meets battle fleet, there is no advantage to the defender analogous to that secured by the entrenchment of infantry. Again, from the time of opening fire, the individual ship is the mark of the gunner, and there is no phase of the battle or range at which areas are searched in a general way. In a naval battle every shot fired is aimed or directed at some definite one of the enemy's ships; there is no firing on the mass or "into the brown." Under the old conditions of the sailing-ship and cannon of some 1,000 or 1,200 yards maximum effective range, advantage could be taken of concentration within limits; and an examination of the latter 18th century tactics makes it apparent that with any ordinary disparity of numbers (probably in no case exceeding 2 to 1) the effect of concentration must have been not far from that indicated by theory. But to whatever extent this was the case, it is certain that with a battle-fleet action at the present day the conditions are still more favourable to the weight of numbers, since with the modern battle range—some 4 to 5 miles—there is virtually no limit to the degree of concentration of fire. Further than this, there is in modern naval warfare practically no chance of coming to close quarters in ship-to-ship combats, as in the old days.

Thus the conditions are to-day almost ideal from the point of view of theoretical treatment. A numerical superiority of ships of individually equal strength will mean definitely that the inferior fleet at the outset has to face the full fire of the superior, and as the battle proceeds and the smaller fleet is knocked to pieces, the initial disparity will become worse and worse, and the fire to which it is subjected more and more concentrated. These are precisely the conditions taken as the basis of the investigation from which the *n-square* law has been derived. The same observations will probably be found to apply to aerial warfare when air fleets engage in conflict, more especially so in view of the fact that aeroplane can attack aeroplane in three dimensions of space instead of being limited to two, as is the case with the battleship. This will mean that even with weapons of moderate range the degree of fire concentration possible will be very great. By attacking from above and below, as well as from all points of the compass, there is, within reason, no limit to the number of machines which can be brought to bear on a given small force of the enemy, and so a numerically superior fleet will be able to reap every ounce of advantage from its numbers.

*Individual Value of Ships or Units.* The factor the most difficult to assess in the evaluation of a fleet as a fighting machine is (apart from the *personnel*) the individual value of its units, when these vary amongst themselves. There is no possibility of entirely obviating this difficulty, since the fighting value of any given ship depends not only upon its gun armament, but also upon its protective armour. One ship may be stronger than another at some one range, and weaker at some longer or shorter range, so that the question of fleet strength can never be reduced quite to a matter of simple arithmetic, nor the design of the battleship to an exact science. In practice the drawing up of a naval programme resolves itself, in great part at least, into the answering of the prospective enemy's programme type by type and ship by ship. It is, however, generally accepted that so long as we are confining our attention to the main battle fleets, and so are dealing with ships of closely comparable gun calibre and range, and armour of approximately equivalent weight, the fighting value of the individual ship may be gauged by the weight of its "broadside," or more accurately, taking into account the speed with which the different guns can be served, by the weight of shot that can be thrown per minute. Another basis, and one that perhaps affords a fairer comparison, is to give the figure for the *energy per minute* for broadside fire, which represents, if we like so to express it, the horsepower of the ship as a fighting machine. Similar means of comparison will probably be found applicable to the fighting aeroplane, though it may be that the *downward fire* capacity will be regarded as of vital importance rather than the broadside fire as pertaining to the battleship.

*Applications of the n-square Law.* The *n-square* law tells us at once the price or penalty that must be paid if elementary principles are outraged by the division of our battle fleet[2] into two or more isolated detachments. In this respect our present disposition—a single battle fleet or "Grand" fleet—is far more economical and strategically preferable as a defensive power to the old-time distribution of the Channel Fleet, Mediterranean Fleet, etc. If it had been really necessary, for any political or geographical reason, to maintain two separate battle fleets at such distance asunder as to preclude their immediate concentration in case of attack, the cost to the country would have been enormously increased. In the case, for example, of our total battle fleet being separated into two equal parts, forming separate fleets or squadrons, the increase would require to be fixed at approximately 40 per cent.—that is to say, in the relation of 1 to $\sqrt{2}$; more generally the solution is given by a right-angled triangle, as in Figure 8. In must not be forgotten that, even with this enormous increase, the security will not be so great as appears on paper, for the enemy's fleet, having met and defeated one section of our fleet, may succeed in falling back on his base for repair and refit, and emerge later with the advantage of strength in his favour. Also one must not overlook the demoralizing effect on the *personnel* of the fleet first to go into action, of the knowledge that they are hopelessly outnumbered and already beaten on paper—that they are, in fact, regarded by their King and country as "cannon fodder." Further than this, presuming two successive fleet actions and the enemy finally beaten, the cost of victory in men and *matériel* will be greater in the case of the divided fleet than in the case of a single fleet of equal total fighting strength, in the proportion of the total numbers engaged—that is to say, in Figure 8, in the proportion that the two sides of the right-angled triangle are greater than the hypotenuse.

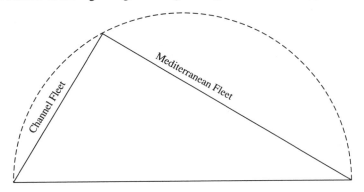

**FIGURE 8**
*Single or "Grand" Fleet of Equal Strength (Lines give numerical values).*

In brief, however potent political or geographical influences or reasons may be, it is questionable whether *under any circumstances* it can be considered sound strategy to divide the main battle fleet on which the defence of a country depends. This is to-day the accepted view of every naval strategist of repute, and is the basis of the present distribution of Great Britain's naval forces.

*Fire Concentration the Basis of Naval Tactics.* The question of fire concentration is again found to be paramount when we turn to the consideration and study of naval tactics. It is worthy of note that the recognition of the value of any definite tactical scheme does not seem to have been universal until quite the latter end of the 18th century. It is even said that the French Admiral Suffren, about the year 1780, went so far as to attribute the reverses suffered by the French at sea to "the introduction of tactics" which he stigmatised as "the veil of timidity";[3] the probability is that the then existing standard of seamanship in the French Navy was so low that anything beyond the simplest of manœuvres led to confusion, not unattended by danger. The subject, however, was, about that date, receiving considerable attention. A writer, Clerk, about 1780, pointed out that in meeting the attack of the English the French had adopted a system of defence consisting of a kind of running fight, in which, initially taking the "lee gage," they would await the English attack in line ahead, and having delivered their broadsides on the leading English ships (advancing usually in line abreast), they would bear away to leeward and take up position, once more waiting for the renewal of the attack, when the same process was repeated. By these tactics the French obtained a concentration of fire on a small portion of the English fleet, and so were able to inflict severe punishment with little injury to themselves.[4] Here we see the beginnings of sound tactical method adapted to the needs of defence.

Up to the date in question there appears to have been no studied attempt to found a scheme of attack on the basis of concentration; the old order was to give battle in parallel columns or lines, ship to ship, the excess of ships, if either force were numerically superior, being doubled on the rear ships of the enemy. It was not till the "Battle of the Saints," in 1782, that a change took place; Rodney (by accident or intention) broke away from tradition, and cutting through the lines of the enemy, was able to concentrate on his centre and rear, achieving thereby a decisive victory.

*British Naval Tactics in 1805. The Nelson "Touch."* The accident or experiment of 1782 had evidently become the established tactics of the British in the course of the twenty years which followed, for not only do we find the method in question carefully laid down in the plan of attack given in the Memorandum issued by Nelson just prior to the Battle of Trafalgar in 1805, but the French Admiral Villeneuve[5] confidently asserted in a note issued to his staff in anticipation of the battle that:—"The British Fleet will not be formed in a line-of-battle parallel to the combined fleet according to the usage of former days. Nelson, assuming him to be, as represented, really in command, will seek to break our line, envelop our rear, and overpower with groups of his ships as many as he can isolate and cut off." Here we have a concise statement of a definite tactical scheme based on a clear understanding of the advantages of fire concentration.

It will be understood by those acquainted with the sailing-ship of the period that the van could only turn to come to the assistance of those in the rear at the cost of a considerable interval of time, especially if the van should happen to be to leeward of the centre and rear. The time taken to "wear ship," or in light winds to "go about" (often only to be effected by manning the boats and rowing to assist the manoeuvre), was by no means an inconsiderable item. Thus it would not uncommonly be a matter of some hours before the leading ships could be brought within decisive range, and take an active part in the fray.

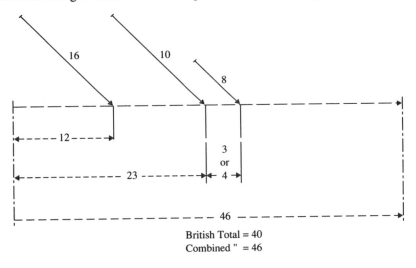

British Total = 40
Combined " = 46

**FIGURE 9**

*Nelson's Memorandum and Tactical Scheme.* In order further to embarrass the enemy's van, and more effectively to prevent it from coming into action, it became part of the scheme of attack that a few ships, a comparatively insignificant force, should be told off to intercept and engage as many of the leading ships as possible; in brief, to fight an independent action on a small scale; we may say admittedly a losing action. In this connection Nelson's memorandum of October 9 is illuminating. Nelson assumed for the purpose of framing his plan of attack that his own force would consist of forty sail of the line, against forty-six of the combined (French and Spanish) fleet. These numbers are considerably greater, as things turned out, than those ultimately engaged; but we are here dealing with the memorandum, and not with the actual battle. The British Fleet was to form in two main columns, comprising sixteen sail of the line each, and a smaller column of eight ships only. The plan of attack prescribed in the event of the enemy being found in line ahead was briefly as follows:—One of the main columns was to cut the enemy's line about the centre, the other to break through about twelve ships from the rear, the smaller column being ordered to engage the rear of the enemy's van three or four ships ahead of the centre, and to frustrate, as far as possible, every effort the van might make to come to the succour of the threatened centre or rear. Its object, in short, was to prevent the van of the combined fleet from taking part in the main action. The plan is shown diagrammatically in Figure 9.

*Nelson's Tactical Scheme Analysed.* An examination of the numerical values resulting from the foregoing disposition is instructive. The force with which Nelson planned to envelop the half—*i.e.*, 23 ships—of the combined fleet amounted to 32 ships in all; this according to the $n^2$ law would give him a superiority of fighting strength of almost exactly two to one,[6] and would mean that if subsequently he had to meet the other half of the combined fleet, without allowing for any injury done by the special eight-ship column, he would have been able to do so on terms of equality. The fact that the van of the combined fleet would most certainly be in some degree crippled by its previous encounter is an indication and measure of the positive advantage of strength provided by the tactical scheme. Dealing with the position arithmetically, we have:—

Strength of British (in arbitrary $n^2$ units),
$$32^2 + 8^2 = 1088$$
And combined fleet,
$$23^2 + 23^2 = \underline{1058}$$
British advantage . . . .     30

Or, the numerical equivalent of the remains of the British Fleet (assuming the action fought to the last gasp), $= \sqrt{30}$ or 5½ ships.

If for the purpose of comparison we suppose the total forces had engaged under the conditions described by Villeneuve as "the usage of former days," we have:—

Strength of combined fleet, $46^2$ . . . . $= 2116$
Strength of British fleet,     $40^2$ . . . . $= \underline{1600}$
Balance in favour of enemy     . . . .     516

Or, the equivalent numerical value of the remainder of the combined fleet, assuming complete annihilation of the British, $= \sqrt{516} = 23$ ships approximately.

Thus we are led to appreciate the commanding importance of a correct tactical scheme. If in the actual battle the old-time method of attack had been adopted, it is extremely doubtful whether the superior seamanship and gunnery of the British could have averted defeat. The actual forces on the day were 27 British sail of the line against the combined fleet numbering 33, a rather less favourable ratio than assumed in the Memorandum. In the battle, as it took place, the British attacked in two columns instead of three, as laid down in the Memorandum; but the scheme of concentration followed the original idea. The fact that the wind was of the lightest was alone sufficient to determine the exclusion of the enemy's van from the action. However, as a study the Memorandum is far more important than the actual event, and in the foregoing analysis it is truly remarkable to find, firstly, the definite statement of the cutting the enemy into two *equal* parts—according to the *n-square* law the exact proportion corresponding to the reduction of his total effective strength to a minimum; and, secondly, the selection of a proportion, the nearest whole-number equivalent to the $\sqrt{2}$ ratio of theory, required to give a fighting strength equal to tackling the two halves of the enemy on level terms, and the detachment of the remainder, the column of eight sail, to weaken and impede the leading half of the enemy's fleet to guarantee the success of the main idea. If, as might fairly be assumed, the foregoing is more than a coincidence,[7] it suggests itself that Nelson, if not actually acquainted with the *n-square* law, must have had some equivalent basis on which to figure his tactical values.

## ENDNOTES

[1]This is not strictly true, since towards the close of the fight the last few men will be attacked by more than their own number. The main principle is, however, untouched.

[2]Capital ships:—Dreadnoughts and Super-Dreadnoughts.

[3]Mahan, "Sea Power," page 425.

[4]Incidentally, also, the scheme in question had the advantage of subjecting the English to a raking fire from the French broadsides before they were themselves able to bring their own broadside fire to bear.

[5]"The Enemy at Trafalgar," Ed. Fraser; Hodder and Stoughton, page 54.

[6]$23 \times \sqrt{2} = 32.5$.

[7]Although we may take it to be a case in which the dictates of experience resulted is a disposition now confirmed by theory, the agreement is remarkable.

# COMMENTARY ON
# OPERATIONS RESEARCH

The sprawling activity known as operations research had its beginning during the Second World War. Science has of course contributed ideas to destruction since the time of Archimedes and in both the great wars of the present century it furnished the technical assistance making possible the development of every major weapon from the machine gun to the atom bomb. Operations research, however, is a different kind of scientific work. It is a conglomerate of methods. It has been defined as "a scientific method of providing executive departments with a quantitative basis for decisions regarding the operations under their control." The definition is a little inflated but it conveys the general outline of the subject.

In the last war operational analysts were to be found at work in strange places and under unlikely circumstances. Mathematicians discussed gunnery problems with British soldiers in Burma; chemists did bomb damage assessment with economist colleagues at Princes Risborough, a "secure" headquarters outside London; generals conferred about tank strategy in the Italian campaign with biochemists and lawyers; a famous British zoölogist was key man in planning the bombardment of Pantellaria; naval officers took statisticians and entomologists into their confidence regarding submarine losses in the Pacific; the high command of the R.A.F. and American Airforce shared its headaches over Rumanian oil fields, French marshaling yards, German ball-bearing and propeller factories and mysterious ski-sites in the Pas-de-Calais with psychologists, architects, paleontologists, astronomers and physicists. It was a lively, informal, paradoxical exchange of ideas between amateur and professional warmakers and it produced some brilliant successes. It led to the solution of important gunnery and bombardment problems; improved the efficiency of our antisubmarine air patrol in the Bay of Biscay and elsewhere; shed light on convoying methods in the North Atlantic; helped our submarines to catch enemy ships and also to avoid getting caught; supplied a quantitative basis for weapons evaluation; altered basic concepts of air to air and naval combat; simplified difficult recurring problems of supply and transport. There were of course many more failures than successes but the over-all record is impressive.

What scientists brought to operational problems—apart from specialized knowledge—was the scientific outlook. This in fact was their major contribution. They tended to think anew, to suspect preconceptions, to act only on evidence. Their indispensable tool was the mathematics of probability and they made use of its subtlest theories and most powerful techniques. They were unhampered by laboratory dogma, but the experimental method was their inseparable guide. A thoughtful student of the subject, the British mathematician J. Bronowski (1908–1974), aptly described their work: "A war or a battle, a mission or a sortie, none is repeatable and none is an experiment. Yet the young scientists brought to them the conviction that in them and nowhere else must be found the empirical evidence for the rightness or wrongness of the assumptions and underlying strategy by which war is made. The passion of these men was to trace in operations involving life and death the tough skeleton of experimental truth."[1]

The material I have selected is from the best book thus far published on the subject. Morse, a physicist, and Kimball, a chemist,[2] had wide experience in operations research in the last war, and subsequently as consultants. The examples are confined to military problems but I should point out that, having got its start in the war, operations research is now being extended to engineering, to communication, to coal mining, to business, to manufacture and to other branches of industry. The new problems are not as easy or as enticing as were many of the military exercises (even the art of war, now that the simple mistakes have been put right, offers a less "creamy surface to skim") and opinions differ as to whether first-class men will find satisfaction in such work. I incline to Bronowski's view: "The heroic age is over; and dropping with a sigh the glamour and the heady sense of power, we have to face the recognition that the field of opportunity will never again be quite so blank, so simple and so lavish. What was new and speculative on the battlefield turns out, in the practical affairs of industry, to become only a painstaking combination of cost accounting, job analysis, time and motion study and the general integration of plant flow. There is an extension of this to the larger economics of whole industries and nations, but it is hardly likely to be rewarding to first-rate scientists and calls at bottom for the immense educational task of interesting economists and administrators in the mathematics of differentials and of prediction."

Sir Charles Darwin has suggested that in the future—not too remote—computing machines will take over the job. This is more likely, at any rate, than that administrators will master differentials.

## ENDNOTES

[1]Review of the Morse and Kimball book (*Methods of Operations Research*, New York, 1951) in *Scientific American*, October 1951, pp. 75–77.

[2]Dr. Phillip M. Morse (1903–1985) was a professor of physics at the Massachusetts Institute of Technology from 1931 until his death. During the war he was director of the U. S. Navy Operations Research Group, and later he held other equally responsible positions in the field. Dr. George E. Kimball (1906–1967) was, among others, Deputy Director, Operations Evaluation Group, U. S. Navy and was professor of chemistry at Columbia University.

# HOW TO HUNT A SUBMARINE

## By Phillip M. Morse and George E. Kimball

. . . Just as with every other field of applied science, the improvement of operations of war by the application of scientific analysis requires a certain flair which comes with practice, but which is difficult to put into words.

It is important first to obtain an overall quantitative picture of the operation under study. One must first see what is similar in operations of a given kind before it will be worthwhile seeing how they differ from each other. In order to make a start in so complex a subject, one must ruthlessly strip away details (which can be taken into account later), and arrive at a few broad, very approximate "constants of the operation." By studying the variations of these constants, one can then perhaps begin to see how to improve the operation.

It is well to emphasize that these constants which measure the operation are useful even though they are extremely approximate; it might almost be said that they are more valuable *because* they are very approximate. This is because successful application of operations research usually results in improvements by factors of 3 or 10 or more. Many operations are ineffectively compared to their theoretical optimum because of a single faulty component: inadequate training of crews, or incorrect use of equipment, or inadequate equipment. Usually, when the "bottleneck" has been discovered and removed, the improvements in effectiveness are measured in hundreds or even thousands of per cent. In our first study of any operation we are looking for these large factors of possible improvement. They can be discovered if the constants of the operation are given only to one significant figure, and any greater accuracy simply adds unessential detail.

One might term this type of thinking "hemibel thinking." A bel is defined as a unit in a logarithmic scale[1] corresponding to a factor of 10. Consequently, a hemibel corresponds to a factor of the square root of 10, or approximately 3.

Ordinarily, in the preliminary analysis of an operation, it is sufficient to locate the value of the constant to within a factor of 3. Hemibel thinking is extremely useful in any branch of science, and most successful scientists employ it habitually. It is particularly useful in operations research.

Having obtained the constants of the operation under study in units of hemibels (or to one significant figure), we take our next step by comparing these constants. We first compare the value of the constants obtained in actual operations with the optimum theoretical value, if this can be computed. If the actual value is within a hemibel (i.e., within a factor of 3) of the theoretical value, then it is extremely unlikely that any improvement in the details of the operation will result in significant improvement. In the usual case, however, there is a wide gap between the actual and theoretical results. In these cases a hint as to the possible means of improvement can usually be obtained by a crude sorting of the operational data to see whether changes in personnel, equipment, or tactics produce a significant change in the constants. In many cases a theoretical study of the optimum values of the constants will indicate possibilities of improvement. . . .

## SWEEP RATES

An important function for some naval forces, particularly for some naval aircraft, is that of scouting or patrol, that is, search for the enemy. In submarine warfare search is particularly important. The submarine must find the enemy shipping before it can fire its torpedoes, and the antisubmarine craft must find the enemy submarine in order to attack it, or to route its convoys evasively, and so on.

Patrol or search is an operation which is peculiarly amenable to operations research. The action is simple, and repeated often enough under conditions sufficiently similar to enable satisfactory data to be accumulated. From these data measures of effectiveness can be computed periodically from which a great deal can be deduced. By comparing the operational values of the constants with the theoretically optimum values, one can obtain an overall picture as to the efficiency of our own forces. Sudden changes in the constants without change in our own tactics will usually mean a change in enemy tactics which, of course, needs investigation and usually counteraction.

## CALCULATION OF CONSTANTS

In the simplest case a number of search units (e.g., aircraft or submarine) are sent into a certain area $A$ of the ocean to search for enemy craft. A total of $T$ units of time (hours or days) is spent by one or another of the search craft in the area, and a number of contacts $C$ with an enemy unit are reported. It is obvious that the total number of contacts obtained in a month is not a significant measure of the effectiveness of the searching craft because it depends on the length of time spent in searching. A more useful constant would be the average number of contacts made in the area per unit of time spent in searching ($C$ divided by $T$).

The number of contacts per unit of searching time is a simple measure which is useful for some purposes and not useful for others. As long as the scene of the search remains the same, the quantity $(C/T)$ depends on the efficiency of the individual searching craft and also on the number $N$ of enemy craft which are in the area on the average. Consequently, any sudden change in this quantity would indicate a change in enemy concealment tactics, or else a change in the number of enemy craft present. Since this quantity depends so strongly on the enemy's actions, it is not a satisfactory one to compare against theoretically optimum values in order to see whether the searching effort can be appreciably improved or not. Nor is it an expedient quantity to use in comparing the search efforts in two different areas.

A large area is more difficult to search over than a small one since it takes more time to cover the larger area with the same density of search. Consequently, the number of contacts per unit searching time should be multiplied by the area searched over in order to compensate for this area effect, and so that the searching effort in two different areas can be compared on a more or less equal basis.

## OPERATIONAL SWEEP RATE

One further particularly profitable step can be taken, if other sources of intelligence allow one to estimate (to within a factor of 3) the average number of enemy craft in the area while the search was going on.

The quantity which can then be computed is the number of contacts per unit search time, multiplied by the area searched over and divided by the estimated number of enemy units in the area. Since the dimensions of this quantity are square miles per hour, it is usually called the effective, or operational, *sweep rate*.

Operational sweep rate:

$$Q_{op} = \left( \frac{CA}{NT} \right) \frac{\text{square miles}}{\text{hour (or day)}}. \tag{1}$$

$C$ = number of contacts;
$A$ = area searched over in square miles;
$T$ = total searching time in hours (or days);
$N$ = probable number of enemy craft in area.

This quantity is a measure of the ability of a single search craft to find a single enemy unit under actual operational conditions. It equals the effective area of ocean swept over by a single search craft in an hour (or day).

Another way of looking at this constant is taken by remembering that $(N/A)$ is the *average density* of target craft, in number per square mile. Since $(C/T)$ is the number of contacts produced per hour (or day) $Q_{op} = (C/T) \div (N/A)$ is the number of contacts which would be obtained per hour (or day) if the density of target craft were one per square mile.

## THEORETICAL SWEEP RATE

Sweep rates can be compared from area to area and from time to time, since the effects of different size of areas and of different numbers of enemy craft are already balanced out. Sweep rates can also be compared with the theoretical optimum for the craft in question. Elsewhere we have shown that the sweep rate is equal to twice the "effective lateral range of detection" of the search craft equipment, multiplied by the speed of the search craft.[2]

Theoretical sweep rate:

$$Q_{th} = 2Rv \frac{\text{square miles}}{\text{hour (or day)}} \tag{2}$$

$R$ = effective lateral range of detection in miles;
$v$ = average speed of search craft in miles per hour (or day).

A comparison of this sweep rate with the operational value will provide us with the criterion for excellence which we need.

The ratio between $Q_{op}$ and $Q_{th}$ is a factor which depends both on the effectiveness of our side in using the search equipment available, and on the effectiveness of the enemy in evading detection. For instance, if the search craft is a plane equipped with radar, and if the radar is in poor operational condition on the average, this ratio will be correspondingly diminished. Similarly, if the enemy craft is a submarine, then a reduction of the average time it spent on the surface would reduce the ratio for search planes using radar or visual sighting. The ratio also would be reduced if the area were covered by the searching craft in a nonuniform manner, and if the enemy craft tended to congregate in those regions which were searched least. Correspondingly, the ratio ($Q_{op}/Q_{th}$) will be increased (and may even be greater than unity) if the enemy craft tend to congregate in one region of the area, and if the searching effort is also concentrated there. It can be seen that a comparison of the two sweep rates constitutes a very powerful means of following the fluctuations in efficacy of the search operation as the warfare develops.

## SUBMARINE PATROL

A few examples will show the usefulness of the quantities mentioned here. The first example comes from data on the sighting of merchant vessels by submarines on patrol. Typical figures are given in Table 1. All numbers are rounded off to one or two significant figures, since the estimate of the number of ships present in the area is uncertain, and there is no need of having the accuracy of the other figures any larger. The operational sweep rate (computed from the data) is also tabulated. Since the ratio of the values of $Q$ for regions $B$

and $E$ is less than 1 hemibel, the difference in the sweep rates for those regions is probably due to the rather wide limits of error of the values of $N$. The difference in sweep rate between areas $B$ and $D$ is probably significant however (it corresponds to a ratio of more than a hemibel). Investigation of this difference shows that the antisubmarine activity in region $B$ was considerably more effective than in $D$, and, consequently, the submarines in region $B$ had to spend more time submerged and had correspondingly less time to make sightings. The obvious suggestion (unless there are other strategic reasons to the contrary) is to transfer some of the effort from region $B$ to region $D$, since the yield per submarine per day is as good, and since the danger to the submarine is considerably less.

TABLE 1. CONTACTS ON MERCHANT VESSELS BY SUBMARINES

| Region | B | D | E |
|---|---|---|---|
| Area, sq. miles, $A$ | 80,000 | 250,000 | 400,000 |
| Avg. No. ships present, $N$ | 20 | 20 | 25 |
| Ship flow through area per day, $F$ | 6 | 3 | 4 |
| Sub-days in area, $T$ | 800 | 250 | 700 |
| Contacts, $C$ | 400 | 140 | 200 |
| Sweep rate, $Q_{op}$ | 2,000 | 7,000 | 4,500 |
| Fraction of ship flow sighted by a sub, $C/FT$ | 0.08 | 0.2 | 0.07 |
| Sightings per sub per day | 0.5 | 0.6 | 0.3 |

For purposes of comparison, we compute the theoretical sweep rate. A submarine on patrol covers about 200 miles a day on the average, and the average range of visibility for a merchant vessel is between 15 and 20 miles. The theoretical sweep rate, therefore, is about 6,000 to 8,000 square miles per day. This corresponds remarkably closely with the operational sweep rate in regions $D$ and $E$. The close correspondence indicates that the submarines are seeing all the shipping they could be expected to see (i.e., with detection equipment having a range of 15 to 20 miles). It also indicates that the enemy has not been at all successful in evading the patrolling submarines, for such evasion would have shown up as a relative diminution in $Q_{op}$. The reduced value of sweep rate in region $B$ has already been explained.

Therefore, a study of the sweep rate for submarines against merchant vessels has indicated (for the case tabulated) that no important amount of shipping is missed because of poor training of lookouts or of failure of detection equipment. It has also indicated that one of the three regions is less productive than the other two; further investigation has revealed the reason. The fact that each submarine in region $D$ sighted one ship in every five that passed through the region is a further indication of the extraordinary effectiveness of the submarines patrolling these areas.

# AIRCRAFT SEARCH FOR SUBMARINES

Another example, not quite so impressive, but perhaps more instructive, can be taken from data on search for submarines by antisubmarine aircraft. Typical values are shown in Table 2, for three successive months, for three contiguous areas. Here the quantity $T$ represents the total time spent by aircraft over the ocean on antisubmarine patrol of all sorts in the region during the month in question. The quantity $C$ represents the total number of verified sightings of a surfaced submarine in the area and during the month in question. From these data the value of the operational sweep rate, $Q_{op}$, can be computed and is expressed also on a hemibel scale. From these figures a number of interesting conclusions can be drawn, and a number of useful suggestions can be made for the improving of the operational results.

TABLE 2. SIGHTINGS OF SUBMARINES BY ANTISUBMARINE AIRCRAFT

| Region | A | | | B | | | C | | |
|---|---|---|---|---|---|---|---|---|---|
| Area, sq. miles, $A$ | 300,000 | | | 600,000 | | | 900,000 | | |
| Month | A | M | J | A | M | J | A | M | J |
| Avg. No. subs, $N$ | 7 | 7 | 6 | 1 | 4 | 3 | 3 | 7 | 5 |
| Total plane time (in thousands of hours), $T$ | 20 | 25 | 24 | 6 | 7 | 9 | 5 | 5 | 6 |
| Contacts, $C$ | 39 | 37 | 30 | 2 | 35 | 14 | 4 | 11 | 9 |
| Sweep rate, $Q_{op}$ | 80 | 60 | 60 | 200 | 750 | 300 | 240 | 280 | 270 |
| Sweep rate in hemibels | 4 | 4 | 4 | 5 | 6 | 5 | 5 | 5 | 5 |

We first compare the operational sweep rate with the theoretically optimum rate. The usual antisubmarine patrol plane flies at a speed of about 150 knots. The average range of visibility of a surfaced U-boat in flyable weather is about 10 miles. Therefore, if the submarines were on the surface all of the time during which the planes were searching, we should expect the theoretical search rate to be 3,000 square miles per hour, according to equation (2). On the hemibel scale this is a value of 7. If the submarines on the average spent a certain fraction of the time submerged, then $Q_{th}$ would be proportionally diminished. We see that the average value of the sweep rate in regions $B$ and $C$ is about one-tenth (2 hemibels) smaller than the maximum theoretical value of 3,000.

Part of this discrepancy is undoubtedly due to the submergence tactics of the submarines. In fact, the sudden rise in the sweep rate in region $B$ from April to May was later discovered to be almost entirely due to a change in tactics on the part of the submarines. During the latter month the submarines carried on an all-out attack, coming closer to shore than before or since, and staying longer on the surface, in order to sight more shipping. This bolder policy exposed the submarines to too many attacks, so they returned to more cautious tactics in June. The episode serves to indicate that at least one-half of the 2 hemibel discrepancy between operational and theoretically maximum sweep rates is probably due to the submergence tactics of the submarine.

The other factor of 3 is partially attributable to a deficiency in operational training and practice in antisubmarine lookout keeping. Antisubmarine patrol is a monotonous duty. The average plane can fly for hundreds of hours (representing an elapsed time of six months or more) before a sighting is made. Experience has shown that, unless special competitive practice exercises are used continuously, performance of such tasks can easily fall below one-third of their maximum effectiveness. Data in similar circumstances, mentioned later in this chapter, show that a diversion of 10 per cent of the operational effort into carefully planned practice can increase the overall effectiveness by factors of two to four.

We have thus partially explained the discrepancy between the operational sweep rate in regions $B$ and $C$ and the theoretically optimum sweep rate; we have seen the reason for the sudden increase for one month in region $B$. We must now investigate the result of region $A$ which displays a consistently low score in spite of (or perhaps because of) the large number of antisubmarine flying hours in the region. Search in region $A$ is consistently 1 hemibel worse (a factor of 3) than in the other two regions. Study of the details of the attacks indicates that the submarines were not more wary in this region; the factor of 3 could thus not be explained by assuming that the submarines spent one-third as much time on the surface in region $A$. Nor could training entirely account for the difference. A number of new squadrons were "broken in" in region $A$, but even the more experienced squadrons turned in the lower average.

## DISTRIBUTION OF FLYING EFFORT

In this case the actual track plans of the antisubmarine patrols in region $A$ were studied in order to see whether the patrol perhaps concentrated the flying effort in regions where the submarines were not likely to be. This indeed proved to be the case, for it was found that a disproportionately large fraction of the total antisubmarine flying in region $A$ was too close to shore to have a very large chance of finding a submarine on the surface. The data for the month of April (and also for other months) was broken down according to the amount of patrol time spent a given distance off shore. The results for the one month are given in Table 3. In this analysis it was not necessary to compute the sweep rate, but only to compare the number of contacts per thousand hours flown in various strips at different distances from the shore. This simplification is possible since different strips of the same region are being compared for the same periods of time; consequently, the areas are equal and the average distribution of submarines is the same. The simplification is desirable since it is not known, even approximately, where the seven submarines, which were present in that region in that month, were distributed among the offshore zones.

TABLE 3. SIGHTINGS OF SUBMARINES BY ANTISUBMARINE PLANES,
OFFSHORE EFFECT

| Distance from shore in miles | 0 to 60 | 60 to 120 | 120 to 180 | 180 to 240 |
|---|---|---|---|---|
| Flying time in sub-region, $T$ (in thousands of hours) | 15.50 | 3.70 | 0.60 | 0.17 |
| Contacts made in sub-region, $C$ | 21 | 11 | 5 | 2 |
| Contacts per 1,000 hours flown, $(C/T)$ | 1.3 | 3 | 8 | 12 |
| Contacts per 1,000 hours flown, in hemibels | 0 | 1 | 2 | 2 |

A comparison of the different values of contacts per 1,000 hours flown for the different offshore bands immediately explains the ineffectiveness of the search effort in region *A*. Flying in the inner zone, where three-quarters of the flying was done, is only one-tenth as effective as flying in the outer zone, where less than 1 per cent of the flying was done. Due perhaps to the large amount of flying in the inner zone, the submarines did not come this close to shore very often, and, when they came, kept well submerged. In the outer zones, however, they appeared to have been as unwary as in region *B* in the month of May.

If a redistribution of flying effort would not have changed submarine tactics, then a shift of 2,000 hours of flying per month from the inner zone to the outer (which would have made practically no change in the density of flying in the inner zone, but which would have increased the density of flying in the outer zone by a factor of 13) would have approximately doubled the number of contacts made in the whole region during that month. Actually, of course, when a more uniform distribution of flying effort was inaugurated in this region, the submarines in the outer zones soon became more wary and the number of contacts per thousand hours flown in the outer region soon dropped to about 4 or 5. This still represented a factor of 3, however, over the inshore flying yield. We therefore can conclude that the discrepancy of one hemibel in sweep rate between region *A* and regions *B* and *C* is primarily due to a maldistribution of patrol flying in region *A*, the great preponderance of flying in that region being in localities where the submarines were not. When these facts were pointed out, a certain amount of redistribution of flying was made (within the limitations imposed by other factors), and a certain amount of improvement was observed.

The case described here is not a unique one; in fact, it is a good illustration of a situation often encountered in operations research. The planning officials did not have the time to make the detailed analysis necessary for the filling in of Table 3. They saw that many more contacts were being made on submarines close inshore than farther out, and they did not have at hand the data to show that this was entirely due to the fact that nearly all the flying was close to shore. The data on contacts, which is more conspicuous, might have actually persuaded the operations officer to increase still further the proportion of flying close to shore. Only a detailed analysis of the amount of flying time in each zone, resulting in a tabulation of the sort given in Table 3, was able to give the officer a true picture of the situation. When this had been done, it was possible for the officer to balance the discernible gains to be obtained by increasing the offshore flying against other possible detriments. In this case, as with most others encountered in this field, other factors enter; the usefulness of the patrol planes could not be measured solely by their collection of contacts, and the other factors favored inshore flying.

## ANTISUBMARINE FLYING IN THE BAY OF BISCAY

An example of the use of sweep rate for following tactical changes in a phase of warfare will be taken from the RAF Coastal Command struggle against German U-boats in the Bay of Biscay. After the Germans had captured France, the Bay of Biscay ports were the principal operational bases for U-boats. Nearly all of the German submarines operating in the Atlantic went out and came back through the Bay of Biscay. About the beginning of 1942, when the RAF began to have enough long range planes, a number of them were assigned to antisubmarine duty in the Bay to harass these transit U-boats. Since the submarines had to be discovered before they could be attacked, and since these planes were out only to attack submarines, a measure of the success of the campaign was the number of U-boat sightings made by the aircraft.

The relevant data for this part of the operation are shown in Figure 1 for the years 1942 and 1943. The number of hours of antisubmarine patrol flying in the Bay per month, the number of sightings of U-boats resulting, and the estimated average number of U-boats in the Bay area during the month are plotted in the upper part of the figure. From these values and from the area of the Bay searched over (130,000 square miles), one can compute the values of the operational sweep rate which are shown in the lower half of the figure.

The graph for $Q_{op}$ indicates that two complete cycles of events have occurred during the two years shown. The first half of 1942 and the first half of 1943 gave sweep rates of the order of 300 square miles per hour, which correspond favorably with the sweep rates obtained in regions $B$ and $C$ in Table 2. The factor of 10 difference between these values and the theoretically maximum value of 3,000 square miles per hour can be explained, as before, partly by the known discrepancy between lookout practice in actual operation and theoretically optimum lookout effectiveness, and mainly by submarine evasive tactics. It was known at the beginning of 1942 that the submarines came to the surface for the most part at night, and stayed submerged during a good part of the day. Since most of the antisubmarine patrols were during daylight, these tactics could account for a possible factor of 5, leaving a factor of 2 to be accounted for (perhaps) by lookout fatigue, etc.

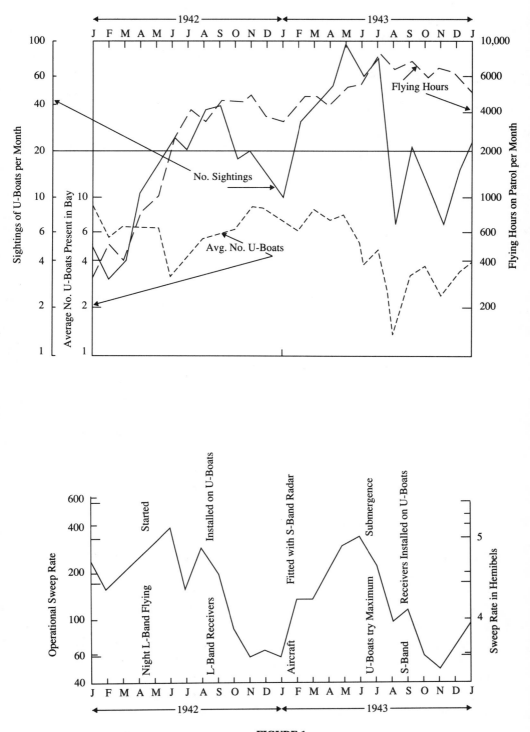

**FIGURE 1**
*Sightings of U-boats by antisubmarine aircraft in the Bay of Biscay in 1942–1943.*

During the early part of 1942, the air cover over the Bay of Biscay increased, and the transit submarines began to experience a serious number of attacks. In the spring a few squadrons of radar planes were equipped for night-flying, with searchlights to enable them to make attacks at night on the submarines. When these went into operation, the effective search rate for all types of planes increased at first. The night-flying planes caught a large number of submarines on the surface at night. These night attacks caused the submarines to submerge more at night and surface more in the daytime; therefore the day-flying planes also found more submarines on the surface.

The consequent additional hazard to the U-boats forced a countermeasure from the Germans; for even though the night-flying was a small percentage of the total air effort in the Bay, the effects of night attack on morale were quite serious. The Germans started equipping their submarines with radar receivers capable of hearing the L-band radar set carried in the British planes. When these sets were operating properly, they would give the submarine adequate warning of the approach of a radar plane, so that it could submerge before the plane could make a sighting or attack. Despite difficulties in getting the sets to work effectively, they became more and more successful, and the operational sweep rate for the British planes dropped abruptly in the late summer of 1942, reaching a value about one-fifth of that previously attained.

When this low value of sweep rate continued for several months, it was obviously necessary for the British to introduce a new measure. This was done by fitting the antisubmarine aircraft with S-band radar which could not be detected by the L-band receivers on the German submarines at that time. Commencing with the first of 1943, the sweep rate accordingly rose again as more and more planes were fitted with the shorter wave radar sets. Again the U-boats proved particularly susceptible to the attacks of night-flying planes equipped with the new radar sets and with searchlights. By midsummer of 1943, the sweep rate was back as high as it had been a year before.

The obvious German countermeasure was to equip the submarines with S-band receivers. This, however, involved a great many design and manufacturing difficulties, and these receivers were not to be available until the fall of 1943. In the interim the Germans sharply reduced the number of submarines sent out, and instructed those which did go out to stay submerged as much as possible in the Bay region. This reduced the operational sweep rate for the RAF planes to some extent, and, by the time the U-boats had been equipped with S-band receivers in the fall, the sweep rate reached the same low values it had reached in the previous fall. The later cycle, which occurred in 1944, involved other factors which we will not have time to discuss here.

This last example shows how it is sometimes possible to watch the overall course of a part of warfare by watching the fluctuations of a measure of effectiveness. One can at the same time see the actual benefits accruing from a new measure and also see how effective are the countermeasures. By keeping a month-to-month chart of the quantity, one can time the introduction of new measures, and also can assess the danger of an enemy measure. A number of other examples of this sort will be given later in this chapter.

# EXCHANGE RATES

A useful measure of effectiveness for all forms of warfare is the exchange rate, the ratio between enemy loss and own loss. Knowledge of its value enables one to estimate the cost of any given operation and to balance this cost against other benefits accruing from the operation. Here again a great deal of insight can be obtained into the tactical trends by comparing exchange rates; in particular, by determining how the rate depends on the relative strength of the forces involved.

When the engagement is between similar units, as in a battle between tanks or between fighter planes, the units of strength on each side are the same, and the problem is fairly straightforward. Data are needed on a large number of engagements involving a range of sizes of forces involved. Data on the strength of the opposed forces at the beginning of each engagement and on the resulting losses to both sides are needed. These can then be subjected to statistical analysis to determine the dependence of the losses on the other factors involved.

Suppose $m$ and $n$ are the number of own and enemy units involved, and suppose $k$ and $l$ are the respective losses in the single engagements. In general, $k$ and $l$ will depend on $m$ and $n$, and the nature of the dependence is determined by the tactics involved in the engagement. For instance, if the engagement consists of a sequence of individual combats between single opposed units, then both $k$ and $l$ are proportional to either $m$ or $n$ (whichever is smaller), and the exchange rate $(l/k)$ is independent of the size of the opposing forces. On the other hand, if each unit on one side gets about an equal chance to shoot at each unit on the other side, then the losses on one side will be proportional to the number of opposing units (that is, $k$ will be proportional to $n$, and $l$ will be proportional to $m$).

## AIR-TO-AIR COMBAT

The engagements between American and Japanese fighter aircraft in the Pacific in 1943–44 seem to have corresponded more closely to the individual-combat type of engagements. The data which have been analyzed indicate that the exchange rate for Japanese against U.S. fighters $(l/k)$ was approximately independent of the size of the forces in the engagement. The percentage of Japanese fighters lost per engagement seems to have been independent of the numbers involved (i.e., $k$ was proportional to $n$); whereas the percentage of U.S. fighters lost per engagement seemed to increase with an increase of Japanese fighters, and decrease with an increase of U.S. fighters (i.e., $l$ was also proportional to $n$).

The exchange rate for U.S. fighters in the Pacific during the years 1943 and 1944 remained at the surprisingly high value of approximately 10. This circumstance contributed to a very high degree to the success of the U.S. Navy in the Pacific. It was, therefore, of importance to analyze as far as possible the reasons for this high exchange rate in order to see the importance of the various contributing factors, such as training and combat experience, the effect of the characteristics of planes, etc. The problem is naturally very complex, and it is possible here only to give an indication of the relative importance of the contributing factors.

Certainly a very considerable factor has been the longer training which the U.S. pilots underwent compared to the Japanese pilots. A thoroughgoing study of the results of training and of the proper balance between primary training and operational practice training has not yet been made, so that a quantitative appraisal of the effects of training is as yet impossible. Later in this chapter we shall give an example which indicates that it sometimes is worth while even to withdraw aircraft from operations for a short time in order to give the pilots increased training.[3] There is considerable need for further operational research in such problems. It is suspected that, in general, the total effectiveness of many forces would be increased if somewhat more time were given to refresher training in the field, and slightly less to operations.

The combat experience of the pilot involved has also had its part in the high exchange rate. The RAF Fighter Command Operations Research Group has studied the chance of a pilot being shot down as a function of the number of combats the pilot has been in. This chance decreases by about a factor of 3 from the first to the sixth combat. A study made by the Operations Research Group, U. S. Army Air Forces, indicates that the chance of shooting down the enemy when once in a combat increases by 50 per cent or more with increasing experience.

The exchange rate will also depend on the types of planes entering the engagement. An analysis of British-German engagements indicates that Spitfire 9 has an exchange rate about twice that of Spitfire 5. The difference is probably mostly due to the difference in speed, about 40 knots. There are indications that the exchange rate for F6F-5 is considerably larger than that for the F6F-3. Since the factors of training, experience, and plane type all appear to have been in the favor of the United States, it is not surprising that the exchange rate turned out to be as large as 10.

## TACTICS TO EVADE TORPEDOES

The last example given in this section will continue the analysis of the submarine versus submarine problem discussed. It has been shown that there was a possibility that our own submarines in the Pacific were being torpedoed by Japanese submarines, and that there was a good chance that several of our casualties were due to this cause. Presumably the danger was greatest when our submarine was traveling on the surface and the enemy submarine was submerged. It was important, therefore, to consider possible measures to minimize this danger. One possibility was to install a simple underwater listening device beneath the hull of the submarine, to indicate the presence of a torpedo headed toward the submarine. Torpedoes driven by compressed air can be spotted by a lookout, since they leave a characteristic wake; electric torpedoes, on the other hand, cannot be spotted by their wake. All types of torpedoes, however, have to run at a speed considerably greater than that of the target, and therefore their propellers generate a great deal of underwater sound. This sound, a characteristic high whine, can be detected by very simple underwater microphones, and the general direction from which the sound comes can be determined by fairly simple means.

Microphone equipment to perform this function had already been developed by NDRC; it remained to determine the value of installing it. In other words, even if the torpedo could be heard and warning given, could it be evaded? The chief possibility, of course, lay in radical maneuvers. A submarine (or a ship) presents a much smaller target to the torpedo end on than it does broadside. Consequently, as soon as a torpedo is heard, and its direction is determined, it is advisable for the submarine to turn toward or away from the torpedo, depending on which is the easier maneuver.

## GEOMETRICAL DETAILS

The situation is shown in Figure 2. Here the submarine is shown traveling with speed $u$ along the dash-dot line. It discovers a torpedo at range $R$ and at angle on the bow $\theta$ headed toward it. For correct firing, the torpedo is not aimed at

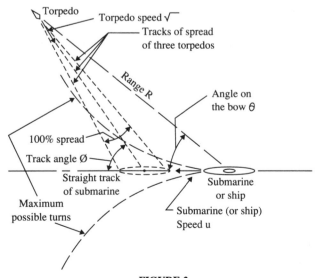

**FIGURE 2**
*Quantities connected with analysis of torpedo attack on submarine or ship.*

where the submarine is, but at where the submarine will be when the torpedo gets there. The relation between the track angle $\phi$, the angle on the bow $\theta$, the speed of torpedo and submarine, and the range $R$ can be worked out from the geometry of triangles. The aim, of course, is never perfect, and operational data indicate that the standard deviation for torpedoes fired from U. S. submarines is about 6 degrees of angle.

In most cases, more than one torpedo is fired. For instance, if three torpedoes are fired in a salvo, the center torpedo is usually aimed at the center of the target. If the other two are aimed to hit the bow and stern of the target, the salvo of three is said to have a 100 per cent spread. Due to the probable error in aim, it turns out to be somewhat better to increase the spread to 150 per cent, so that, if the aim were perfect, the center torpedo would hit amidships, and the other two would miss ahead and astern. Analysis of the type to be given later shows that a salvo of three with 150 per cent spread gives a somewhat greater probability of hit than does a salvo with 100 per cent spread.[4]

A glance at Figure 2 shows that if the *track angle* $\phi$ is less than 90° the submarine should turn as sharply as possible toward the torpedoes in order to present as small a target as possible; if the track angle is greater than 90° the turn should be away from the torpedoes. Assuming a three torpedo salvo, with 150 per cent spread and 7° standard deviation in aim, and knowing the maximum rate of turn of the submarine and the speed of the submarine and torpedo, it is then possible to compute the probability of hit of the salvo, as a function of the angle-on-the-bow $\theta$ and the range $R$ at which the submarine starts its turn. If the range is large enough, the submarine can turn completely toward or away from the torpedoes (this is called "combing the tracks") and may even move completely outside of the track of the salvo. If the torpedoes are not discovered until at short range, however, very little improvement can be obtained by turning.

One can therefore compute the probability of hitting the submarine if it starts to turn when it hears a torpedo at some range and angle-on-the-bow. This can be plotted on a diagram showing contours of equal probability of sinking, and these can be compared with contours for probability of sinking if the submarine takes no evasive action, but continues on a straight course. A typical set of contours is shown in Figure 3.

The solid contours show the probabilities of a hit when the submarine takes correct evasive action. The dotted contours give the corresponding chances when a submarine continues on a steady course. One sees that the dotted contour for 30 per cent chance of hit covers a much greater area than a solid contour for the same chance. In other words, at these longer ranges the evasive action of the submarine has a greater effect. The contours for 60 per cent chance of hit do not show the corresponding improvement, since, by the time the torpedo is so close to the submarine, maneuver has little chance of helping the situation. One sees that, if one can hear the torpedo as far away as 2,000 yards, a very large reduction in the chance of being hit can be produced by the correct evasive maneuvers.

Since these contours represent, in effect, vulnerability diagrams for torpedo attack, they suggest the directions in which lookout activity should be emphasized. The greatest danger exists at a relative bearing corresponding to a 90 degree track angle, and the sector from about 30 degrees to 105 degrees on the bow should receive by far the most attention. The narrow separation of the contours corresponding to evasive action emphasizes the extreme importance of the range of torpedo detection. In many instances a reduction of 500 yards in detection range may cut in half the target probability of escaping.

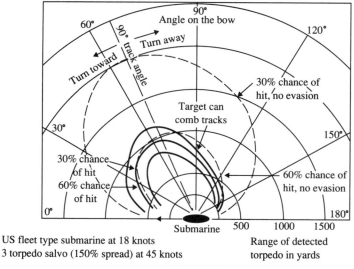

US fleet type submarine at 18 knots
3 torpedo salvo (150% spread) at 45 knots

Range of detected
torpedo in yards

**FIGURE 3**

*Chance of surviving torpedo salvo by sharp turns as soon as torpedo is detected, as function of torpedo range and bearing when detected, compared to chance of survival when no evasive action is taken.*

Another factor vital to the efficacy of evasive turning is the promptness with which it is initiated. For 45-knot torpedoes, every 10 seconds delay in execution of the turn corresponds approximately to a reduction of 250 yards in the distance from the torpedo to the target. Thus it is apparent that a 20 seconds delay in beginning the evasive turning will probably halve the chances of successful evasion.

These same calculations, with different speeds and different dimensions for the target vessel, may be used to indicate to the submarine where it is best to launch its torpedoes in order to minimize the effect of evasive turning of the target ship. One sees that it is best to launch torpedoes, if possible, with a track angle of approximately 90 degrees. One sees also the importance of coming close to the target before firing the salvo, since evasive action is much less effective when begun with the torpedo less than 2,000 yards away.

This study showed the value of good torpedo-detection microphones, with ranges of at least 2,000 yards, and supported the case for their being installed on fleet submarines. Publication of the study to the fleet indicating the danger from Japanese submarines and of the usefulness of evasive turns, produced an alertness which saved at least four U.S. submarines from being torpedoed, according to the records.

## THE SQUID PROBLEM

As a somewhat more complicated example, we shall now consider the problem of determining the effectiveness of the antisubmarine device known as *Squid*. This is a device which throws three proximity-fuzed depth charges ahead of the launching ship in a triangular pattern. In order to simplify the problem we shall make the assumption that the heading of the submarine is known, and also the assumption that the aiming errors are distributed in a circular normal fashion, with the same standard deviation for all depths. We shall also assume that, if a single depth charge passes within a lethal radius $R$

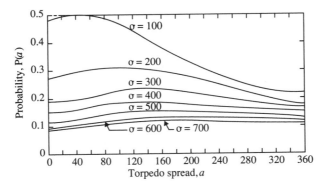

**FIGURE 4**
*Probability of sinking ship with spread of three torpedoes.*

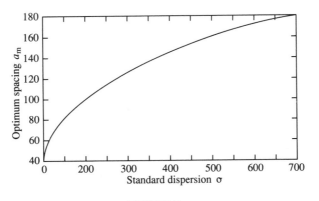

**FIGURE 5**
*Optimum spread as a function of dispersion of aiming errors.*

of the submarine, the submarine will be sunk. We wish to determine the best pattern for the depth charges.

For any given pattern, the pattern damage function depends on two variables, $x$ and $y$, the aiming errors along and perpendicular to the course of the submarine. For any pair of values of $x$ and $y$, $D_p(x, y)$ is 1 if the submarine is sunk, and 0 otherwise. A typical case is shown in Figure 6. The origin is the point of aim, and the positions of the depth charges in the pattern are indicated by crosses. Each possible position of the center of the submarine is represented by a point in this plane. (Note that $x$ and $y$ are actually the *negatives* of the aiming errors.) The three shaded regions represent the positions at which the submarine is destroyed by each of the three depth charges. The pattern damage function is 1 inside the shaded regions, and 0 in the unshaded regions.

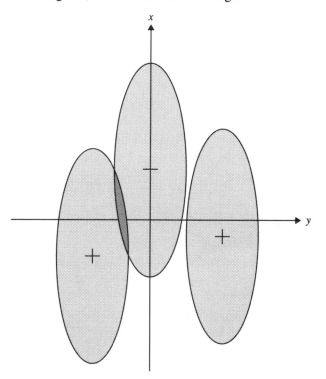

**FIGURE 6**
*Damage function for "Squid" pattern.*

Let $f(x, y)\, dxdy$ be the probability that the center of the submarine be in the area element $dxdy$. Then the probability of destroying the submarine is

$$P = \int D(x,y) f(x,y) dxdy$$

$$= \int_D f(x,y) dxdy \,. \tag{1}$$

In the last equation the region $D$ of integration is just the shaded area in Figure 6. Because of the irregular shape of this region, analytical evaluation of this integral is impractical, and graphical methods must be used. In problems of this type a very convenient aid is a form of graph paper known as "circular probability paper." This paper is divided into cells in such a way that, if a point is chosen from a circular normal distribution, the point is equally likely to fall in any of the cells. If an area is drawn on the paper, the chance of a point falling inside the area is proportional to the number of cells in the area. It follows that the integral of equation (1) can be easily evaluated by drawing the damage function to the proper scale on circular probability paper, and counting the cells in the shaded area.

This method gives a rapid means of finding the probability of destroying the submarine with any given pattern. To find the best pattern, we note that changing the position of any one of the depth charges amounts to shifting the corresponding shaded area in Figure 6 parallel to itself to a new position. If three templates are made by cutting the outline of the shaded area for a single depth charge out of a sheet of transparent material, to the correct scale to go with the circular probability paper, then the best pattern can be found by moving the templates around on a sheet of circular probability paper until the number of cells within the lethal area is a maximum.

## ENDNOTES

[1]This suggests the advantage of using logarithmic graph paper in plotting data. Unity is zero hemibels, 3 is 1 hemibel, 10 is 2 hemibels, 30 is 3 hemibels, and 10,000 is 8 hemibels. A hemibel is 5 decibels. An appropriate abbreviation would be hb, corresponding to db for decibel.

[2][The reference is to another section of the book from which this excerpt has been taken. ED.]

[3][Not included in this selection. ED.]

[4][Not included in this selection. ED.]

# PART XXI

# A MATHEMATICAL THEORY OF ART

1. Mathematics of Aesthetics
   *by George David Birkhoff*

# COMMENTARY ON
# GEORGE DAVID BIRKHOFF

George David Birkhoff (1884–1944) was a leading figure among the mathematicians of the present century. He stood out for the powerful faculty which he brought to bear on complex and fundamental problems, and for the diversity of his researches. He cared about many things and took fruitful thought to all of them.

Birkhoff studied at Chicago and at Harvard. He wrote his Ph.D. thesis on differential equations and continued to work in this field and in group theory during his early teaching years at the University of Wisconsin and at Princeton. A memoir published in 1911[1] on the theory of difference equations attracted wide notice, as did an extremely ingenious verification, two years later, of a famous conjecture by Poincaré as to the topological properties of a ring-shaped region bounded by concentric circles. This proof was important because it uncovered an interesting relationship between analysis situs and dynamics. Poincaré had predicted this result by pointing out that, if his conjecture were verified, "it would lead to concluding the existence of infinitely many periodic motions in the restricted problem of three bodies and similar problems."[2] Birkhoff continued to win recognition for papers on general dynamics and the theory of orbits, turning his attention in these studies from the purely mathematical topics with which he had dealt at the outset of his career. He was thus led to his investigations of the ergodic hypothesis.

Some eighty years ago the German physicist Boltzmann gave the name ergodic "to those mechanical systems which had the property that each particular motion, when continued indefinitely, passes through every configuration and state of motion of the system which is compatible with the value of the total energy."[3] Clerk Maxwell and Boltzmann put forward the hypothesis that "the systems considered in the kinetic theory of gases are ergodic," and Birkhoff produced a beautiful theorem to prove that these intuitions were justified. He showed that an "idealized billiard ball" moving on an "idealized convex billiard table" tends "in the limit to lie in any assigned area of the table a definite proportion of the time."[4] The theorem is not of vital significance to billiard players but turns out to be applicable to numerous deep problems of analysis in applied mathematics. Among them is the celebrated celestial mechanics problem of three bodies—astronomical counterparts of the billiard ball.[5]

Birkhoff contributed to the development of point-set theory, to the study of n-dimensional space and to mathematical physics. He wrote two books on relativity which were "widely read and characteristically original in treatment."[6] In the latter part of the 1920s he began to formulate his views on a relatively novel subject, the mathematical treatment of aesthetics. Pythagoras, it will be recalled, had some time earlier applied mathematics to music by showing that certain simple arithmetical ratios of lengths of strings determined the musical intervals: the octave, fifth and fourth.[7] Birkhoff was similarly attracted to aesthetics by an interest in the formal structure of Western music, but he later conceived the more ambitious goal of creating a "general mathematical theory of the fine arts, which would do for aesthetics what had been achieved in another philosophical subject, logic, by the symbolisms of Boole, Peano, and Russell."[8] He describes the aesthetic feeling as "intuitive" and "sui generis," but holds nevertheless that the attributes upon which aesthetic value depends are accessible to measure. Three main variables constitute "the typical aesthetic experience": the *complexity* (C) of the object, the feeling of value or *aesthetic measure* (M), and the property of harmony, symmetry or *order* (O). These yield the basic formula

$$M = \frac{O}{C}$$

expressing the hypothesis that the aesthetic measure is determined "by the density of order relations in the aesthetic object." The formula may be regarded as a symbolic restatement of the famous definition of the Beautiful by the eighteenth-century Dutch philosopher Frans Hemsterhuis: "that which gives us the greatest number of ideas in the shortest space of time."[9] Birkhoff proceeded from this basic definition to a consideration of what appeals to us in polygonal forms, ornaments, vases, diatonic chords and harmony, melody, and the musical quality in poetry.[10]

Having applied his methods to aesthetics, he was emboldened to tackle ethics. The question presented itself "almost irresistibly" to his mind: "Is not a similar treatment of analytic ethics possible?" Here too he had been anticipated by Pythagoras who had asserted that "justice is represented by a square number." This is not a very heavy contribution to science or to morals, but it has the considerable merit of not contradicting any other definition of justice; also it is pleasingly mystical. Birkhoff's approach is not mystical, though he applauds Pythagoras. He outlines a rational program to "clarify and codify the vast ethical domain." The formula of ethical measure is analogous in many respects to the earlier undertaking in aesthetics.

The two selections following exhibit the inventiveness and the stimulating play of Birkhoff's mind. His views are modestly put forward; no claim is made that they form a well-rounded system. For my part, they are unconvincing but never tedious. In any case it may be argued that mathematicians should have a turn at examining the beautiful and the good; philosophers, theologians, writers on aesthetics and other experts have been probing these matters for more than 2000 years without making any notable advance.

# ENDNOTES

[1]"General theory of linear difference equations," *Trans. Amer. Math. Soc.*, 12 (1911), 242–284. Much of the information in this note is taken from the obituary on Birkhoff by Sir Edmund T. Whittaker, *Journal of the London Mathematical Society*, Vol. 20, Part 2, April 1945, pp. 121–128.

[2]Whittaker, *op. cit.*, p. 124.

[3]Whittaker, *op. cit.*, p. 125.

[4]G. D. Birkhoff, "What is the Ergodic Theorem?" *American Mathematical Monthly*, Vol. 49, April 1942.

[5]"Thus in G. W. Hill's celebrated idealization of the earth-sun-moon problem (the restricted problem of three bodies) we can at once assert (with probability 1) that the moon possesses a true mean angular state of rotation about the earth (measured from the epoch), the same in both directions of time." Birkhoff, *loc. cit.*

[6]Birkhoff, *Relativity and Modern Physics*, with the co-operation of R. E. Langer, Cambridge, 1923. Birkhoff, *The Origin, Nature, and Influence of Relativity*, New York, 1925.

[7]"The fact that pitch is numerically measurable was known to the early Greek philosopher Pythagoras who observed that if the length of a musical string be divided in the ratio of 1 to 2, then the note of the shorter string is an octave higher," G. D. Birkhoff, *Aesthetic Measure*, Cambridge, 1933, p. 90.

[8]Whittaker, *op. cit.*, p. 127.

[9]*Lettre sur la sculpture*, 1769.

[10]The first account of Birkhoff's theory of aesthetic measure was given in 1928 at Bologna: *Atti Congressi Bologna*, 1 (1928), 315–333; the most complete statement appears in his book *Aesthetic Measure*, Cambridge, 1933.

*The business of a poet is to examine not the individual but the species; to remark general properties and large appearances. He does not number the streaks of the tulip, or describe the different shades of verdure of the forest; he is to exhibit . . . such prominent and striking features as recall the original to every mind.*

—SAMUEL JOHNSON (*Imlac in Rasselas*)

# MATHEMATICS OF AESTHETICS

## By George David Birkhoff

## THE BASIC FORMULA

### 1. THE AESTHETIC PROBLEM

Many auditory and visual perceptions are accompanied by a certain intuitive feeling of value, which is clearly separable from sensuous, emotional, moral, or intellectual feeling. The branch of knowledge called aesthetics is concerned primarily with this aesthetic feeling and the aesthetic objects which produce it.

There are numerous kinds of aesthetic objects, and each gives rise to aesthetic feeling which is *sui generis*. Such objects fall, however, in two categories: some, like sunsets, are found in nature, while others are created by the artist. The first category is more or less accidental in quality, while the second category comes into existence as the free expression of aesthetic ideals. It is for this reason that art rather than nature provides the principal material of aesthetics.

Of primary significance for aesthetics is the fact that the objects belonging to a definite class admit of direct intuitive comparison with respect to aesthetic value. The artist and the connoisseur excel in their power to make discriminations of this kind.

To the extent that aesthetics is successful in its scientific aims, it must provide some rational basis for such intuitive comparisons. In fact it is the fundamental problem of aesthetics to determine, within each class of aesthetic objects, those specific attributes upon which the aesthetic value depends.

### 2. NATURE OF THE AESTHETIC EXPERIENCE

The typical aesthetic experience may be regarded as compounded of three successive phases: (1) a preliminary effort of attention, which is necessary for the act of perception, and which increases in proportion to what we shall call the *complexity (C)* of the object; (2) the feeling of value or *aesthetic measure (M)* which rewards this effort; and finally (3) a realization that the object is characterized by a certain harmony, symmetry, or *order (O)*, more or less concealed, which seems necessary to the aesthetic effect.

## 3. MATHEMATICAL FORMULATION OF THE PROBLEM

This analysis of the aesthetic experience suggests that the aesthetic feelings arise primarily because of an unusual degree of harmonious interrelation within the object. More definitely, if we regard *M, O,* and *C* as measurable variables, we are led to write

$$M = \frac{O}{C}$$

and thus to embody in a basic formula the conjecture that the aesthetic measure is determined by the density of order relations in the aesthetic object.

The well known aesthetic demand for 'unity in variety' is evidently closely connected with this formula. The definition of the beautiful as that which gives us the greatest number of ideas in the shortest space of time (formulated by Hemsterhuis in the eighteenth century) is of an analogous nature.

If we admit the validity of such a formula, the following mathematical formulation of the fundamental aesthetic problem may be made: *Within each class of aesthetic objects, to define the order* O *and the complexity* C *so that their ratio* M = O/C *yields the aesthetic measure of any object of the class.*

It will be our chief aim to consider various simple classes of aesthetic objects, and in these cases to solve as best we can the fundamental aesthetic problem in the mathematical form just stated. Preliminary to such actual application, however, it is desirable to indicate the psychological basis of the formula and the conditions under which it can be applied.

## 4. THE FEELING OF EFFORT IN AESTHETIC EXPERIENCE

From the physiological-psychological point of view, the act of perception of an aesthetic object begins with the stimulation of the auditory or visual organs of sense, and continues until this stimulation and the resultant cerebral excitation terminate. In order that the act of perception be successfully performed, there is also required the appropriate field of attention in consciousness. The attentive attitude has of course its physiological correlative, which in particular ensures that the motor adjustments requisite to the act of perception are effected when required. These adjustments are usually made without the intervention of motor ideas such as accompany all voluntary motor acts, and in this sense are 'automatic.' In more physiological terms, the stimulation sets up a nerve current which, after reaching the cerebral cortex, in part reverts to the periphery as a motor nerve current along a path of extreme habituation, such as corresponds to any automatic act.

Now, although these automatic adjustments are made without the intervention of motor ideas, nevertheless there is a well known feeling of effort or varying tension while the successive adjustments are called for and performed. This constitutes a definite and important part of the general feeling characteristic of the state of attention. The fact that interest of some kind is almost necessary for sustained attention would seem to indicate that this feeling has not a positive (pleasurable) tone but rather a negative one. Furthermore, if we bear in mind that the so-called automatic acts are nothing but the outcome of unvarying voluntary acts habitually performed, we may reasonably believe that there remain vestiges of the motor ideas originally involved, and that it is these which make up this feeling of effort.

From such a point of view, the feeling of effort always attendant upon perception appears as a summation of the feelings of tension which accompany the various automatic adjustments.

## 5. THE PSYCHOLOGICAL MEANING OF 'COMPLEXITY'

Suppose that $A, B, C, \ldots$ are the various automatic adjustments required, with respective indices of tension $a, b, c, \ldots$, and that these adjustments $A, B, C, \ldots$ take place $r, s, t, \ldots$ times respectively. Now it is the feeling of effort or tension which is the psychological counterpart of what has been referred to as the complexity $C$ of the aesthetic object. In this manner we are led to regard the sum of the various indices as the measure of complexity, and thus to write

$$C = ra + sb + tc + \cdots.$$

A simple illustration may serve to clarify the point of view. Suppose that we fix attention upon a convex polygonal tile. The act of perception involved is so quickly performed as to seem nearly instantaneous. The feeling of effort is almost negligible while the eye follows the successive sides of the polygon and the corresponding motor adjustments are effected automatically. Nevertheless, according to the point of view advanced above, there is a slight feeling of tension attendant upon each adjustment, and the complexity $C$ will be measured by the number of sides of the polygon.

Perhaps a more satisfying illustration is furnished by any simple melody. Here the automatic motor adjustments necessary to the act of perception are the incipient adjustments of the vocal cords to the successive tones. Evidently in this case the complexity $C$ will be measured by the number of notes in the melody.

## 6. ASSOCIATIONS AND AESTHETIC FEELING

Up to this point we have only considered the act of perception of an aesthetic object as involving a certain effort of attention. This feeling of effort is correlated with the efferent part of the nerve current which gives rise to the required automatic motor adjustments, and has no direct reference to aesthetic feeling.

For the cause (physiologically speaking) of aesthetic feeling, we must look to that complementary part of the nerve current which, impinging on the auditory and visual centers, gives rise to sensations derived from the object, and, spreading from thence, calls various associated ideas with their attendant feelings into play. These sensations, together with the associated ideas and their attendant feelings, constitute the full perception of the object. It is in these associations rather than in the sensations themselves that we shall find the determining aesthetic factor.

In many cases of aesthetic perception there is more or less complete identification of the percipient with the aesthetic object. This feeling of 'empathy,' whose importance has been stressed by the psychologist Lipps,[1] contributes to the enhancement of the aesthetic effect. Similarly, actual participation on the part of the percipient, as in the case of singing a tune as well as hearing it, will enhance the effect.

## 7. THE INTUITIVE NATURE OF SUCH ASSOCIATIONS

Mere verbal associations are irrelevant to the aesthetic experience. In other words, aesthetic associations are *intuitive* in type.

When, for instance, I see a symmetrical object, I feel its pleasurable quality, but do not need to assert explicitly to myself, "How symmetrical!" This characteristic feature may be explained as follows. In the course of individual experience it is found generally that symmetrical objects possess exceptional and desirable qualities. Thus our own bodies are not regarded as perfectly formed unless they are symmetrical. Furthermore, the visual and tactual technique by which we perceive the symmetry of various objects is uniform, highly developed, and almost instantaneously applied. It is this technique which forms the associative 'pointer.' In consequence of it, the perception of any symmetrical object is accompanied by an intuitive aesthetic feeling of positive tone.

It would even seem to be almost preferable that no verbal association be made. The unusual effectiveness of more or less occult associations in aesthetic experience is probably due to the fact that such associations are never given verbal reference.

## 8. THE RÔLE OF SENSUOUS FEELING

The typical aesthetic perception is primarily of auditory or visual type, and so is not accompanied by stimulation of the end-organs of the so-called lower senses. Thus the sensuous feeling which enters will be highly refined. Nevertheless, since sensuous feeling with a slight positive tone ordinarily accompanies sensations of sight and of sound, it might appear that such sensuous feeling requires some consideration as part of the aesthetic feeling. Now, in my opinion, this component can be set aside in the cases of most interest just because the positive tone of sensuous feeling is always present, and in no way differentiates one perception from another.

For example, all sequences of pure musical tones are equally agreeable as far as the individual sensations are concerned. Yet some of these sequences are melodic in quality, while others are not. Hence, although the agreeableness of the individual sounds forms part of the tone of feeling, we may set aside this sensuous component when we compare the melodic quality of various sequences of musical tones.

To support this opinion further, I will take up briefly certain auditory facts which at first sight appear to be in contradiction with it.

If a dissonant musical interval, such as a semitone, is heard, the resultant tone of feeling is negative. Similarly, if a consonant interval like the perfect fifth is heard, the resultant tone of feeling is positive. But is not the sensation of a dissonant interval to be considered a single auditory sensation comparable with that of a consonant interval, and is it not necessary in this case at least to modify the conclusion as to the constancy of the sensuous factor?

In order to answer this question, let us recall that musical tones, as produced either mechanically or by the human voice, contain a pure fundamental tone of a certain frequency of vibration and pure overtones of double the frequency (the octave), of triple the frequency (the octave of the perfect fifth), etc.; here, with Helmholtz, we regard a pure tone as the true individual sensation of sound. Thus 'association by contiguity' operates to connect any tone with its overtones.

If such be the case, a dissonant interval, being made up of two dissociated tones, may possess a negative tone of feeling on account of this dissociation; while the two constituent tones of a consonant interval, being connected by association through their overtones, may possess a positive tone of feeling in consequence. Hence the obvious difference in the aesthetic effect of a consonant and a dissonant musical interval can be explained on the basis of association alone.

## 9. FORMAL AND CONNOTATIVE ASSOCIATIONS

It is necessary to call attention to a fundamental division of the types of associations which enter into the aesthetic experience.

Certain kinds of associations are so simple and unitary that they can be at once defined and their rôle can be ascertained with accuracy. On the other hand, there are many associations, of utmost importance from the aesthetic point of view, which defy analysis because they touch our experience at so many points. The associations of the first type are those such as symmetry; an instance of the second type would be the associations which are stirred by the *meaning* of a beautiful poem.

For the purpose of convenient differentiation, associations will be called 'formal' or 'connotative' according as they are of the first or second type. There will of course be intermediate possibilities.

More precisely, formal associations are such as involve reference to some simple physical property of the aesthetic object. Two simple instances of these are the following:

> rectangle in vertical position → symmetry about vertical;
> interval of note and its octave → consonance.

There is no naming of the corresponding property, which is merely pointed out, as it were, by the visual or auditory technique involved. All associations which are not of this simple formal type will be called connotative.

## 10. FORMAL AND CONNOTATIVE ELEMENTS OF ORDER

The property of the aesthetic object which corresponds to any association will be called an 'element of order' in the object; and such an element of order will be called formal or connotative according to the nature of the association. Thus a formal element of order arises from a simple physical property such, for instance, as that of consonance in the case of a musical interval or of symmetry in the case of a geometrical figure.

It is not always the case that the elements of order and the corresponding associations are accompanied by a positive tone of feeling. For example, sharp dissonance is to be looked upon as an element of order with a negative tone of feeling.

## 11. TYPES OF FORMAL ELEMENTS OF ORDER

The actual types of formal elements of order which will be met with are mainly such obvious positive ones as repetition, similarity, contrast, equality, symmetry, balance, and sequence, each of which takes many forms. These are in general to be reckoned as positive in their effect.

Furthermore there is a somewhat less obvious positive element of order, due to suitable centers of interest or repose, which plays a rôle. For example, a painting should have one predominant center of interest on which the eye can rest; similarly in Western music it is desirable to commence in the central tonic chord and to return to this center at the end.

On the other hand, ambiguity, undue repetition, and unnecessary imperfection are formal elements of order which are of strongly negative type. A rectangle nearly but not quite a square is unpleasantly ambiguous; a poem overburdened with alliteration and assonance fatigues by undue repetition; a musical performance in which a single wrong note is heard is marred by the unnecessary imperfection.

## 12. THE PSYCHOLOGICAL MEANING OF 'ORDER'

We are now prepared to deal with the order $O$ of the aesthetic object in a manner analogous to that used in dealing with the complexity $C$.

Let us suppose that associations of various types $L, M, N, \ldots$ take place with respective indices of tone of feeling $l, m, n, \ldots$ In this case the indices may be positive, zero, or negative, according as the corresponding tones of feeling are positive, indifferent, or negative. If the associations, $L, M, N, \ldots$ occur $u, v, w, \ldots$ times respectively, then we may regard the total tone of feeling as a summational effect represented by the sum $ul + vm + \cdots$.

This effect is the psychological counterpart of what we have called the order $O$ of the aesthetic object, inasmuch as $L, M, N, \ldots$ correspond to what have been termed the elements of order in the aesthetic object. Thus we are led to write

$$O = ul + vm + wn + \cdots.$$

By way of illustration, let us suppose that we have before us various polygonal tiles in vertical position. What are the elements of order and the corresponding associations which determine the feeling of aesthetic value accompanying the act of perception of such a tile? Inasmuch as a detailed study of polygonal form is made in the next chapter, we shall merely mention three obvious positive elements of order, without making any attempt to choose indices. If a tile is symmetric about a vertical axis, the vertical symmetry is felt pleasantly. Again, a tile may have symmetry of rotation; a square tile, for example, has this property, for it can be rotated through a right angle without affecting its position. Such symmetry of rotation is also appreciated immediately. Lastly, if the sides of a tile fall along a rectangular network, as in the case of a Greek cross, the relation to the network is felt agreeably.

## 13. THE CONCEPT OF AESTHETIC MEASURE

The aesthetic measure $M$ of a class of aesthetic objects is primarily any quantitative index of their comparative aesthetic effectiveness.

It is impossible to compare objects of different types, as we observed at the outset. Who, for instance, would attempt to compare a vase with a melody? In fact, for comparison to be possible, such classes must be severely restricted. Thus it is futile to compare a painting in oils with one in water colors, except indirectly, by the comparison of each with the best examples of its type; to be sure, the two paintings might be compared, in respect to composition alone, by means of photographic reproduction. On the other hand, photographic portraits of the same person are readily compared and arranged in order of preference.

But even when the class is sufficiently restricted, the preferences of different individuals will vary according to their taste and aesthetic experience. Moreover the preference of an individual will change somewhat from time to time. Thus such aesthetic comparison, of which the aesthetic measure $M$ is the determining index, will have substantial meaning only when it represents the normal or average judgment of some selected group of observers. For example, in the consideration of Western music it would be natural to abide by the consensus of opinion of those who are familiar with it.

Consequently the concept of aesthetic measure $M$ is applicable only if the class of objects is so restricted that direct intuitive comparison of the different objects becomes possible, in which case the arrangement in order of aesthetic measure represents the aesthetic judgment of an idealized 'normal observer.'

## 14. THE BASIC FORMULA

If our earlier analysis be correct, it is the intuitive estimate of the amount of order $O$ inherent in the aesthetic object, as compared with its complexity $C$, from which arises the derivative feeling of the aesthetic measure $M$ of the different objects of the class considered. We shall first make an argument to this effect on the basis of an analogy, and then proceed to a more purely mathematical argument.

The analogy will be drawn from the economic field. Among business enterprises of a single definite type, which shall be held the most successful? The usual answer would take the following form. In each business there is involved a certain investment $i$ and a certain annual profit $p$. The ratio $p/i$, which represents the percentage of interest on the investment, is regarded as the economic measure of success.

Similarly in the perception of aesthetic objects belonging to a definite class, there is involved a feeling of effort of attention, measured by $C$, which is rewarded by a certain positive tone of feeling, measured by $O$. It is natural that reward should be proportional to effort, as in the case of a business enterprise. By analogy, then, it is the ratio $O/C$ which best represents the aesthetic measure $M$.

## 15. A MATHEMATICAL ARGUMENT

More mathematically, but perhaps not more convincingly, we can argue as follows. In the first place it must be supposed that if two objects of the class have the same order $O$ and the same complexity $C$, their aesthetic measures are to be regarded as the same. Hence we may write

$$M = f(O, C)$$

and thus assert that the aesthetic measure depends functionally upon $O$ and $C$ alone.

It is obvious that if we increase the order without altering the complexity, or if we diminish the complexity without altering the order, the value of $M$ should be increased. But these two laws do not serve to determine the function $f$.

In order to do so, we imagine the following hypothetical experiment. Suppose that we have before us a certain set of $k$ objects of the class, all having the same order $O$ and the same complexity $C$, and also a second set of $k'$ objects of the class, all having the order $O'$ and complexity $C'$. Let us choose $k$ and $k'$ so that $k'C'$ equals $kC$.

Now proceed as follows. Let all of the first set of objects be observed, one after the other; the total effort will be measured by $kC$ of course, and the total tone of aesthetic feeling by $kO$. Similarly let all of the second set be observed. The effort will be the same as before, since $k'C'$ equals $kC$; and the total tone of feeling will be measured by $k'O'$.

If the aesthetic measure of the individual objects of the second class is the same as of the first, it would appear inevitable that the total tone of feeling must be the same in both cases, so that $k'O'$ equals $kO$. With this granted, we conclude at once that the ratios $O'/C'$ and $O/C$ are the same. In consequence the aesthetic measure only depends upon the ratio $O$ to $C$:

$$M = f\left(\frac{O}{C}\right).$$

The final step can now be taken. Since it is not the actual numerical magnitude of $f$ that is important but only the relative magnitude when we order according to aesthetic measure, and since $M$ must increase with $O/C$, we can properly define $M$ as equal to the ratio of $O$ to $C$.

It is obvious that the aesthetic measure $M$ as thus determined is zero ($M = 0$) when the tone of feeling due to the associated ideas is indifferent.

## 16. THE SCOPE OF THE FORMULA

As presented above, the basic formula admits of theoretic application to any properly restricted class of aesthetic objects.

Now it would seem not to be difficult in any case to devise a reasonable and simple measure of the complexity $C$ of the aesthetic objects of the class. On the other hand, the order $O$ must take account of all types of associations induced by the objects, whether formal or connotative; and a suitable index is to be assigned to each. Unfortunately the connotative elements of order cannot be so treated, since they are of inconceivable variety and lie beyond the range of precise analysis.

It is clear then that complete quantitative application of the basic formula can only be effected when the elements of order are mainly formal. Of course it is always possible to consider the formula only in so far as the formal elements of order are concerned, and to arrive in this way at a partial application.

Consequently our attention will be directed almost exclusively towards the formal side of art, to which alone the basic formula of aesthetic measure can be quantitatively applied. Our first and principal aim will be to effect an analysis in typical important cases of the utmost simplicity. From the vantage point so reached it will be possible to consider briefly more general questions. In following this program, there is of course no intention of denying the transcendent importance of the connotative side in all creative art.

## 17. A DIAGRAM

The diagram with the attached legend in Figure 1 may be of assistance in recalling the above analysis of the aesthetic experience and the basic aesthetic formula to which it leads.

## 18. THE METHOD OF APPLICATION

Even in the most favorable cases, the precise rules adopted for the determination of $O$, $C$, and thence of the aesthetic measure $M$, are necessarily empirical. In fact the symbols $O$ and $C$ represent social values, and share in the uncertainty common to such values. For example, the 'purchasing power of money' can only be determined approximately by means of empirical rules, and yet the concept involved is of fundamental economic importance.

At the same time it should be added that this empirical method seems to be the only one by which concepts of this general category can be approached scientifically.

We shall endeavor at all times to choose formal elements of order having unquestionable aesthetic importance, and to define indices in the most simple and reasonable manner possible. The underlying facts have to be ascertained by the method of direct introspection.

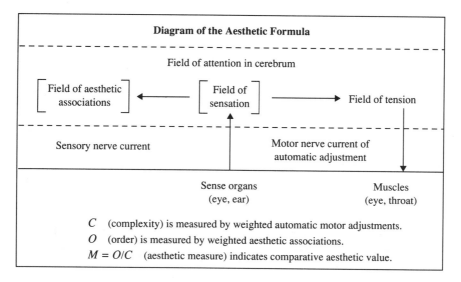

**FIGURE 1**

In particular we shall pay attention to the two following *desiderata*:

As far as possible these indices are to be taken as equal, or else in the simplest manner compatible with the facts.

The various elements of order are to be considered only in so far as they are logically independent. If, for example, $a = b$ is an equality which enters in $O$, and if $b = c$ is another such equality, then the equality $a = c$ will not be counted separately.

## ENDNOTE

[1]*Ästhetik: Psychologie des Schönen und der Kunst,* Hamburg and Leipzig, vol. 1 (1903), vol. 2 (1906).

# PART XXII

# MATHEMATICS OF THE GOOD

## 1

*Philosophers count about two hundred and eighty-eight views of the sovereign good.*

—Pascal

# A Mathematical Approach to Ethics

## By George David Birkhoff

Since the time of the German philosopher, Immanuel Kant, it has been clear that, for certain purposes, philosophic thought may be treated separately in its logical, aesthetic, and ethical aspects, concerned respectively with the true, the beautiful, and the good.

In the last century logic has developed into an independent discipline—the edifice of syllogistic thought—of which all of mathematics appears as the grandiose superstructure.

The concept of "aesthetic measure" which I laid before you in 1932 made possible a more or less mathematical treatment of aesthetics giving promise of taking the subject of analytical aesthetics out of the domain of philosophic speculation into the region of common sense thought. The question thus presents itself almost irresistibly to the mind: Is not a similar treatment of analytic ethics possible? My aim here is to show that such a program seems to be feasible.

To most mathematicians the tendency towards increasing mathematization in these three fundamental aspects of philosophic thought—logic, aesthetics, and ethics—is only what was to be expected; for they are likely to agree with the dictum of the great French philosopher and mathematician, René Descartes, *omnia apud me mathematica fiunt*—with me everything turns into mathematics!

Even in early Greek times the philosopher Pythagoras tried to bring mathematical order into the ethical field by asserting that justice is represented by a square number. This must be looked upon as a mystical conjecture of real importance for ethics. Similarly Plato and Aristotle were always desirous of showing the close relationship of the good and the beautiful, if not their essential identity; and they regarded the beautiful as characterized by unity in variety. Thus, there has always been observable in ethics, as well as in aesthetics, a tendency towards quantitative formulation. The supreme goal of the *summum bonum* or highest good, adopted by the Greeks, is suggestive of this; and the modern utilitarian principle of "the greatest good of the greatest number" reveals still more clearly the same tendency.

A very interesting analogy between aesthetics and ethics is the following. Individuals of so-called artistic temperament often look upon their personal experiences as a succession of aesthetic adventures from which they try to extract the greatest possible enjoyment. Similarly, persons of predominantly moralistic type strive for a maximum of moral satisfaction by making in their daily lives such ethical decisions as will best promote the material and spiritual well-being of their fellows.

Just as the analysis of experience from the aesthetic point of view yields the concept of "aesthetic measure"—the ratio of aesthetic reward to effort of attention—as basic in the evaluation of aesthetic pleasure, so the consideration of experience in its ethical aspects leads to an analogous concept of "ethical measure"—the amount of moral satisfaction based on good accomplished.

The simple ethical formula evidently suggested is:

*M (ethical measure) = G (total good achieved).*

From this point of view the ethically-minded person[1] endeavors always to select that one of the possible courses of action which *maximizes* the ethical measure $G$, just as the aesthetically-minded person continually compares aesthetic objects and prefers those which maximize the aesthetic measure $O/C$.[2] The utilitarian calculus of Jeremy Bentham represents a suggestive semi-philosophical attempt in the same direction.[3]

Let us consider a little more in detail this general parallelism between the aesthetic and ethical domains. In order to do this the use of parallel columns is convenient.

### Aesthetics

Some of the principal aesthetic 'factors' are (+, of positive type) repetition, similarity, contrast, balance, sequence, centers of interest or repose; (−, of negative type) complexity, ambiguity, undue repetition, unnecessary imperfection. These factors enter into the terms $O$ and $C$ of the aesthetic formula,

$$M = O/C$$

The factors involved in the order $O$ may be divided into formal and connotative elements of order, while the complexity $C$ is formal. Only the formal type of elements in $O$ admits of quantitative treatment.

### Ethics

Some of the principal ethical 'factors' are: (+, of positive type) material good, sensuous enjoyment, happiness, intellectual and spiritual achievement; (−, of negative type) material waste and destruction, pain, sorrow, intellectual and spiritual deterioration. These enter into the term $G$ of the ethical formula,

$$M = G$$

The factors involved in the good, $G$, may be divided into the material and the immaterial elements of the good. Only the material type of elements admits of quantitative treatment by the formula.

In aesthetics, objects of a definite class are to be compared in regard to their relative aesthetic measures $M$. Such classes are of extraordinary variety. The theory of aesthetic measure is best exemplified by certain simple formal visual and auditory fields, provided by art rather than by nature.

Artists, connoisseurs, and critics of all kinds are considered to be especially competent judges in their special aesthetic fields. But the aggregate opinion of ordinary lay observers plays a vital rôle.

Aesthetic tastes vary from one individual to another, and are relative to the period and culture concerned. Nevertheless there is a certain grand parallelism to be discerned, due to the presence of certain absolute elements of order, as, for instance, rhythm in music. Cultivated human beings are generally able to understand and appreciate aesthetic objects of all kinds and periods.

Finally, the main phases in the history of aesthetic ideas and literary criticism of special artistic forms can be concisely interpreted by use of the concept of aesthetic measure.

In ethics, each single definite problem is to be considered by itself, and the possible solutions are compared as to their ethical measures, $M$. These problems are also of extraordinary variety. The main interest in ethics is provided by problems arising in practice rather than by artificial problems.

Religious leaders, statesmen, judges, and the socially elect are regarded as the best judges in their several ethical fields. But the general intuitive opinion of mankind often has decisive weight.

Ethical values and ideals vary in a similar manner. Nevertheless, there are always to be found certain absolute elements of the good as, for instance, bravery and loyalty in their socially validated forms. Careful study of the development of such specific forms serves to explain them acceptably to men everywhere as varied manifestations of these absolute elements of the good.

Similarly, the main phases in the history of ethical ideas and of their many special social manifestations admit of concise interpretation through the concept of ethical measure.

Having called attention to this significant general parallelism between aesthetics and ethics, I propose to consider some specific problems which illustrate how the concept of ethical measure can be used. I make no apologies for the simple character of the ideas involved, since it is inevitable that the initial results obtained be rudimentary. As possibly suggestive in this connection, it may be recalled that the first classification of matter as solid, liquid, or gaseous provided a crude trifurcation of nature, which ultimately led to the mathematical theories of elasticity and hydrodynamics.

*Problem I.* A bus driver regularly takes passengers from the starting point $A$ to their destinations along the main road from $A$ to $M$ and along certain side roads on one side of the main road. The majority of the passengers live along the main road, and the side roads are short. The driver wishes to be as

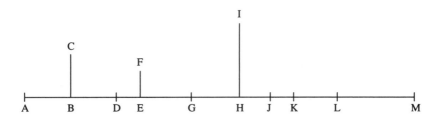

accommodating as possible and to give all the passengers equal consideration. In what order should he take the passengers to their destinations?

His decision is always to deliver the passengers in the natural order of going from $A$ to $M$ along the main road. To justify this decision he might argue as follows:

Suppose first that all the passengers on some trip wished to alight at points on the main route, as not infrequently was the case. If he took them to their destinations in other than the natural order, the series of passengers (as a series) would be less quickly delivered one by one than otherwise, i.e., the first passenger would alight later, the second passenger also, etc. Since all the passengers are to be treated as well as possible, this would be regarded as extremely objectionable by them. But in the event that some of the passengers wish to alight along the short side roads, the additional times required are very small and in the driver's judgment do not need to be considered. Hence he finds that the delivery of passengers should always be as stated.

Let us attempt to formalize this simple reasoning. The underlying good here, $G$, may be regarded here as negative $(-)$, if we reckon upon the unrealizable good of immediate delivery of the passengers as the neutral point $(0)$ from which the reckoning starts. Thus we write

$$G = -\text{(sum of all the trip-durations for the passengers)}.$$

The possible solutions to be considered are the various ways of taking the passengers to their destinations.

The two basic assumptions of the driver are almost but not entirely in agreement with this definition of $G$; they are: (1) the individual trip-durations along the main road are to be diminished as far as possible; (2) the trip-durations along the side roads need not be considered. On this basis his decision is obviously as stated and in general will maximize the good, $G$, as just defined.

However, there might occasionally arise situations in which this solution was not actually the best one by the formula written above. Suppose, for example, there were six passengers, one to be delivered at $C$, and five at $D$, with equal distances $AB$, $BC$, and $BD$ (see the figure on page 2177). Clearly the 'best' solution in this exceptional case would be to deliver the five passengers at $D$ along the main road, and then to return along the main road and deliver the remaining passenger at $C$. In fact, if the driver follows his general rule, we have

$$G = -22a,$$

where $a$ stands for the time required for the bus to go over any one of the equal distances, while if the driver were to deliver his passengers in the reverse order, we would have

$$G = -14a,$$

so that $8a$ units of time would be thereby saved to the passengers.

Nevertheless, the driver decides to deliver the passengers in the usual way. In doing so he goes directly against a perfectly natural postulate referred to above, namely, that if he can shorten (or in no case lengthen) the trip-durations of the successive passengers, he certainly should do so.

Obviously the naturalness and uniformity of the solution adopted by him operates as an important factor in its favor. For the rule of procedure chosen by the driver is readily understood by the passengers and any modification of it in the direction of increased complication might lead to dissatisfaction, especially because the time-schedules of the trip would become even more unpredictable.

*Thus we are led to realize that there are instances in which the simplicity and elegance of the solution of an ethical problem must itself be regarded as one of the imponderable elements of the good which enter into G.*

There is a kind of counterpart to this phenomenon in the aesthetic field: Apparently the intuitive aesthetic judgment tends through an inner necessity to prefer formally simple elements of order in the aesthetic object.

Our second problem is intended to present a very different type of ethical situation which is of high significance, and which involves both material and immaterial elements of the good. Though presented in a specialized form, I believe that the problem selected embodies a situation characteristic of critical moments in the lives of many human beings—moments when the choice must be made between material good with attendant failure in loyalty, on the one hand, or the sacrifice of this material gain with preservation of loyalty, on the other. As has been indicated previously, a complete quantitative treatment cannot be hoped for in such a problem.

*Problem II.* One or the other of two friends of long standing, *A* and *B,* is to be advanced to an opening in the organization in which they hold positions of the same rank. *A* happens to learn that the actual selection will hinge upon the judgment of a certain person *L* belonging to the same organization. Ought *A* to pass this information on to *B?*

The answer of course is that in the circumstances stated *A* ought to inform his friend *B*.

*A*'s reasons for this decision might be formalized as follows: The material goods $g_A$ and $g_B$ which will accrue to him or to his friend through such an advancement are the same: $g_A = g_B = g$. If *A* informs *B*, the immaterial good of his friendship, $f$, with *A* is retained. Therefore, we have simply

$$M = g.$$

On the other hand, if *A* does not tell *B* what he has learned, the friendship between them is destroyed, even if *B* never learns of *A*'s unfriendly act; and so we have

$$M = g - f.$$

Since $g$ exceeds $g - f$, *A* ought to tell *B*, although he realizes that by doing so he gives up a definite personal advantage. In the above reckoning the unfavorable effect upon $A$'s character of not informing $B$ is intentionally disregarded although it might really be the most important consideration of all.

$A$'s decision to pass on the information to $B$ is here assumed to be made on the utilitarian basis. On a hedonistic basis, $A$ might conclude that if he fails to inform $B$, then

$$M = g - f,$$

since he will be certain to win $L$'s special favor, whereas, in the contrary case,

$$M = \frac{g}{2},$$

inasmuch as he would then only have an equal chance with $B$. In this event, he would have to balance the prospect of material advancement against his friendship with $B$.

Again, according to the extent that $A$ believes himself inferior to $B$, he will feel that his chances are lessened by telling $B$. If $A$ is a loyal friend, however, he will not be moved from his decision by such thoughts.

The basic hypothesis has been made here that the information about $L$ is of legitimate practical advantage to $A$ and $B$. It is also assumed that the friendship between $A$ and $B$ is a sincere one, founded upon mutual esteem. For clearly if there were no real friendship, $A$ would not be under any obligation to inform $B$, any more than he would consider it an obligation on $B$'s part to tell him. Of course if $A$ believes that $B$ would not tell him if circumstances were reversed, or that $B$ would employ unfair or unscrupulous tactics to gain $L$'s favor, the bond of friendship between them is already weak; and so the situation would not be the one envisaged in the problem under consideration.

A somewhat similar type of problem, also not infrequently exemplified in human experience, is the following:

*Problem III.* Two men, $A$ and $B$, among six, $A$, $B$, $C$, $D$, $E$, $F$ in control of a certain business, have orally agreed to exchange all relevant information before entering into any arrangement with the others. $A$ and $B$ do this in order to protect their interests in the business. $A$ is approached confidentially by $C$, $D$, $E$, and $F$, and asked if he will concur in a vote giving him important special privileges which are to be withheld from $B$. Actually $A$ does not feel he is entitled to these special privileges any more than $B$ is. How ought $A$ to act?

The ethical course for $A$ to follow is clearly to refuse to connive with $C$, $D$, $E$, and $F$. He should further inform $C$, $D$, $E$, and $F$ that in his opinion to do otherwise would not be fair to $B$.

If $A$ acts in this manner we may write

$$M = 0,$$

meaning thereby that the *status quo ante* is not altered. If $A$ consents to their proposal, we may write for $A$

$$M = g_A - f - e_B,$$

meaning that $A$ gains the privileges mentioned ($g_A$), loses $B$'s friendship ($f$), and possibly incurs $B$'s positive enmity ($e_B$), fraught with danger to him—for instance, the enmity of $B$ might lead to his loss of a valuable reputation for business integrity. Here we find in $G$ two elements of mainly material nature ($g_A$, $e_B$) and one of immaterial nature ($f$).

The two preceding problems have been taken from the field of social ethics. It is of interest that similar problems can be drawn from the field of international ethics. In the problem about to be stated there is no intention to parallel closely any actual problem. The intention is rather to suggest that there may exist somewhat analogous problems which admit of clarification when approached from the point of view of ethical measure.

*Problem IV.* As the result of a war, $B$ has lost a colony $C$ to the nation $A$. This colony $C$ has subsequently been given nearly complete independence by $A$. This action leaves $C$ well satisfied with her status and favorably disposed towards $A$. However, $B$ has an economic need for her former colony $C$, by reason of lack of raw materials which $C$ had formerly supplied; and for this and other more political reasons, $B$ demands the cession of $C$ back to her by $A$. How is $A$ to reply to the demand?

A reasonable analysis on $A$'s part might be the following: $A$ concludes that to return the colony $C$ would not only be objectionable to $C$ but extremely detrimental to $A$'s international standing and prestige as a concession under duress. Furthermore, $A$ feels that if she did agree to $B$'s demand, other similar demands reinforced by further military threats would soon follow. Thus $A$ (and $C$) might write in the event of return of $C$ to $B$

$$M = g_B - h_{A,C}$$

(where $g_B$ = material good to $B$; $h_{A,C}$ = *material and immaterial harm to $A$ and $C$*) and, in the contrary case,

$$M = 0,$$

since there is no reason to believe that the prospect of ultimate war is effectively lessened. Hence $A$ and $C$ would have to balance $B$'s good against their own harm; and so $A$ would almost certainly refuse to cede $C$ back to $B$.

From $B$'s standpoint, however, the analysis in the case of cession would more correctly be

$$M = g_B + p,$$

(where $p$ = good of peace), since $B$ would not admit that $A$ or $C$ would suffer much economic loss thereby, nor that there could be enduring peace without cession; and $B$'s analysis in the contrary case would be

$$M = 0.$$

Thus the balance in favor of cession is $g_B + p$ in $B$'s estimate, and at least $h_{A,C} - g_B$ against cession in $A$ and $C$'s estimate. Thus there is a very serious conflict of ethical judgment. Such a situation naturally raises the question of possible compromise.

*In this and similar cases of apparent conflict in ethical judgments the thorough exploration of all possibilities of compromise is absolutely essential.*

The following is a suggestion of a possibility of such a compromise in this particular case: $A$ notifies $B$ and $C$ that in recognition of $B$'s economic needs and of her claims, she will henceforth not accept from her colony $C$ any more favorable trade status than $C$ accords to $B$.

There is then the possibility that despite $A$'s refusal of $B$'s demand for the return of $C$ to her, $B$ can recover a substantial portion of her former trade with $C$. Thus one might write, in behalf of all the three parties $A$, $B$, and $C$, a formula such as

$$M = \tfrac{1}{2}g_B - \tfrac{1}{4}g_A + \tfrac{1}{2}p,$$

in the event of such a compromise, as against $M = 0$ if the *status quo ante* is preserved. The sole loss for $A$ would be loss of an estimated quarter of $B$'s trade with her ($\tfrac{1}{4}g_A$); $B$ would recover an estimated half of her former trade with $C$ ($\tfrac{1}{2}g_B$) without loss to $C$; and it might be that the resultant improvement in the friendliness of relations between $A$ and $B$ would increase the likelihood of a permanent peace and so slow down the expensive armament race between $A$ and $B$ ($\tfrac{1}{2}p$).

The question of compromise is extremely important in many ethical problems. Is it reasonable to suppose that such compromises can generally be reached? In this connection I recall a conversation with Dean Roscoe Pound and Count Korzybski some years ago. Count Korzybski had expressed the opinion that many conflicts of points of view had their origin mainly in misunderstandings as to the meaning of terms, so that the conflict would disappear as soon as these meanings were agreed upon. I replied that in many disputes the situation resembled rather that arising between two boys contending for a single piece of pie; and Dean Pound was inclined to agree with me. In the tragic condition of the world today, the suggestion might be made that if the division of the single piece of pie into two equal pieces were made (a reasonable compromise), both boys could be induced to accept their portion! . . .

*In my judgment it would be a very constructive program for analytic ethics, to catalogue systematically various significant problems in the three fields of procedural ethics, social ethics, and international ethics, and to classify the main types of solutions on the basis of the formula for ethical measure.*

This has been attempted to a very rudimentary extent in the special problems above.

More specifically, social customs and systems of law and of religion contain a vast mass of ethical data, embodying the accepted ethical solutions of innumerable practical problems of analytic ethics; and the inductive method can generally be applied to treat new problems when they arise. In so far as these solutions are not purely empirical, they could be codified by means of the ethical formula. Such a codification would list and classify the very extensive variety of ethical intuitions (postulates), in part the cause of, and in part the result of, specific social interactions. There is little doubt of the basic rôle which the sentiments of love, goodwill, loyalty, and other feelings of sensuous, aesthetic, or intellectual type play in such intuitions. These provide a substratum of absolute elements, of which the specific manifestation depends on the particular culture and period concerned.

Another useful service of such a program might be to treat the extremely interesting history of ethical ideas by use of the same ethical formula. Thus the early Greek conception of ethical behavior as directed towards the attainment of the *summum bonum* is evidently in consonance with the ethical formula. The customary threefold division of ethical theories into those of hedonistic (or egoistic) type, of utilitarian (or universalistic) type, and of altruistic type is immediately explained in the same way; for if $G_I$ denotes the good of an individual and $G_F$ the good of his fellows, then the three types of ethical theory correspond to the respective formulas: $M = G_I$, $M = G_I + G_F$, $M = G_F$ respectively. It is hardly necessary to say that in promulgating a theory which is (supposedly) of the first or last non-utilitarian type, it is frequently necessary to rob Peter to pay Paul!

Many ethical theorists have tended to take the good and the pleasurable as synonymous; thus, according to Bentham, pleasures differ quantitatively but not qualitatively. From our point of view, this is a necessary assumption if all the constituents of the good in $G$ are looked upon as comparable quantities, as required by the ethical formula.

Some have regarded the striving for perfection as supremely important, thereby emphasizing the achievement of potential good as the final goal; this reaches far into the domain of the *qualitative* application of the ethical formula.

Still others, like Kant, insist upon the dominating rôle of the sense of duty as the "categorical imperative." This validates the innumerable ethical intuitions on which concrete decisions concerning the immaterial good must always depend. Through the sense of duty we feel that it is possible to distinguish clearly between right and wrong, independently of our particular backgrounds, although careful analysis reveals that this independence is by no means complete. In fact the formalization of such intuitions, combined with the use of the general ethical formula, leads to the analytic solution of ethical problems by means of reasoning—a point of view going back to Socrates, Plato, and Aristotle.

There is a further reason why the systematic codification of ethical notions might be of genuine service. Ordinary language provides a vast storehouse of convenient symbols, which (as has been recently emphasized) often bring together under one name a number of quite different entities. For example, we speak of "fatigue" with a good deal of certainty. But what is fatigue? There are specific conditions of fatigue of the muscles, of special nerves, etc.; but what have they in common? Careful experimentation, systematic analysis, and detailed classification are necessary for the proper elucidation of this question. This has indeed been accomplished recently by Professor L. J. Henderson and his colleagues in the Harvard Fatigue Laboratory; and the final upshot is that "fatigue" designates not one thing but many, grouped together largely because of intellectual convenience. Of course the notion of fatigue has no immediate ethical import.

Now many of the terms used constantly in ethical discussions have an even less definite meaning, and frequently provide a convenient emotional support for ethical or unethical action. Certain of these general terms, such as "wisdom" and "justice" seem to be mainly constructive in their effect, but others, like "racial superiority," for example, are positively destructive and dangerous, unless their various meanings have been made very specific. For instance, in speaking of racial superiority, which of the qualities listed below do we regard as really characteristic? Physical prowess and beauty; racial purity; descent from divine ancestors; intellectual capacity and achievement; aesthetic sensibility and artistic creativeness; unselfish idealism; unlimited devotion to the state; economic efficiency; military might; high potentiality of further development? Evidently there are many consistent points of view as to what constitutes "racial superiority"; and so wherever the idea is used it needs to be properly defined and accurately applied in the selected sense.

Our thought here is akin to that of Korzybski, that when human beings realize that certain important general terms have a variety of distinct meanings, the effect of this realization is definitely prophylactic against misunderstanding, prejudice, and intolerance. If the clarification of such important but multiple-valued ideas is not effected soon, the present tragic confusion among men may end in the destruction of civilization.

Such is the general program for ethics to which I desire to direct attention. It is analogous in many respects to that which I have attempted to carry out provisionally in aesthetics. The program involves the introduction of elementary quantitative ideas based on a simple formula for "ethical measure" in order to clarify and codify the vast ethical domain. Conceivably such a program might perform the same kind of useful service for ethics as elementary logic performs for mathematics, and grammar for language.

## ENDNOTES

[1]Or corporate body or state.

[2]$O$ = order, $C$ = complexity.

[3]In this connection, Mr. P. A. Samuelson of the Society of Fellows at Harvard University calls my attention to F. Y. Southworth's very interesting volume on *Mathematical Psychics* (1881).

# PART XXIII

# MATHEMATICS IN LITERATURE

# COMMENTARY ON
# THE ISLAND OF LAPUTA

C aptain Lemuel Gulliver's voyage to the Flying Island, Laputa, is the
longest part of Swift's marvelous book, but in originality, inventiveness
and persuasiveness of satire, it is inferior to the other voyages. At times,
indeed, the reader of *Laputa* wishes that Gulliver might just once complete a
voyage without mishap.

*Gulliver's Travels* was published in October, 1726. The "chief end" of all his
labors, Swift had written to Pope the year before, was "to vex the world rather
than divert it."[1] The world chose to be diverted. From the first the *Travels* were
enormously successful; the first edition sold out in a week, and the book
"received the immediate compliment of wide imitation and translation."[2]
*Laputa,* however, was condemned as uninspired and even imitative. Sharp-
eyed readers noticed that some of its details and situations were copied from
other books, in particular from one of the classic "fantastic voyages," the *True
History* of the Roman writer Lucian, whose works Swift had read in translation.[3] But the fact that Swift borrowed is generally known and has earned him
little criticism in connection with the other voyages. The real objection in the
case of *Laputa* is that he borrowed to poor effect. In a joint letter to Swift a
month after *Gulliver* appeared, Pope and Gay reported a widely held opinion:
"As to other critics they think the flying island is the least entertaining; and so
great an opinion the town have of the impossibility of Gulliver's writing below
himself, it is agreed that part was not writ by the same hand."[4] Many modern
critics have echoed this view, declaring that *Laputa* is not to be compared with
the rest of the *Travels,* that it lacks unity and that its satire is forced and of only
contemporary interest.

This disparagement is too severe. *Laputa* is not as fascinating as *Lilliput* or
*Brobdingnag* or as powerful as the voyage to the Houyhnhnms; but it is by no
means devoid of the superlative literary and philosophic qualities that have won
for the *Travels* a place among the masterpieces of imagination and social
commentary. "There is a pedantry in manners, as in all arts and sciences; and
sometimes in trades," Swift wrote; "Pedantry is properly the over-rating any
kind of knowledge we pretend to."[5] *Laputa* is aimed at pedantry in science and
learning. The Laputians, at least those of "better Quality," are mathematicians,
astrologers and musicians, so wrapped up in their speculations as to be clumsily
unfit for the simplest practical tasks. Their food is served in geometrical shapes
and tastes as flat as a plane; their houses are "very ill built" because of "the
Contempt they bear for practical Geometry, which they despise as vulgar and
mechanick"; their clothes, measured by tailors who use a quadrant, ruler and

compass, are "very ill-made and quite out of shape." Altogether, they are an "awkward and unhandy people," opinionated, "very bad Reasoners" and live in perpetual fear that the earth will collide with a comet or be swallowed by the sun or suffer some other celestial calamity. On the neighboring continent of Balnibarbi, in the metropolis, Lagado, is an Academy of Projectors whose scientists and philosophers spend their time in such pursuits as writing treatises on the malleability of fire, "contriving a new method for building Houses, by beginning at the Roof, and working downwards to the Foundation," and extracting sunbeams out of cucumbers. There is also a mathematical school where Gulliver sees the master teach his pupils "after a Method scarce imaginable to us in *Europe*. The Proposition and Demonstration were fairly written on a thin Wafer, with Ink composed of a Cephalick Tincture. This the Student was to swallow upon a fasting Stomach, and for three Days following eat nothing but Bread and Water. As the Wafer digested, the Tincture mounted to his Brain, bearing the Proposition along with it." For various reasons this method failed of its purpose.

So far as the other voyages are concerned, it is clear whom Swift meant to ridicule, the foibles and follies he wished to expose. *Laputa*'s target is less obvious. Why was Swift so acutely roiled by mathematicians and natural scientists in a period when mathematics was being enriched by major discoveries, when mathematical physics, a field first opened up by Galileo, Huygens and Newton, was being systematically extended, and when the experimental sciences were flourishing in almost every department? It is hard to think of a time when science was more lively and creative, more widely and justly esteemed.

*Laiputa* cannot be explained merely by ascribing it to the "great foundation of misanthropy" upon which Swift said he had erected "the whole building" of his *Travels*; to his "perfect rage and resentment" (his own words) against practically everyone for practically everything. Nor can it be asserted that Swift, though he admitted "little acquaintance with mathematical knowledge," was either ignorant of the contemporary ferment in scientific thought or indifferent to it. Interest in scientific discoveries was widespread among English and French men of letters during the seventeenth and eighteenth centuries. Swift, as recent researches show, not only shared this fashionable interest but was impelled by scientific curiosity to dig deeper.[6] Through his friendship with men engaged in scientific work, notably the admirable Dr. Arbuthnot, and by carefully reading the *Transactions of the Royal Society,* he learned to distinuish the honest men from the fakers and dilettantes.

The *Transactions* were a remarkable mixture of nonsense and sober ideas; of magic, mathematics, fantasy, experimental fact, foolishness, logic and pedantry. Swift scanned the contributions with a "humorously critical eye," and with the deliberate purpose of gathering material for his writings. Diligent and ingenious scholars have traced almost everything in *Laputa* to contemporary science and to certain prevalent popular reactions to scientific doctrines and discoveries. There were discussions in the *Transactions* of comets and strange voyages; the analogy between music and mathematics was treated by the famous English mathematician John Wallis (1616–1703) and by one of the virtuosos, the Rev. T. Salmon, in a paper called "The Theory of Musick Reduced to Arithmetical and Geometrical Progressions." The Rev. T. Salmon was, I believe, an Englishman, but his interests were evidently identical with those of the wise men of Laputa.[7] The Laputian dread of the sun and of comets was also widespread in the eighteenth century. Having begun to accept Newton's clockwork universe, men were haunted by fear that the clockwork might get out of order. Halley had predicted the return of his comet in 1758; no one could be certain he was right, much less that it might not succumb to a churlish whim and bump into the earth.[8] Realization that the stability of the earth's orbit depended "on a nice balance between the velocity with which the earth is falling toward the sun and its tangential velocity at right angles to that fall" did not unduly disturb the handful of mathematicians and astronomers who understood what this equilibrium meant. But there were many others, of weaker faith in arithmetic and geometry, who would have preferred a less precarious arrangement. The Grand Academy in the metropolis of Lagado has been identified as Gresham College in London, where the Royal Society for many years held its meetings, had its library and museum. The Projectors of the Academy dominated their nation as the members of the Royal Society dominated England.[9] Many more examples could be given of the derivation of *Laputa* from circumstances of the eighteenth-century scientific scene, as well as from earlier works of fantasy.

It is true, as Whitehead has said, that Swift lived at a time "peculiarly unfitted for gibes at contemporary mathematics."[10] It is also true that in *Laputa,* as in the other voyages, Swift exhibits complex and to some extent contradictory emotions. He detested fakers and fools whether their specialty was geometry or politics. He lashed out at pride in every sphere of human affairs. At the same time he treasured human excellence and with characteristic fervor valued high moral and rational qualities wherever he found them.[11] It is fair to say, I think, that Swift was both enormously impressed by science and contemptuous of it; aware of its importance in human affairs and insensitive to the usefulness of useless knowledge. Like other writers of his age—Butler, Shadwell, Addison—he ridiculed the impracticality of mathematics. This was partly justified, partly the result of his essential ignorance of the subject itself. If he had been more of a mathematician he would perhaps have been less contemptuous, or, at any rate, more discriminating in his ridicule. But if he had been more of a mathematician it is doubtful he would have written *Gulliver.*

# ENDNOTES

[1]September 29, 1725. This is the famous letter that continues: "I have ever hated all nations, professions, and communities, and all my love is toward individuals: for instance I hate the tribe of lawyers, but I love Counsellor Such-a-one, and Judge Such-a-one: so with physicians—I will not speak of my own trade—soldiers, English, Scotch, French, and the rest. But principally I hate and detest that animal called man, although I heartily love John, Peter, Thomas, and so forth."

[2]*Gulliver's Travels, etc.,* edited by William Alfred Eddy, New York, 1933; from the introduction by Eddy, p. 1.

[3]William A. Eddy, *Gulliver's Travels, A Critical Study,* Princeton, 1923, pp. 53–57.

[4]November 17, 1726. *Correspondence,* III, 360; quoted by Eddy, *op. cit.,* preceding note.

[5]*A Treatise on Good Manners and Good Breeding in the Prose Works of Jonathan Swift, DD.,* XI, 81. (Bohn's Standard Library, 1897–1911, ed. by Temple Scott.)

[6]Two leading papers on this subject are Marjorie Nicholson and Nora Mohler, "The Scientific Background of Swift's Voyages to Laputa," *Annals of Science,* Vol. 2, 1937; and George Reuben Potter, "Swift and Natural Science," *Philological Quarterly,* Vol. 20, April, 1941.

[7]Nicholson and Mohler, *op. cit.,* Note 6 above.

[8]One of the Laputians' anxieties was that since the "Earth very narrowly escaped a brush from the tail of the last comet, which would have infallibly reduced it to ashes," the next, "which they have calculated for One and Thirty years hence, will probably destroy us." Note that "one and thirty years" from 1726, the date of publication of the *Travels,* is 1757; Halley predicted the return of the comet in 1758, See Nicholson and Mohler, *op. cit.,* p. 310.

[9]Nicholson and Mohler, *op. cit.,* pp. 318–320.

[10]Alfred North Whitehead, *An Introduction to Mathematics,* New York and London, 1948, p. 3.

[11]For an interesting analysis of this aspect of Swift's satire, see John M. Bullitt, *Jonathan Swift and the Anatomy of Satire,* Cambridge (Mass.), 1953, Chapter I; and for the exact point under discussion, pp. 12–13.

1

*The proof of the pudding is in the eating.*

—ENGLISH PROVERB (*14th century*)

*It's a very odd thing—*
*As odd as can be—*
*That whatever Miss T eats*
*Turns into Miss T.*

—WALTER DE LA MARE

# CYCLOID PUDDING

## *By Jonathan Swift*

## (EXCERPTS: "VOYAGE TO LAPUTA" FROM *Gulliver's Travels*)

*The Humours and Dispositions of the Laputians described. An Account of their Learning. Of the King and his Court. The Author's reception there. The Inhabitants subject to fears and disquietudes. An Account of the Women.*

At my alighting I was surrounded by a Crowd of People, but those who stood nearest seemed to be of better Quality. They beheld me with all the Marks and Circumstances of Wonder; neither indeed was I much in their Debt; having never till then seen a Race of Mortals so singular in their Shapes, Habits, and Countenances. Their Heads were all reclined to the Right, or the Left; one of their Eyes turned inward, and the other directly up to the Zenith. Their outward Garments were adorned with the Figures of Suns, Moons, and Stars, interwoven with those of Fiddles, Flutes, Harps, Trumpets, Guittars, Harpsicords, and many more Instruments of Musick, unknown to us in *Europe*. I observed here and there many in the Habit of Servants, with a blown Bladder fastned like a Flail to the End of a short Stick, which they carried in their Hands. In each Bladder was a small Quantity of dried Pease, or little Pebbles, (as I was afterwards informed). With these Bladders they now and then flapped the Mouths and Ears of those who stood near them, of which Practice I could not then conceive the Meaning. It seems, the Minds of these People are so taken up with intense Speculations, that they neither can speak, or attend to the Discourses of others, without being rouzed by some external Taction upon the Organs of Speech and Hearing; for which Reason, those Persons who are able to afford it, always keep a *Flapper*, (the Original is *Climenole*) in their Family, as one of their Domesticks; nor even walk abroad or make Visits without him. And the Business of this Officer is, when two or more Persons are in Company, gently to strike with his Bladder the Mouth of him who is to speak, and the Right Ear of him or them to whom the Speaker addresseth himself. This *Flapper* is likewise employed diligently to attend his Master in his Walks, and upon Occasion to give him a soft Flap on his Eyes; because he is always so wrapped up in Cogitation, that he is in manifest Danger of falling down every Precipice, and bouncing his Head against every Post; and in the Streets, of jostling others, or being jostled himself into the Kennel.

It was necessary to give the Reader this Information, without which he would be at the same Loss with me, to understand the Proceedings of these People, as they conducted me up the Stairs, to the Top of the Island, and from thence to the Royal Palace. While we were ascending they forgot several Times what they were about, and left me to my self, till their Memories were again rouzed by their *Flappers*; for they appeared altogether unmoved by the Sight of my foreign Habit and Countenance, and by the Shouts of the Vulgar, whose Thoughts and Minds were more disengaged.

At last we entered the Palace, and proceeded into the Chamber of Presence; where I saw the King seated on his Throne, attended on each Side by Persons of prime Quality. Before the Throne, was a large Table filled with Globes and Spheres, and Mathematical Instruments of all Kinds. His Majesty took not the least Notice of us, although our Entrance were not without sufficient Noise, by the Concourse of all Persons belonging to the Court. But, he was then deep in a Problem, and we attended at least an Hour, before he could solve it. There stood by him on each Side, a young Page, with Flaps in their Hands; and when they saw he was at Leisure, one of them gently struck his Mouth, and the other his Right Ear; at which he started like one awaked on the sudden, and looking towards me, and the Company I was in, recollected the Occasion of our coming, whereof he had been informed before. He spoke some Words; where-upon immediately a young Man with a Flap came up to my Side, and flapt me gently on the Right Ear; but I made Signs as well as I could, that I had no Occasion for such an Instrument; which as I afterwards found, gave his Majesty and the whole Court a very mean Opinion of my Understanding. The King, as far as I could conjecture, asked me several Questions, and I addressed my self to him in all the Languages I had. When it was found, that I could neither understand nor be understood, I was conducted by his Order to an Apartment in his Palace, (this Prince being distinguished above all his Predecessors for his Hospitality to Strangers,) where two Servants were appointed to attend me. My Dinner was brought, and four Persons of Quality, whom I remembered to have seen very near the King's Person, did me the Honour to dine with me. We had two Courses, of three Dishes each. In the first Course, there was a Shoulder of Mutton, cut into an Æquilateral Triangle; a Piece of Beef into a Rhomboides; and a Pudding into a Cycloid. The second Course was two Ducks, trussed up into the Form of Fiddles; Sausages and Puddings resembling Flutes and Haut-boys, and a Breast of Veal in the Shape of a Harp. The Servants cut our Bread into Cones, Cylinders, Parallelograms, and several other Mathematical Figures. While we were at Dinner, I made bold to ask the Names of several Things in their Language; and those noble Persons, by the Assistance of their *Flappers,* delighted to give me Answers, hoping to raise my Admiration of their great Abilities, if I could be brought to converse with them. I was soon able to call for Bread, and Drink, or whatever else I wanted.

After Dinner my Company withdrew, and a Person was sent to me by the King's Order, attended by a *Flapper.* He brought with him Pen, Ink, and Paper, and three or four Books; giving me to understand by Signs, that he was sent to teach me the Language. We sat together four Hours, in which Time I wrote down a great Number of Words in Columns, with the Translations over against them. I likewise made a Shift to learn several short Sentences. For my Tutor would order one of my Servants to fetch something, to turn about, to make a Bow, to sit, or stand, or walk, and the like. Then I took down the Sentence in Writing. He shewed me also in one of his Books, the Figures of the Sun, Moon, and Stars, the Zodiack, the Tropics and Polar Circles, together with the Denominations of many Figures of Planes and Solids. He gave me the Names and Descriptions of all the Musical Instruments, and the general Terms of Art in playing on each of them. After he had left me, I placed all my Words with their Interpretations in alphabetical Order. And thus in a few Days, by the Help of a very faithful Memory, I got some Insight into their Language.

The Word, which I interpret the *Flying* or *Floating Island,* is in the Original *Laputa*; whereof I could never learn the true Etymology. *Lap* in the old obsolete Language signifieth *High,* and *Untuh* a *Governor*; from which they say by Corruption was derived *Laputa* from *Lapuntuh.* But I do not approve of this Derivation, which seems to be a little strained. I ventured to offer to the Learned among them a Conjecture of my own, that *Laputa* was *quasi Lap outed*; *Lap* signifying properly the dancing of the Sun Beams in the Sea; and *outed* a Wing, which however I shall not obtrude, but submit to the judicious Reader.

Those to whom the King had entrusted me, observing how ill I was clad, ordered a Taylor to come next Morning, and take my Measure for a Suit of Cloths. This Operator did his Office after a different Manner from those of his Trade in *Europe.* He first took my Altitude by a Quadrant, and then with Rule and Compasses, described the Dimensions and Out-Lines of my whole Body; all which he entred upon Paper, and in six Days brought my Cloths very ill made, and quite out of Shape, by happening to mistake a Figure in the Calculation. But my Comfort was, that I observed such Accidents very frequent, and little regarded.

During my Confinement for want of Cloaths, and by an Indisposition that held me some Days longer, I much enlarged my Dictionary; and when I went next to Court, was able to understand many Things the King spoke, and to return him some Kind of Answers. His Majesty had given Orders, that the Island should move North-East and by East, to the vertical Point over *Lagado*, the Metropolis of the whole Kingdom, below upon the firm Earth. It was about Ninety Leagues distant, and our Voyage lasted four Days and an Half. I was not in the least sensible of the progressive Motion made in the Air by the Island. On the second Morning, about Eleven o'Clock, the King himself in Person, attended by his Nobility, Courtiers, and Officers, having prepared all their Musical Instruments, played on them for three Hours without Intermission; so that I was quite stunned with the Noise; neither could I possibly guess the Meaning, till my Tutor informed me. He said, that the People of their Island had their Ears adapted to hear the Musick of the Spheres, which always played at certain Periods; and the Court was now prepared to bear their Part in whatever Instrument they most excelled.

In our Journey towards *Lagado* the Capital City, his Majesty ordered that the Island should stop over certain Towns and Villages, from whence he might receive the Petitions of his Subjects. And to this Purpose, several Packthreads were let down with small Weights at the Bottom. On these Packthreads the People strung their Petitions, which mounted up directly like the Scraps of Paper fastned by School-boys at the End of the String that holds their Kite. Sometimes we received Wine and Victuals from below, which were drawn up by Pullies.

The Knowledge I had in Mathematicks gave me great Assistance in acquiring their Phraseology, which depended much upon that Science and Musick; and in the latter I was not unskilled. Their Ideas are perpetually conversant in Lines and Figures. If they would, for Example, praise the Beauty of a Woman, or any other Animal, they describe it by Rhombs, Circles, Parallelograms, Ellipses, and other Geometrical Terms; or else by Words of Art drawn from Musick, needless here to repeat. I observed in the King's Kitchen all Sorts of Mathematical and Musical Instruments, after the Figures of which they cut up the Joynts that were served to his Majesty's Table.

Their Houses are very ill built, the Walls bevil, without one right Angle in any Apartment; and this Defect ariseth from the Contempt they bear for practical Geometry; which they despise as vulgar and mechanick, those Instructions they give being too refined for the Intellectuals of their Workmen; which occasions perpetual Mistakes. And although they are dextrous enough upon a Piece of Paper, in the Management of the Rule, the Pencil, and the Divider, yet in the common Actions and Behaviour of Life, I have not seen a more clumsy, awkward, and unhandy People, nor so slow and perplexed in their Conceptions upon all other Subjects, except those of Mathematicks and Musick. They are very bad Reasoners, and vehemently given to Opposition, unless when they happen to be of the right Opinion, which is seldom their Case. Imagination, Fancy, and Invention, they are wholly Strangers to, nor have any Words in their Language by which those Ideas can be expressed; the whole Compass of their thoughts and Mind, being shut up within the two forementioned Sciences.

Most of them, and especially those who deal in the Astronomical Part, have great Faith in judicial Astrology, although they are ashamed to own it publickly. But, what I chiefly admired, and thought altogether unaccountable, was the strong Disposition I observed in them towards News and Politicks; perpetually enquiring into publick Affairs, giving their Judgments in Matters of State; and passionately disputing every Inch of a Party Opinion. I have indeed observed the same Disposition among most of the Mathematicians I have known in *Europe*; although I could never discover the least Analogy between the two Sciences; unless those People suppose, that because the smallest Circle hath as many Degrees as the largest, therefore the Regulation and Management of the World require no more Abilities than the handling and turning of a Globe. But, I rather take this Quality to spring from a very common Infirmity of human Nature, inclining us to be more curious and conceited in Matters where we have least Concern, and for which we are least adapted either by Study or Nature.

These People are under continual Disquietudes, never enjoying a Minute's Peace of Mind; and their Disturbances proceed from Causes which very little affect the rest of Mortals. Their Apprehensions arise from several Changes they dread in the Celestial Bodies. For Instance; that the Earth by the continual Approaches of the Sun towards it, must in Course of Time be absorbed or swallowed up. That the Face of the Sun will by Degrees be encrusted with its own Effluvia, and give no more Light to the World. That, the Earth very narrowly escaped a Brush from the Tail of the last Comet, which would have infallibly reduced it to Ashes; and that the next, which they have calculated for One and Thirty Years hence, will probably destroy us. For, if in its Perihelion it should approach within a certain Degree of the Sun, (as by their Calculations they have Reason to dread) it will conceive a Degree of Heat ten Thousand Times more intense than that of red hot glowing Iron; and in its Absence from the Sun, carry a blazing Tail Ten Hundred Thousand and Fourteen Miles long; through which if the Earth should pass at the Distance of one Hundred Thousand Miles from the *Nucleus,* or main Body of the Comet, it must in its Passage be set on Fire, and reduced to Ashes. That the Sun daily spending its Rays without any Nutriment to supply them, will at last be wholly consumed and annihilated; which must be attended with the Destruction of this Earth, and of all the Planets that receive their Light from it.

They are so perpetually alarmed with the Apprehenisons of these and the like impending Dangers, that they can neither sleep quietly in their Beds, nor have any Relish for the common Pleasures or Amusements of Life. When they meet an Acquaintance in the Morning, the first Question is about the Sun's Health; how he looked at his Setting and Rising, and what Hopes they have to avoid the Stroak of the approaching Comet. This Conversation they are apt to run into with the same Temper that Boys discover, in delighting to hear terrible Stories of Sprites and Hobgoblins, which they greedily listen to, and dare not go to Bed for fear.

The Women of the Island have Abundance of Vivacity; they contemn their Husbands, and are exceedingly fond of Strangers, whereof there is always a considerable Number from the Continent below, attending at Court, either upon Affairs of the several Towns and Corporations, or their own particular Occasions; but are much despised, because they want the same Endowments. Among these the Ladies chuse their Gallants: But the Vexation is, that they act with too much Ease and Security; for the Husband is always so wrapped in Speculation, that the Mistress and Lover may proceed to the greatest Familiarities before his Face, if he be but provided with Paper and Implements, and without his *Flapper* at his Side.

The Wives and Daughters lament their Confinement to the Island, although I think it the most delicious Spot of Ground in the World; and although they live here in the greatest Plenty and Magnificence, and are allowed to do whatever they please: They long to see the World, and take the Diversions of the Metropolis, which they are not allowed to do without a particular Licence from the King; and this is not easy to be obtained, because the People of Quality have found by frequent Experience, how hard it is to persuade their Women to return from below. I was told, that a great Court Lady, who had several Children, is married to the prime Minister, the richest Subject in the Kingdom, a very graceful Person, extremely fond of her, and lives in the finest Palace of the Island; went down to *Lagado,* on the Pretence of Health, there hid her self for several Months, till the King sent a Warrant to search for her; and she was found in an obscure Eating-House all in Rags, having pawned her Cloths to maintain an old deformed Footman, who beat her every Day, and in whose Company she was taken much against her Will. And although her Husband received her with all possible Kindness, and without the least Reproach; she soon after contrived to steal down again with all her Jewels, to the same Gallant, and hath not been heard of since.

This may perhaps pass with the Reader rather for an *European* or *English* Story, than for one of a Country so remote. But he may please to consider, that the Caprices of Womankind are not limited by any Climate or Nation; and that they are much more uniform than can be easily imagined.

In about a Month's Time I had made a tolerable Proficiency in their Language, and was able to answer most of the King's Questions, when I had the Honour to attend him. His Majesty discovered not the least Curiosity to enquire into the Laws, Government, History, Religion, or Manners of the Countries where I had been; but confined his Questions to the State of Mathematicks, and received the Account I gave him, with great Contempt and Indifference, though often rouzed by his *Flapper* on each Side.

. . . I was at the Mathematical School, where the Master taught his Pupils after a Method scarce imaginable to us in *Europe.* The Proposition and Demonstration were fairly written on a thin Wafer, with Ink composed of a Cephalick Tincture. This the Student was to swallow upon a fasting Stomach, and for three Days following eat nothing but Bread and Water. As the Wafer digested, the Tincture mounted to his Brain, bearing the Proposition along with it. But the Success hath not hitherto been answerable, partly by some Error in the *Quantum* or Composition, and partly by the Perverseness of Lads; to whom this Bolus is so nauseous, that they generally steal aside, and discharge it upwards before it can operate; neither have they been yet persuaded to use so long an Abstinence as the Prescription requires.

# COMMENTARY ON
# ALDOUS HUXLEY

Aldous Huxley (1894–1963) was the son of Leonard Huxley, the grandson of Thomas Henry Huxley and Thomas Arnold, the nephew of Mrs. Humphrey Ward and the grandnephew of Matthew Arnold. He was himself a novelist and an essayist of high gifts; his writings are marked by imagination, insight and feeling, unfailing intelligence and a superb command of language. The rational, skeptical and scientific strain of T. H. Huxley is in his work; also defeatism, a lack of faith in humanity, anger and disgust. His grandfather had urged upon Charles Kingsley "the merits of life on 'this narrow ledge of uncertainty.'" Like others of his generation, the grandson came to regard these merits as insufficient to offset the discomfort and the anxiety of living on a ledge. He moved, therefore, to the higher and broader shelf of mysticism, a strange sanctuary for one to whom a rigorous nineteenth-century naturalism was still so persuasive. The side-by-sideness of a rational and a mystical outlook was one of the most striking features of his work.

Huxley was born in Surrey in 1894, and was educated at a preparatory school and at Eton. Science, he said, was a "gospel and exhortation" in his family and he planned to become a doctor. Unfortunately he contracted keratitis and became within a few months almost completely blind. "I learned to read books and music in Braille and to use a typewriter, and continued my education with tutors. At this period, when I was about eighteen, I wrote a complete novel which I was never able to read, as it was written by touch on the typewriter without the help of eyes. By the time I could read again, the manuscript was lost."[1] After two years he had recovered sufficiently to read with one eye, aided by a magnifying glass. Though the limitations of his vision made a scientific career impossible, Huxley was able to attend courses at Oxford, and take his degree in English literature and philology.

In 1916 appeared Huxley's first book, *The Burning Wheel,* a collection of symbolist poetry. After the war he joined the staff of the *Athenaeum,* under the editorship of John Middleton Murry, and for several years "did a great variety of literary journalism for many periodicals." During the 1920s he and his wife spent much time in Italy, and in close association with D. H. Lawrence. He lived also in France, in Central America and, since 1938, in Southern California. The early novels of "skeptical brilliance" were *Crome Yellow* (1921), *Antic Hay* (1923) and *Point Counter Point* (1926). They are still regarded by many critics as the best of Huxley's work. Among the better known of his later writings are the novels *Brave New World* (1932), *Eyeless in Gaza* (1936), *After Many a Summer Dies the Swan* (1940), *The Devils of Loudon* (1952); his essays

*Jesting Pilate* (1926), *Brief Candles* (1930), *Ends and Means* (1937), *Literature and Science* (1963); and his biography of Father Joseph, a seventeenth-century French mystic, entitled *Grey Eminence* (1941). He also worked as a screen writer. His death followed a long period of recurrent illness.

J. W. N. Sullivan (1886–1937), who interviewed Huxley in the thirties, recorded some clarifying glimpses into his mind. "My chief motive in writing has been the desire to express a point of view. Or, rather, the desire to clarify a point of view to myself. I do not write for my readers; in fact, I don't like thinking about my readers." Sullivan asked whether he thought mankind had progressed. Huxley replied: "Yes, since Neanderthal times. But it is very difficult to say that mankind is now progressing. The question is, what do we want to aim at. Progress in one direction hinders progress in some other direction. For instance, our great mechanical progress has hindered intellectual progress. And it seems clear to me that intellectual development often hinders emotional development. If we evolved a race of Isaac Newtons, that would not be progress. For the price Newton had to pay for being a supreme intellect was that he was incapable of friendship, love, fatherhood, and many other desirable things. As a man he was a failure; as a monster he was superb. I admit that mathematical science is a good thing. But excessive devotion to it is a bad thing. Excessive, or rather exclusive, devotion to anything is bad."[2] This statement is an admirable epitome of his outlook on life.

"Young Archimedes" is the title story of a collection published in 1924. It is a moving, graceful example of Huxley's artistry.

### ENDNOTES

[1]Stanley J. Kunitz and Howard Haycraft, *Twentieth Century Authors,* article on Huxley, New York, 1942. I have followed Huxley's own account of his life as given in this biographical dictionary.
[2]J. W. N. Sullivan, *Contemporary Mind,* London, 1934, pp. 141–143.

2

*To see a World in a Grain of Sand,*
*And a Heaven in a Wild Flower,*
*Hold Infinity in the palm of your hand,*
*And Eternity in an hour.*

—WILLIAM BLAKE

# YOUNG ARCHIMEDES

## *By Aldous Huxley*

It was the view which finally made us take the place. True, the house had its disadvantages. It was a long way out of town and had no telephone. The rent was unduly high, the drainage system poor. On windy nights, when the ill-fitting panes were rattling so furiously in the window frames that you could fancy yourself in an hotel omnibus, the electric light, for some mysterious reason, used invariably to go out and leave you in the noisy dark. There was a splendid bathroom; but the electric pump, which was supposed to send up water from the rainwater tanks in the terrace, did not work. Punctually every autumn the drinking well ran dry. And our landlady was a liar and a cheat.

But these are the little disadvantages of every hired house, all over the world. For Italy they were not really at all serious. I have seen plenty of houses which had them all and a hundred others, without possessing the compensating advantages of ours—the southward-facing garden and terrace for the winter and spring, the large cool rooms against the midsummer heat, the hilltop air and freedom from mosquitoes, and finally the view.

And what a view it was! Or rather, what a succession of views. For it was different every day; and without stirring from the house one had the impression of an incessant change of scene: all the delights of travel without its fatigues. There were autumn days when all the valleys were filled with mist and the crest of the Apennines rose darkly out of a flat white lake. There were days when the mist invaded even our hilltop and we were enveloped in a soft vapor in which the mist-colored olive trees, that sloped away below our windows towards the valley, disappeared as though into their own spiritual essence; and the only firm and definite things in the small, dim world within which we found ourselves confined were the two tall black cypresses growing on a little projecting terrace a hundred feet down the hill. Black, sharp, and solid, they stood there, twin pillars of Hercules at the extremity of the known universe; and beyond them there was only pale cloud and round them only the cloudy olive trees.

These were the wintry days; but there were days of spring and autumn, days unchallengingly cloudless, or—more lovely still—made various by the huge floating shapes of vapor that, snowy above the faraway, snowcapped mountains, gradually unfolded, against the pale bright blue, enormous heroic gestures. And in the height of the sky the bellying draperies, the swans, the aerial marbles, hewed and left unfinished by gods grown tired of creation almost before they had begun, drifted sleeping along the wind, changing form as they moved. And the sun would come and go behind them; and now the town in the valley would fade and almost vanish in the shadow, and now, like an immense fretted jewel between the hills, it would glow as though by its own light. And looking across the nearer tributary valley that wound from below our crest down towards the Arno, looking over the low dark shoulder of hill on whose extreme promonotory stood the towered church of San Miniato, one saw the huge dome airily hanging on its ribs of masonry, the square campanile, the sharp spire of Santa Croce, and the canopied tower of the Signoria, rising above the intricate maze of houses, distinct and brilliant, like small treasures carved out of precious stones. For a moment only, and then their light would fade away once more, and the traveling beam would pick out, among the indigo hill beyond, a single golden crest.

There were days when the air was wet with passed or with approaching rain, and all the distances seemed miraculously near and clear. The olive trees detached themselves one from another on the distant slopes; the faraway villages were lovely and pathetic like the most exquisite small toys. There were days in summertime, days of impending thunder when, bright and sunlit against huge bellying masses of black and purple, the hills and the white houses shone as it were precariously, in a dying splendor, on the brink of some fearful calamity.

How the hills changed and varied! Every day and every hour of the day, almost, they were different. There would be moments when, looking across the plans of Florence, one would see only a dark blue silhouette against the sky. The scene had no depth; there was only a hanging curtain painted flatly with the symbols of the mountains. And then, suddenly almost, with the passing of a cloud, or when the sun had declined to a certain level in the sky, the flat scene transformed itself; and where there had been only a painted curtain, now there were ranges behind ranges of hills, graduated tone after tone from brown, or gray, or a green gold to faraway blue. Shapes that a moment before had been fused together indiscriminately into a single mass now came apart into their constituents. Fiesole, which had seemed only a spur of Monte Morello, now revealed itself as the jutting headland of another system of hills, divided from the nearest bastions of its greater neighbor by a deep and shadowy valley.

At noon, during the heats of summer, the landscape became dim, powdery, vague and almost colorless under the midday sun; the hills disappeared into the trembling fringes of the sky. But as the afternoon wore on the landscape emerged again, it dropped its anonymity, it climbed back out of nothingness into form and life. And its life, as the sun sank and slowly sank through the long afternoon, grew richer, grew more intense with every moment. The level light, with its attendant long, dark shadows, laid bare, so to speak, the anatomy of the land; the hills—each western escarpment shining, and each slope averted from the sunlight profoundly shadowed—became massive, jutty, and solid. Little folds of dimples in the seemingly even ground revealed themselves. Eastward from our hilltop, across the plain of the Ema, a great bluff casts its ever-increasing shadow; in the surrounding brightness of the valley a whole town lay eclipsed within it. And as the sun expired on the horizon, the further hills flushed in its warm light, till their illumined flanks were the color of tawny roses; but the valleys were already filled with the blue mist of the evening. And it mounted, mounted; the fire went out of the western windows of the populous slopes; only the crests were still alight, and at last they too were all extinct. The mountains faded and fused together again into a flat painting of mountains against the pale evening sky. In a little while it was night; and if the moon were full, a ghost of the dead scene still haunted the horizons.

Changed in its beauty, this wide landscape always preserved a quality of humanness and domestication which made it, to my mind at any rate, the best of all landscapes to live with. Day by day one traveled through its different beauties; but the journey, like our ancestors' Grand Tour, was always a journey through civilization. For all its mountains, its deep slopes and deep valleys, the Tuscan scene is dominated by its inhabitants. They have cultivated every rood of ground that can be cultivated; their houses are thickly scattered even over the hills, and the valleys are populous. Solitary on the hilltop, one is not alone in a wilderness. Man's traces are across the country, and already—one feels it with satisfaction as one looks out across it—for centuries, for thousands of years; it has been his, submissive, tamed, and humanized. The wide, blank moorlands, the sands, the forests of innumerable trees—these are places for occasional visitation, healthful to the spirit which submits itself to them for not too long. But fiendish influences as well as divine haunt these total solitudes. The vegetative life of plants and things is alien and hostile to the human. Men cannot live at ease except where they have mastered their surroundings and where their accumulated lives outnumber and outweigh the vegetative lives about them. Stripped of its dark wood, planted, terraced and tilled almost to the mountains' tops, the Tuscan landscape is humanized and safe. Sometimes upon those who live in the midst of it there comes a longing for some place that is solitary, inhuman, lifeless, or peopled only with alien life. But the longing is soon satisfied, and one is glad to return to the civilized and submissive scene.

I found that house on the hilltop the ideal dwelling place. For there, safe in the midst of a humanized landscape, one was yet alone; one could be as solitary as one liked. Neighbors whom one never sees at close quarters are the ideal and perfect neighbors.

Our nearest neighbors, in terms of physical proximity, lived very near. We had two sets of them, as a matter of fact, almost in the same house with us. One was the peasant family, who lived in a long, low building, part dwelling house, part stables, storerooms and cow sheds, adjoining the villa. Our other neighbors—intermittent neighbors, however, for they only ventured out of town every now and then, during the most flawless weather—were the owners of the villa, who had reserved for themselves the smaller wing of the huge L-shaped house—a mere dozen rooms or so —leaving the remaining eighteen or twenty to us.

They were a curious couple, our proprietors. An old husband, gray, listless, tottering, seventy at least; and a signora of about forty, short, very plump, with tiny fat hands and feet and a pair of very large, very dark black eyes, which she used with all the skill of a born comedian. Her vitality, if you could have harnessed it and made it do some useful work, would have supplied a whole town with electric light. The physicists talk of deriving energy from the atom; they would be more profitably employed nearer home—in discovering some way of tapping those enormous stores of vital energy which accumulate in unemployed women of sanguine temperament and which, in the present imperfect state of social and scientific organization, vent themselves in ways that are generally so deplorable in interfering with other people's affairs, in working up emotional scenes, in thinking about love and making it, and in bothering men till they cannot get on with their work.

Signora Bondi got rid of her superfluous energy, among other ways, by "doing in" her tenants. The old gentleman, who was a retired merchant with a reputation for the most perfect rectitude, was allowed to have no dealings with us. When we came to see the house, it was the wife who showed us round. It was she who, with a lavish display of charm, with irresistible rollings of the eyes, expatiated on the merits of the place, sang the praises of the electric pump, glorified the bathroom (considering which, she insisted, the rent was remarkably moderate), and when we suggested calling in a surveyor to look over the house, earnestly begged us, as though our well-being were her only consideration, not to waste our money unnecessarily in doing anything so superfluous. "After all," she said, "we are honest people. I wouldn't dream of letting you the house except in perfect condition. Have confidence." And she looked at me with an appealing, pained expression in her magnificent eyes, as though begging me not to insult her by my coarse suspiciousness. And leaving us no time to pursue the subject of surveyors any further, she began assuring us that our little boy was the most beautiful angel she had ever seen. By the time our interview with Signora Bondi was at an end, we had definitely decided to take the house.

"Charming woman," I said, as we left the house. But I think that Elizabeth was not quite so certain of it as I.

Then the pump episode began.

On the evening of our arrival in the house we switched on the electricity. The pump made a very professional whirring noise; but no water came out of the taps in the bathroom. We looked at one another doubtfully.

"Charming woman?" Elizabeth raised her eyebrows.

We asked for interviews; but somehow the old gentleman could never see us, and the Signora was invariably out or indisposed. We left notes; they were never answered. In the end, we found that the only method of communicating with our landlords, who were living in the same house with us, was to go down into Florence and send a registered express letter to them. For this they had to sign two separate receipts and even, if we chose to pay forty centimes more, a third incriminating document, which was then returned to us. There could be no pretending, as there always was with ordinary letters or notes, that the communication had never been received. We began at last to get answers to our complaints. The Signora, who wrote all the letters, started by telling us that, naturally, the pump didn't work, as the cisterns were empty, owing to the long drought. I had to walk three miles to the post office in order to register my letter reminding her that there had been a violent thunderstorm only last Wednesday, and that the tanks were consequently more than half full. The answer came back: bath water had not been guaranteed in the contract; and if I wanted it, why hadn't I had the pump looked at before I took the house? Another walk into town to ask the Signora next door whether she remembered her adjurations to us to have confidence in her, and to inform her, that the existence in a house of a bathroom was in itself an implicit guarantee of bath water. The reply to that was that the Signora couldn't continue to have communications with people who wrote so rudely to her. After that I put the matter into the hands of a lawyer. Two months later the pump was actually replaced. But we had to serve a writ on the lady before she gave in. And the costs were considerable.

One day, towards the end of the episode, I met the old gentleman in the road, taking his big Maremman dog for a walk—or being taken, rather, for a walk by the dog. For where the dog pulled the old gentleman had perforce to follow. And when it stopped to smell, or scratch the ground, or leave against a gatepost its visiting card or an offensive challenge, patiently, at his end of the leash, the old man had to wait. I passed him standing at the side of the road, a few hundred yards below our house. The dog was sniffing at the roots of one of the twin cypresses which grew one on either side of the entry to a farm; I heard the beast growling indignantly to itself, as though it scented an intolerable insult. Old Signor Bondi, leashed to his dog, was waiting. The knees inside the tubular gray trousers were slightly bent. Leaning on his cane, he stood gazing mournfully and vacantly at the view. The whites of his old eyes were discolored, like ancient billiard balls. In the gray, deeply wrinkled face, his nose was dyspeptically red. His white mustache, ragged and yellowing at the fringes, drooped in a melancholy curve. In his black tie he wore a very large diamond; perhaps that was what Signora Bondi had found so attractive about him.

I took off my hat as I approached. The old man stared at me absently, and it was only when I was already almost past him that he recollected who I was.

"Wait," he called after me, "wait!" And he hastened down the road in pursuit. Taken utterly by surprise and at a disadvantage—for it was engaged in retorting to the affront imprinted on the cypress roots—the dog permitted itself to be jerked after him. Too much astonished to be anything but obedient, it followed its master. "Wait!"

I waited.

"My dear sir," said the old gentleman, catching me by the lapel of my coat and blowing most disagreeably in my face, "I want to apologize." He looked around him, as though afraid that even here he might be overheard. "I want to apologize," he went on, "about the wretched pump business. I assure you that, if it had been only my affair, I'd have put the thing right as soon as you asked. You were quite right: a bathroom is an implicit guarantee of bath water. I saw from the first that we should have no chance if it came to court. And besides, I think one ought to treat one's tenants as handsomely as one can afford to. But my wife"—he lowered his voice—"the fact is that she likes this sort of thing, even when she knows that she's in the wrong and must lose. And besides, she hoped, I dare say, that you'd get tired of asking and have the job done yourself. I told her from the first that we ought to give in; but she wouldn't listen. You see, she enjoys it. Still, now she sees that it must be done. In the course of the next two or three days you'll be having your bath water. But I thought I'd just like to tell you how . . ." But the Maremmano, which had recovered by this time from its surprise of a moment since, suddenly bounded, growling, up the road. The old gentleman tried to hold the beast, strained at the leash, tottered unsteadily, then gave way and allowed himself to be dragged off. ". . . how sorry I am," he went on, as he receded from me, "that this little misunderstanding . . ." But it was no use. "Good-by." He smiled politely, made a little deprecating gesture, as though he had suddenly remembered a pressing engagement, and had no time to explain what it was. "Good-by." He took off his hat and abandoned himself completely to the dog.

A week later the water really did begin to flow, and the day after our first bath Signora Bondi, dressed in dove-gray satin and wearing all her pearls, came to call.

"Is it peace now?" she asked, with a charming frankness, as she shook hands.

We assured her that, so far as we were concerned, it certainly was.

"But why did you write me such dreadfully rude letters?" she said, turning on me a reproachful glance that ought to have moved the most ruthless malefactor to contrition. "And then that writ. How could you? To a lady. . ."

I mumbled something about the pump and our wanting baths.

"But how could you expect me to listen to you while you were in that mood? Why didn't you set about it differently—politely, charmingly?" She smiled at me and dropped her fluttering eyelids.

I thought it best to change the conversation. It is disagreeable, when one is in the right, to be made to appear in the wrong.

A few weeks later we had a letter—duly registered and by express messenger—in which the Signora asked us whether we proposed to renew our lease (which was only for six months), and notifying us that, if we did, the rent would be raised twenty-five per cent, in consideration of the improvements which had been carried out. We thought ourselves lucky, at the end of much bargaining, to get the lease renewed for a whole year with an increase in the rent of only fifteen per cent.

It was chiefly for the sake of the view that we put up with these intolerable extortions. But we had found other reasons, after a few days' residence, for liking the house. Of these the most cogent was that, in the peasant's youngest child, we had discovered what seemed the perfect playfellow for our own small boy. Between little Guido—for that was his name—and the youngest of his brothers and sisters there was a gap of six or seven years. His two older brothers worked with their father in the fields; since the time of the mother's death, two or three years before we knew them, the eldest sister had ruled the house, and the younger, who had just left school, helped her and in between-whiles kept an eye on Guido, who by this time, however, needed very little looking after; for he was between six and seven years old and as precocious, self-assured, and responsible as the children of the poor, left as they are to themselves almost from the time they can walk, generally are.

Though fully two and a half years older than little Robin—and at that age thirty months are crammed with half a lifetime's experience—Guido took no undue advantage of his superior intelligence and strength. I have never seen a child more patient, tolerant, and untyrannical. He never laughed at Robin for his clumsy efforts to imitate his own prodigious feats; he did not tease or bully, but helped his small companion when he was in difficulties and explained when he could not understand. In return, Robin adored him, regarded him as the model and perfect Big Boy, and slavishly imitated him in every way he could.

These attempts of Robin's to imitate his companion were often exceedingly ludicrous. For by an obscure psychological law, words and actions in themselves quite serious become comic as soon as they are copied; and the more accurately, if the imitation is a deliberate parody, the funnier—for an overloaded imitation of someone we know does not make us laugh so much as one that is almost indistinguishably like the original. The bad imitation is only ludicrous when it is a piece of sincere and earnest flattery which does not quite come off. Robin's imitations were mostly of this kind. His heroic and unsuccessful attempts to perform the feats of strength and skill, which Guido could do with ease, were exquisitely comic. And his careful, long-drawn imitations of Guido's habits and mannerisms were no less amusing. Most ludicrous of all, because most earnestly undertaken and most incongruous in the imitator, were Robin's impersonations of Guido in a pensive mood. Guido was a thoughtful child given to brooding and sudden abstractions. One would find him sitting in a corner by himself, chin in hand, elbow on knee, plunged, to all appearances, in the profoundest meditation. And sometimes, even in the midst of his play, he would suddenly break off, to stand, his hands behind his back, frowning and staring at the ground. When this happened, Robin became overawed and a little disquieted. In a puzzled silence he looked at his companion. "Guido," he would say softly, "Guido." But Guido was generally too much preoccupied to answer; and Robin, not venturing to insist, would creep near him, and throwing himself as nearly as possible into Guido's attitude—standing Napoleonically, his hands clasped behind him, or sitting in the posture of Michelangelo's Lorenzo the Magnificent—would try to meditate too. Every few seconds he would turn his bright blue eyes towards the elder child to see whether he was doing it quite right. But at the end of a minute he began to grow impatient; meditation wasn't his strong point. "Guido," he called again and, louder, "Guido!" And he would take him by the hand and try to pull him away. Sometimes Guido roused himself from his reverie and went back to the interrupted game. Sometimes he paid no attention. Melancholy, perplexed, Robin had to take himself off to play by himself. And Guido would go on sitting or standing there, quite still; and his eyes, if one looked into them, were beautiful in their grave and pensive calm.

They were large eyes, set far apart and, what was strange in a dark-haired Italian child, of a luminous pale blue-gray color. They were not always grave and calm, as in these pensive moments. When he was playing, when he talked or laughed, they lit up; and the surface of those clear, pale lakes of thought seemed, as it were, to be shaken into brilliant sun-flashing ripples. Above those eyes was a beautiful forehead, high and steep and domed in a curve that was like a subtle curve of a rose petal. The nose was straight, the chin small and rather pointed, the mouth drooped a little sadly at the corners.

I have a snapshot of the two children sitting together on the parapet of the terrace. Guido sits almost facing the camera, but looking a little to one side and downwards; his hands are crossed in his lap and his expression, his attitude are thoughtful, grave, and meditative. It is Guido in one of those moods of abstraction into which he would pass even at the height of laughter and play— quite suddenly and completely, as though he had all at once taken it into his head to go away and left the silent and beautiful body behind, like an empty house, to wait for his return. And by its side sits little Robin, turning to look up at him, his face half averted from the camera, but the curve of his cheek showing that he is laughing; one little raised hand is caught at the top of a gesture, the other clutches at Guido's sleeves, as though he were urging him to come away and play. And the legs dangling from the parapet have been seen by the blinking instrument in the midst of an impatient wriggle; he is on the point of slipping down and running off to play hide-and-seek in the garden. All the essential characteristics of both the children are in that little snapshot.

"If Robin were not Robin," Elizabeth used to say, "I could almost wish he were Guido."

And even at that time, when I took no particular interest in the child, I agreed with her. Guido seemed to me one of the most charming little boys I had ever seen.

We were not alone in admiring him. Signora Bondi when, in those cordial intervals between our quarrels, she came to call, was constantly speaking of him. "Such a beautiful, beautiful child!" she would exclaim with enthusiasm. "It's really a waste that he should belong to peasants who can't afford to dress him properly. If he were mine, I should put him into black velvet; or little white knickers and a white knitted silk jersey with a red line at the collar and cuffs; or perhaps a white sailor suit would be pretty. And in winter a little fur coat, with a squirrelskin cap, and possibly Russian boots . . ." Her imagination was running away with her. "And I'd let his hair grow, like a page's, and have it just curled up a little at the tips. And a straight fringe across his forehead. Everyone would turn round and stare after us if I took him out with me in Via Tornabuoni."

What you want, I should have liked to tell her, is not a child; it's a clockwork doll or a performing monkey. But I did not say so—partly because I could not think of the Italian for a clockwork doll and partly because I did not want to risk having the rent raised another fifteen per cent.

"Ah, if I only had a little boy like that!" She sighed and modestly dropped her eyelids. "I adore children. I sometimes think of adopting one—that is, if my husband would allow it."

I thought of the poor old gentleman being dragged along at the heels of his big white dog and inwardly smiled.

"But I don't know if he would," the signora was continuing, "I don't know if he would." She was silent for a moment, as though considering a new idea.

A few days later, when we were sitting in the garden after luncheon, drinking our coffee, Guido's father instead of passing with a nod and the usual cheerful good day, halted in front of us and began to talk. He was a fine handsome man, not very tall, but well proportioned, quick and elastic in his movements, and full of life. He had a thin brown face, featured like a Roman's and lit by a pair of the most intelligent-looking gray eyes I ever saw. They exhibited almost too much intelligence when, as not infrequently happened, he was trying, with an assumption of perfect frankness and a childlike innocence, to take one in or get something out of one. Delighting in itself, the intelligence shone there mischievously. The face might be ingenuous, impassive, almost imbecile in its expression; but the eyes on these occasions gave him completely away. One knew, when they glittered like that, that one would have to be careful.

Today, however, there was no dangerous light in them. He wanted nothing out of us, nothing of any value—only advice, which is a commodity, he knew, that most people are only too happy to part with. But he wanted advice on what was, for us, rather a delicate subject: on Signora Bondi. Carlo had often complained to us about her. The old man is good, he told us, very good and kind indeed. Which meant, I dare say, among other things, that he could easily be swindled. But his wife. . . . Well, the woman was a beast. And he would tell us stories of her insatiable rapacity: she was always claiming more than the half of the produce which, by the laws of the *métayage* systems, was the proprietor's due. He complained of her suspiciousness: she was forever accusing him of sharp practices, of downright stealing—him, he struck his breast, the soul of honesty. He complained of her shortsighted avarice: she wouldn't spend enough on manure, wouldn't buy him another cow, wouldn't have electric light installed in the stables. And we had sympathized, but cautiously, without expressing too strong an opinion on the subject. The Italians are wonderfully noncommittal in their speech; they will give nothing away to an interested person until they are quite certain that it is right and necessary and, above all, safe to do so. We had lived long enough among them to imitate their caution. What we said to Carlo would be sure, sooner or later, to get back to Signora Bondi. There was nothing to be gained by unnecessarily embittering our relations with the lady—only another fifteen per cent, very likely, to be lost.

Today he wasn't so much complaining as feeling perplexed. The Signora had sent for him, it seemed, and asked him how he would like it if she were to make an offer—it was all very hypothetical in the cautious Italian style—to adopt little Guido. Carlo's first instinct had been to say that he wouldn't like it at all. But an answer like that would have been too coarsely committal. He had preferred to say that he would think about it. And now he was asking for our advice.

Do what you think best, was what in effect we replied. But we gave it distantly but distinctly to be understood that we didn't think that Signora Bondi would make a very good foster mother for the child. And Carlo was inclined to agree. Besides he was very fond of the boy.

"But the thing is," he concluded rather gloomily, "that if she has really set her heart on getting hold of the child, there's nothing she won't do to get him—nothing."

He too, I could see, would have liked the physicists to start on unemployed childless woman of sanguine temperament before they tried to tackle the atom. Still, I reflected, as I watched him striding away along the terrace, singing powerfully from a brazen gullet as he went, there was force there, there was life enough in those elastic limbs, behind those bright gray eyes, to put up a good fight even against the accumulated vital energies of Signora Bondi.

It was a few days after this that my gramophone and two or three boxes of records arrived from England. They were a great comfort to us on the hilltop, providing as they did the only thing in which that spiritually fertile solitude—otherwise a perfect Swiss Family Robinson's island—was lacking: music. There is not much music to be heard nowadays in Florence. The times when Dr. Burney could tour through Italy, listening to an unending succession of new operas, symphonies, quartets, cantatas, are gone. Gone are the days when a learned musician, inferior only to the Reverend Father Martini of Bologna, could admire what the peasants sang and the strolling players thrummed and scraped on their instruments. I have traveled for weeks through the peninsula and hardly heard a note that was not *Salome* or the Fascists' song. Rich in nothing else that makes life agreeable or even supportable, the northern metropolises are rich in music. That is perhaps the only inducement that a reasonable man can find for living there. The other attractions—organized gaiety, people, miscellaneous conversation, the social pleasures—what are those, after all, but an expense of spirit that buys nothing in return? And then the cold, the darkness, the moldering dirt, the damp and squalor. . . . No, where there is no necessity that retains, music can be the only inducement. And that, thanks to the ingenious Edison, can now be taken about in a box and unpacked in whatever solitude one chooses to visit. One can live at Benin, or Nuneaton, or Tozeur in the Sahara, and still hear Mozart quartets, and selections from *The Well-Tempered Clavichord,* and the Fifth Symphony, and the Brahms clarinet quintet, and motets by Palestrina.

Carlo, who had gone down to the station with his mule and cart to fetch the packing case, was vastly interested in the machine.

"One will hear some music again," he said, as he watched me unpacking the gramophone and the disks. "It is difficult to do much oneself."

Still, I reflected, he managed to do a good deal. On warm nights we used to hear him, where he sat at the door of his house, playing his guitar and softly singing; the eldest boy shrilled out the melody on the mandolin and sometimes the whole family would join in, and the darkness would be filled with their passionate, throaty singing. Piedigrotta songs they mostly sang; and the voices drooped slurringly from note to note, lazily climbed or jerked themselves with sudden sobbing emphases from one tone to another. At a distance and under the stars the effect was not unpleasing.

"Before the war," he went on, "in normal times" (and Carlo had a hope, even a belief, that the normal times were coming back and that life would soon be as cheap and easy as it had been in the days before the flood), "I used to go and listen to the operas at the Politeama. Ah, they were magnificent. But it costs five lire now to get in."

"Too much," I agreed.

"Have you got *Trovatore*?" he asked.

I shook my head.

*"Rigoletto?"*

"I'm afraid not."

*"Bohème? Fanciulla del West? Pagliacci?"*

I had to go on disappointing him.

"Not even *Norma*? Or the *Barbiere*?"

I put on Battistini in *"Là ci darem"* out of *Don Giovanni*. He agreed that the singing was good; but I could see that he didn't much like the music, Why not? He found it difficult to explain.

"It's not like *Pagliacci*," he said at last.

"Not palpitating?" I suggested, using a word with which I was sure he would be familiar; for it occurs in every Italian political speech and patriotic leading article.

"Not palpitating," he agreed.

And I reflected that it is precisely by the difference between *Pagliacci* and *Don Giovanni,* between the palpitating and the nonpalpitating, that modern music taste is separated from the old. The corruption of the best, I thought, is the worst. Beethoven taught music to palpitate with his intellectual and spiritual passion. It has gone on palpitating ever since, but with the passion of inferior men. Indirectly, I thought, Beethoven is responsible for *Parsifal, Pagliacci,* and the *Poem of Fire*; still more indirectly for *Samson and Delilah* and "Ivy, cling to me." Mozart's melodies may be brilliant, memorable, infectious; but they don't palpitate, don't catch you between wind and water, don't send the listener off into erotic ecstasies.

Carlo and his elder children found my gramophone, I am afraid, rather a disappointment. They were too polite, however, to say so openly; they merely ceased, after the first day or two, to take any interest in the machine and the music it played. They preferred the guitar and their own singing.

Guido, on the other hand, was immensely interested. And he liked, not the cheerful dance tunes, to whose sharp rhythms our little Robin loved to go stamping round and round the room, pretending that he was a whole regiment of soldiers, but the genuine stuff. The first record he heard, I remember, was that of the slow movement of Bach's Concerto in D Minor for two violins. That was the disk I put on the turntable as soon as Carlo had left me. It seemed to me, so to speak, the most musical piece of music with which I would refresh my long-parched mind—the coolest and clearest of all draughts. The movement had just got under way and was beginning to unfold its pure and melancholy beauties in accordance with the laws of the most exacting intellectual logic, when the two children, Guido in front and little Robin breathlessly following, came clattering into the room from the loggia.

Guido came to a halt in front of the gramophone and stood there, motionless, listening. His pale blue-gray eyes opened themselves wide; making a little nervous gesture that I had often noticed in him before, he plucked at his lower lip with his thumb and forefinger. He must have taken a deep breath; for I noticed that, after listening for a few seconds, he sharply expired and drew in a fresh gulp of air. For an instant he looked at me—a questioning, astonished, rapturous look—gave a little laugh that ended in a kind of nervous shudder, and turned back towards the source of the incredible sounds. Slavishly imitating his elder comrade, Robin had also taken up his stand in front of the gramophone, and in exactly the same position, glancing at Guido from time to time to make sure that he was doing everything, down to plucking at his lip, in the correct way. But after a minute or so he became bored.

"Soldiers," he said, turning to me; "I want soldiers. Like in London." He remembered the ragtime and the jolly marches round and round the room.

I put my fingers to my lips. "Afterwards," I whispered.

Robin managed to remain silent and still for perhaps another twenty seconds. Then he seized Guido by the arm, shouting, "*Vieni,* Guido! Soldiers. *Soldati. Vieni giuocare soldati.*"

It was then, for the first time, that I saw Guido impatient. *"Vai!"* he whispered angrily, slapped at Robin's clutching hand and pushed him roughly away. And he leaned a little closer to the instrument, as though to make up by yet intenser listening for what the interruption had caused him to miss.

Robin looked at him, astonished. Such a thing had never happened before. Then he burst out crying and came to me for consolation.

When the quarrel was made up—and Guido was sincerely repentant, was as nice as he knew how to be when the music had stopped and his mind was free to think of Robin once more—I asked him how he liked the music. He said he thought it was beautiful. But *bello* in Italian is too vague a word, too easily and frequently uttered, to mean very much.

"What did you like best?" I insisted. For he had seemed to enjoy it so much that I was curious to find out what had really impressed him.

He was silent for a moment, pensively frowning. "Well," he said at last, "I liked the bit that went like this." And he hummed a long phrase. "And then there's the other thing singing at the same time—but what are those things," he interrupted himself, "that sing like that?"

"They're called violins," I said.

"Violins." He nodded. "Well, the other violin goes like this." He hummed again. "Why can't one sing both at once? And what is in that box? What makes it make that noise?" The child poured out his questions.

I answered him as best I could, showing him the little spirals on the disk, the needle, the diaphragm. I told him to remember how the string of the guitar trembled when one plucked it; sound is a shaking in the air, I told him, and I tried to explain how those shakings get printed on the black disk. Guido listened to me very gravely, nodding from time to time. I had the impression that he understood perfectly well everything I was saying.

By this time, however, poor Robin was so dreadfully bored that in pity for him I had to send the two children out into the garden to play. Guido went obediently; but I could see that he would have preferred to stay indoors and listen to more music. A little while later, when I looked out, he was hiding in the dark recesses of the big bay tree, roaring like a lion, and Robin laughing, but a little nervously, as though he were afraid that the horrible noise might possibly turn out, after all, to be the roaring of a real lion, was beating the bush with a stick, and shouting, "Come out, come out! I want to shoot you."

After lunch, when Robin had gone upstairs for his afternoon sleep, he reappeared. "May I listen to music now?" he asked. And for an hour he sat there in front of the instrument, his head cocked slightly on one side, listening while I put on one disk after another. Thenceforward he came every afternoon. Very soon he knew all my library of records, had his preferences and dislikes, and could ask for what he wanted by humming the principal theme.

"I don't like that one," he said of Strauss's *Till Eulenspiegel.* "It's like what we sing in our house. Not really like, you know. But somehow rather like, all the same. You understand?" He looked at us perplexedly and appealingly, as though begging us to understand what he meant and so save him from going on explaining. We nodded. Guido went on. "And then," he said, "the end doesn't seem to come properly out of the beginning. It's not like the one you played the first time." He hummed a bar or two from the slow movement of Bach's D Minor Concerto.

"It isn't," I suggested, "like saying: All little boys like playing. Guido is a little boy. Therefore Guido likes playing."

He frowned. "Yes, perhaps that's it," he said at last. "The one you played first is more like that. But, you know," he added, with an excessive regard for truth, "I don't like playing as much as Robin does."

Wagner was among his dislikes; so was Debussy. When I played the record of one of Debussy's arabesques, he said, "Why does he say the same thing over and over again? He ought to say something new, or go on, or make the thing grow. Can't he think of anything different?" But he was less censorious about the *Après-midi d'un faune.* "The things have beautiful voices," he said.

Mozart overwhelmed him with delight. The duet from *Don Giovanni,* which his father had found insufficiently palpitating, enchanted Guido. But he preferred the quartets and the orchestral pieces.

"I like music," he said, "better than singing."

Most people, I reflected, like singing better than music; are more interested in the executant than in what he executes, and find the impersonal orchestra less moving than the soloist. The touch of the pianist is the human touch, and the soprano's high C is the personal note. It is for the sake of this touch, that note, that audiences fill the concert halls.

Guido, however, preferred music. True, he liked *"Là ci darem"*; he liked *"Deh vieni alla finestra"*; he thought *"Che soave zefiretto"* so lovely that almost all our concerts had to begin with it. But he preferred the other things. The *Figaro* overture was one of his favorites. There is a passage not far from the beginning of the piece, when the first violins suddenly go rocketing up into the heights of loveliness; as the music approached that point, I used always to see a smile developing and gradually brightening on Guido's face, and when, punctually, the thing happened, he clapped his hands and laughed aloud with pleasure.

On the other side of the same disk, it happened, was recorded Beethoven's *Egmont* overture. He liked that almost better than *Figaro.*

"It has more voices," he explained. And I was delighted by the acuteness of the criticism; for it is precisely in the richness of its orchestration that *Egmont* goes beyond *Figaro*.

But what stirred him almost more than anything was the *Coriolan* overture. The third movement of the Fifth Symphony, the second movement of the Seventh, the slow movement of the Emperor Concerto—all these things ran it pretty close. But none excited him so much as *Coriolan*. One day he made me play it three or four times in succession; then he put it away.

"I don't think I want to hear that any more," he said.

"Why not?"

"It's too . . . too . . ." he hesitated, "too big," he said at last. "I don't really understand it. Play me the one that goes like this." He hummed the phrase from the D Minor Concerto.

"Do you like that one better?" I asked.

He shook his head. "No, it's not that exactly. But it's easier."

"Easier?" It seemed to me rather a queer word to apply to Bach.

"I understand it better."

One afternoon, while we were in the middle of our concert, Signora Bondi was ushered in. She began at once to be overwhelmingly affectionate towards the child; kissed him, patted his head, paid him the most outrageous compliments on his appearance. Guido edged away from her.

"And do you like music?" she asked.

The child nodded.

"I think he has a gift," I said. "At any rate, he has a wonderful ear and a power of listening and criticizing such as I've never met with in a child of that age. We're thinking of hiring a piano for him to learn on."

A moment later I was cursing myself for my undue frankness in praising the boy. For Signora Bondi began immediately to protest that, if she could have the upbringing of the child, she would give him the best masters, bring out his talent, make an accomplished maestro of him—and, on the way, an infant prodigy. And at that moment, I am sure, she saw herself sitting maternally, in pearls and black satin, in the lea of the huge Steinway, while an angelic Guido, dressed like little Lord Fauntleroy, rattled out Liszt and Chopin to the loud delight of a thronged auditorium. She saw the bouquets and all the elaborate floral tributes, heard the clapping and the few well-chosen words with which the veteran maestri, touched almost to tears, would hail the coming of the little genius. It became more than ever important for her to acquire the child.

"You've sent her away fairly ravening," said Elizabeth, when Signora Bondi had gone. "Better tell her next time that you made a mistake, and that the boy's got no musical talent whatever."

In due course a piano arrived. After giving him the minimum of preliminary instruction, I let Guido loose on it. He began by picking out for himself, the melodies he had heard, reconstructing the harmonies in which they were embedded. After a few lessons, he understood the rudiments of musical notation and could read a simple passage at sight, albeit very slowly. The whole process of reading was still strange to him; he had picked up his letters somehow, but nobody had yet taught him to read whole words and sentences.

I took occasion, next time I saw Signora Bondi, to assure her that Guido had disappointed me. There was nothing in his musical talent, really. She professed to be very sorry to hear it; but I could see that she didn't for a moment believe me. Probably she thought that we were after the child too, and wanted to bag the infant prodigy for ourselves, before she could get in her claim, thus depriving her of what she regarded almost as her feudal right. For, after all, weren't they her peasants? If anyone was to profit by adopting the child it ought to be herself.

Tactfully, diplomatically, she renewed her negotiations with Carlo. The boy, she put it to him, had genius. It was the foreign gentleman who had told her so, and he was the sort of man, clearly, who knew about such things. If Carlo would let her adopt the child, she'd have him trained. He'd become a great maestro and get engagements in the Argentine and the United States, in Paris and London. He'd earn millions and millions. Think of Caruso, for example. Part of the millions, she explained, would of course come to Carlo. But before they began to roll in, those millions, the boy would have to be trained. But training was very expensive. In his own interest, as well as that of his son, he ought to let her take charge of the child. Carlo said he would think it over, and again applied to us for advice. We suggested that it would be best in any case to wait a little and see what progress the boy made.

He made, in spite of my assertions to Signora Bondi, excellent progress. Every afternoon, while Robin was asleep, he came for his concert and his lesson. He was getting along famously with his reading; his small fingers were acquiring strength and agility. But what to me was more interesting was that he had begun to make up little pieces on his own account. A few of them I took down as he played them and I have them still. Most of them, strangely enough, as I thought then, are canons. He had a passion for canons. When I explained to him the principles of the form he was enchanted.

"It is beautiful," he said, with admiration. "Beautiful, beautiful. And so easy!"

Again the word surprised me. The canon is not, after all, so conspicuously simple. Thenceforward he spent most of his time at the piano in working out little canons for his own amusement. They were often remarkably ingenious. But in the invention of other kinds of music he did not show himself so fertile as I had hoped. He composed and harmonized one or two solemn little airs like hymn tunes, with a few sprightlier pieces in the spirit of the military march. They were extraordinary, of course, as being the inventions of a child. But a great many children can do extraordinary things; we are all geniuses up to the age of ten. But I had hoped that Guido was a child who was going to be a genius at forty; in which case what was extraordinary for an ordinary child was not extraordinary enough for him. "He's hardly a Mozart," we agreed, as we played his little pieces over. I felt, it must be confessed, almost aggrieved. Anything less than a Mozart, it seemed to me, was hardly worth thinking about.

He was not a Mozart. No. But he was somebody, as I was to find out, quite as extraordinary. It was one morning in the early summer that I made the discovery. I was sitting in the warm shade of our westward-facing balcony, working. Guido and Robin were playing in the little enclosed garden below. Absorbed in my work, it was only, I suppose, after the silence had prolonged itself a considerable time that I became aware that the children were making remarkably little noise. There was no shouting, no running about; only a quiet talking. Knowing by experience that when children are quiet it generally means that they are absorbed in some delicious mischief, I got up from my chair and looked over the balustrade to see what they were doing. I expected to catch them dabbling in water, making a bonfire, covering themselves with tar. But what I actually saw was Guido, with a burnt stick in his hand, demonstrating on the smooth paving stones of the path, that the square on the hypotenuse of a right-angled triangle is equal to the sum of the squares on the other two sides.

Kneeling on the floor, he was drawing with the point of his blackened stick on the flagstones. And Robin, kneeling imitatively beside him, was growing, I could see, rather impatient with this very slow game.

"Guido," he said. But Guido paid no attention. Pensively frowning, he went on with his diagram. "Guido!" The younger child bent down and then craned round his neck so as to look up into Guido's face. "Why don't you draw a train?"

"Afterwards," said Guido. "But I just want to show you this first. It's *so* beautiful," he added cajolingly.

"But I want a train," Robin persisted.

"In a moment. Do just wait a moment." The tone was almost imploring. Robin armed himself with renewed patience. A minute later Guido had finished both his diagrams.

"There!" he said triumphantly, and straightened himself up to look at them. "Now I'll explain."

And he proceeded to prove the theorem of Pythagoras—not in Euclid's way, but by the simpler and more satisfying method which was, in all probability, employed by Pythagoras himself. He had drawn a square and dissected it, by a pair of crossed perpendiculars, into two squares and two equal rectangles. The equal rectangles he divided up by their diagonals into four equal right-angled triangles. The two squares are then seen to be the squares on the two sides of any of these triangles other than the hypotenuse. So much for the first diagram. In the next he took the four right-angled triangles into which the rectangles had been divided and rearranged them round the original square so that their right angles filled the corners of the square, the hypotenuses looked inwards, and the greater and less sides of the triangles were in continuation along the sides of the squares (which are each equal to the sum of these sides). In this way the original square is redissected into four right-angled triangles and the square on the hypotenuse. The four triangles are equal to the two rectangles of the original dissection. Therefore the square on the hypotenuse is equal to the sum of the two squares—the squares on the two other sides—into which, with the rectangles, the original square was first dissected.

In very untechnical language, but clearly and with a relentless logic, Guido expounded his proof. Robin listened, with an expression on his bright, freckled face of perfect incomprehension.

"*Treno*," he repeated from time to time. "*Treno*. Make a train."

"In a moment," Guido implored. "Wait a moment. But do just look at this. Do." He coaxed and cajoled. "It's so beautiful. It's so easy."

So easy . . . The theorem of Pythagoras seemed to explain for me Guido's musical predilections. It was not an infant Mozart we had been cherishing; it was a little Archimedes with, like most of his kind, an incidental musical twist.

"*Treno, treno!*" shouted Robin, growing more and more restless as the exposition went on. And when Guido insisted on going on with his proof, he lost his temper. "*Cattivo Guido,*" he shouted, and began to hit out at him with his fists.

"All right," said Guido resignedly. "I'll make a train." And with his stick of charcoal he began to scribble on the stones.

I looked on for a moment in silence. It was not a very good train. Guido might be able to invent for himself and prove the theorem of Pythagoras: but he was not much of a draftsman.

"Guido!" I called. The two children turned and looked up. "Who taught you to draw those squares?" It was conceivable, of course, that somebody might have taught him.

"Nobody." He shook his head. Then, rather anxiously, as though he were afraid there might be something wrong about drawing squares, he went on to apologize and explain. "You see," he said, "it seemed to me so beautiful. Because those squares"—he pointed at the two small squares in the first figure—"are just as big as this one." And, indicating the square on the hypotenuse in the second diagram, he looked up at me with a deprecating smile.

I nodded. "Yes, it's very beautiful," I said—"it's very beautiful indeed."

An expression of delighted relief appeared on his face; he laughed with pleasure. "You see it's like this," he went on, eager to initiate me into the glorious secret he had discovered. "You cut these two long squares"—he meant the rectangles—"into two slices. And then there are four slices, all just the same, because, because—Oh, I ought to have said that before—because these long squares are the same because those lines, you see . . ."

"But I want a train," protested Robin.

Leaning on the rail of the balcony, I watched the children below. I thought of the extraordinary thing I had just seen and of what it meant.

I thought of the vast differences between human beings. We classify men by the color of their eyes and hair, the shape of their skulls. Would it not be more sensible to divide them up into intellectual species? There would be even wider gulfs between the extreme mental types than between a Bushman and a Scandinavian. This child, I thought, when he grows up, will be to me, intellectually, what a man is to his dog. And there are other men and women who are, perhaps, almost as dogs to me.

Perhaps the men of genius are the only true men. In all the history of the race there have been only a few thousand real men. And the rest of us—what are we? Teachable animals. Without the help of the real man, we should have found out almost nothing at all. Almost all the ideas with which we are familiar could never have occurred to minds like ours. Plant the seeds there and they will grow; but our minds could never spontaneously have generated them.

There have been whole nations of dogs, I thought; whole epochs in which no Man was born. From the dull Egyptians the Greeks took crude experience and rules of thumb and made sciences. More than a thousand years passed before Archimedes had a comparable successor. There has been only one Buddha, one Jesus, only one Bach that we know of, one Michelangelo.

Is it by a mere chance, I wondered, that a Man is born from time to time? What causes a whole constellation of them to come contemporaneously into being and from out of a single people? Taine thought that Leonardo, Michelangelo, and Raphael were born when they were because the time was ripe for great painters and the Italian scene congenial. In the mouth of a rationalizing nineteenth-century Frenchman the doctrine is strangely mystical; it may be none the less true for that. But what of those born out of time? Blake, for example. What of those?

This child, I thought, has had the fortune to be born at a time when he will be able to make good use of his capacities. He will find the most elaborate analytical methods lying ready to his hand; he will have a prodigious experience behind him. Suppose him born while Stonehenge was building; he might have spent a lifetime discovering the rudiments, guessing darkly where now he might have had a chance of proving. Born at the time of the Norman Conquest, he would have had to wrestle with all the preliminary difficulties created by an inadequate symbolism; it would have taken him long years, for example, to learn the art of dividing MMMCCCCLXXXVIII by MCMXIX. In five years, nowadays, he will learn what it took generations of Men to discover.

And I thought of the fate of all the Men born so hopelessly out of time that they could achieve little or nothing of value. Beethoven born in Greece, I thought, would have had to be content to play thin melodies on the flute or lyre; in those intellectual surroundings it would hardly have been possible for him to imagine the nature of harmony.

From drawing trains, the children in the garden below had gone onto playing trains. They were trotting round and round; with blown round cheeks and pouting mouth like the cherubic symbol of a wind, Robin puff-puffed and

Guido, holding the skirt of his smock, shuffled behind him, tooting. They ran forward, backed, stopped at imaginary stations, shunted, roared over bridges, crashed through tunnels, met with occasional collisions and derailments. The young Archimedes seemed to be just as happy as the little towheaded barbarian. A few minutes ago he had been busy with the theorem of Pythagoras. Now, tooting indefatigably along imaginary rails, he was perfectly content to shuffle backwards and forwards among the flower beds, between the pillars of the loggia, in and out of the dark tunnels of the laurel tree. The fact that one is going to be Archimedes does not prevent one from being an ordinary cheerful child meanwhile. I thought of this strange talent distinct and separate from the rest of the mind, independent, almost, of experience. The typical child prodigies are musical and mathematical; the other talents ripen slowly under the influence of emotional experience and growth. Till he was thirty Balzac gave proof of nothing but ineptitude; but at four the young Mozart was already a musician, and some of Pascal's most brilliant work was done before he was out of his teens.

In the weeks that followed, I alternated the daily piano lessons with lessons in mathematics. Hints rather than lessons they were; for I only made suggestions, indicated methods, and left the child himself to work out the ideas in detail. Thus I introduced him to algebra by showing him another proof of the theorem of Pythagoras. In this proof one drops a perpendicular from the right angle on to the hypotenuse, and arguing from the fact that the two triangles thus created are similar to one another and to the original triangle, and that the proportions which their corresponding sides bear to one another are therefore equal, one can show in algebraical form that $C^2 + D^2$ (the squares on the other two sides) are equal to $A^2 + B^2$ (the squares on the two segments of the hypotenuses) $+ 2AB$; which last, it is easy to show geometrically, is equal to $(A + B)^2$, or the square on the hypotenuse. Guido was as much enchanted by the rudiments of algebra as he would have been if I had given him an engine worked by steam, with a methylated spirit lamp to heat the boiler; more enchanted, perhaps—for the engine would have got broken, and, remaining always itself, would in any case have lost its charm, while the rudiments of algebra continued to grow and blossom in his mind with an unfailing luxuriance. Every day he made the discovery of something which seemed to him exquisitely beautiful; the new toy was inexhaustible in its potentialities.

In the intervals of applying algebra to the second book of Euclid, we experimented with circles; we stuck bamboos into the parched earth, measured their shadows at different hours of the day, and drew exciting conclusions from our observations. Sometimes, for fun, we cut and folded sheets of paper so as to make cubes and pyramids. One afternoon Guido arrived carrying carefully between his small and rather grubby hands a flimsy dodecahedron.

"*E tanto bello!*" he said, as he showed us his paper crystal; and when I asked him how he had managed to make it, he merely smiled and said it had been so easy. I looked at Elizabeth and laughed. But it would have been more symbol-ically to the point, I felt, if I had gone down on all fours, wagged the spiritual outgrowth of my os coccyx, and barked my astonished admiration.

It was an uncommonly hot summer. By the beginning of July our little Robin, unaccustomed to these high temperatures, began to look pale and tired; he was listless, had lost his appetite and energy. The doctor advised mountain air. We decided to spend the next ten or twelve weeks in Switzerland. My parting gift to Guido was the first six books of Euclid in Italian. He turned over the pages, looking ecstatically at the figures.

"If only I knew how to read properly," he said. "I'm so stupid. But now I shall really try to learn."

From our hotel near Grindelwald we sent the child, in Robin's name, various postcards of cows, alphorns, Swiss chalets, edelweiss, and the like. We re-ceived no answers to these cards; but then we did not expect answers. Guido could not write, and there was no reason why his father or his sisters should take the trouble to write for him. No news, we took it, was good news. And then one day, early in September, there arrived at the hotel a strange letter. The manager had it stuck up on the glass-fronted notice board in the hall, so that all the guests might see it, and whoever conscientiously thought that it belonged to him might claim it. Passing the board on the way in to lunch, Elizabeth stopped to look at it.

"But it must be from Guido," she said.

I came and looked at the envelope over her shoulder. It was unstamped and black with postmarks. Traced out in pencil, the big uncertain capital letters sprawled across its face. In the first line was written: *Al Babbo di Robin,* and there followed a travestied version of the name of the hotel and the place. Round the address bewildered postal officials had scrawled suggested emenda-tions. The letter had wandered for a fortnight at least, back and forth across the face of Europe.

"Al Babbo di Robin. To Robin's father." I laughed. "Pretty smart of the postmen to have got it here at all." I went to the manager's office, set forth the justice of my claim to the letter and, having paid the fifty-centime surcharge for the missing stamp, had the case unlocked and the letter given me. We went into lunch.

"The writing's magnificent," we agreed, laughing, as we examined the address at close quarters. "Thanks to Euclid," I added. "That's what comes of pandering to the ruling passion."

But when I opened the envelope and looked at its contents I no longer laughed. The letter was brief and almost telegraphic in style. *Sono Dalla Padrona,* it ran, *non mi Piace ha Rubato il mio libro non Voglio Suonare piu Voglio Tornare a casa Venga Subito Guido.*

"What is it?"

I handed Elizabeth the letter. "That blasted woman's got hold of him," I said.

Busts of men in Homburg hats, angels bathed in marble tears extinguishing torches, statues of little girls, cherubs, veiled figures, allegories and ruthless realisms—the strangest and most diverse idols beckoned and gesticulated as we passed. Printed indelibly on tin and embedded in the living rock, the brown photographs looked out, under glass, from the humbler crosses, headstones, and broken pillars. Dead ladies in cubistic geometrical fashions of thirty years ago—two cones of black satin meeting point to point at the waist, and the arms: a sphere to the elbow, a polished cylinder below—smiled mournfully out of their marble frames; the smiling frames; the smiling faces, the white hands, were the only recognizably human things that emerged from the solid geometry of their clothes. Men with black mustaches, men with white beards, young cleanshaven men, stared or averted their gaze to show a Roman profile. Children in their stiff best opened wide their eyes, smiled hopefully in anticipation of the little bird that was to issue from the camera's muzzle, smiled skeptically in the knowledge that it wouldn't, smiled laboriously and obediently because they had been told to. In spiky Gothic cottages of marble the richer dead reposed; through grilled doors one caught a glimpse of pale Inconsolables weeping, of distraught Geniuses guarding the secret of the tomb. The less prosperous sections of the majority slept in communities, close-crowded but elegantly housed under smooth continuous marble floors, whose every flagstone was the mouth of a separate grave.

These Continental cemeteries, I thought, as Carlo and I made our way among the dead, are more frightful than ours, because these people pay more attention to their dead than we do. That primordial cult of corpses, that tender solicitude for their material well-being, which led the ancients to house their dead in stone, while they themselves lived between wattles and under thatch, still lingers here; persists, I thought, more vigorously than with us. There are a hundred gesticulating statues here for every one in an English graveyard. There are more family vaults, more "luxuriously appointed" (as they say of liners and hotels) than one would find at home. And embedded in every tombstone there are photographs to remind the powdered bones within what form they will have to resume on the Day of Judgment; beside each are little hanging lamps to burn optimistically on All Souls' Day. To the Man who built the Pyramids they are nearer, I thought, than we.

"If I had known," Carlo kept repeating, "if only I had known." His voice came to me through my reflections as though from a distance. "At the time he didn't mind at all. How should I have known that he would take it so much to heart afterwards? And she deceived me, she lied to me."

I assured him yet once more that it wasn't his fault. Though, of course, it was, in part. It was mine too, in part; I ought to have thought of the possibility and somehow guarded against it. And he shouldn't have let the child go, even temporarily and on trial, even though the woman was bringing pressure to bear on him. And the pressure had been considerable. They had worked on the same holding for more than a hundred years, the men of Carlo's family; and now she had made the old man threaten to turn him out. It would be a dreadful thing to leave the place; and besides, another place wasn't so easy to find. It was made quite plain, however, that he could stay if he let her have the child. Only for a little to begin with; just to see how he got on. There would be no compulsion whatever on him to stay if he didn't like it. And it would be all to Guido's advantage; and to his father's, too, in the end. All that the Englishman had said about his not being such a good musician as he had thought at first was obviously untrue—mere jealousy and little-mindedness: the man wanted to take credit for Guido himself, that was all. And the boy, it was obvious, would learn nothing from him. What he needed was a real good professional master.

All the energy that, if the physicists had known their business, would have been driving dynamos, went into this campaign. It began the moment we were out of the house, intensively. She would have more chance of success, the Signora doubtless thought, if we weren't there. And besides, it was essential to take the opportunity when it offered itself and get hold of the child before we could make our bid—for it was obvious to her that we wanted Guido just as much as she did.

Day after day she renewed the assault. At the end of a week she sent her husband to complain about the state of the vines: they were in a shocking condition; he had decided, or very nearly decided, to give Carlo notice. Meekly, shamefacedly, in obedience to higher orders, the old gentleman uttered his threats. Next day Signora Bondi returned to the attack. The padrone, she declared, had been in a towering passion; but she'd do her best, her very best, to mollify him. And after a significant pause she went on to talk about Guido.

In the end Carlo gave in. The woman was too persistent and she held too many trump cards. The child could go and stay with her for a month or two on trial. After that, if he really expressed a desire to remain with her, she would formally adopt him.

At the idea of going for a holiday to the seaside—and it was to the seaside, Signora Bondi told him, that they were going—Guido was pleased and excited. He had heard a lot about the sea from Robin. *"Tanta acqua!"* It had sounded almost too good to be true. And now he was actually to go and see this marvel. It was very cheerfully that he parted from his family.

But after the holiday by the sea was over, and Signora Bondi had brought him back to her town house in Florence, he began to be homesick. The Signora, it was true, treated him exceedingly kindly, bought him new clothes, took him out to tea in the Via Tornabuoni and filled him up with cakes, iced strawberry-ade, whipped cream, and chocolates. But she made him practice the piano more than he liked, and what was worse, she took away his Euclid, on the score that he wasted too much time with it. And when he said that he wanted to go home, she put him off with promises and excuses and downright lies. She told him that she couldn't take him at once, but that next week, if he were good and worked hard at his piano meanwhile, next week . . . And when the time came she told him that his father didn't want him back. And she redoubled her petting, gave him expensive presents, and stuffed him with yet unhealthier foods. To no purpose. Guido didn't like his new life, didn't want to practice scales, pined for his book, and longed to be back with his brothers and sisters. Signora Bondi, meanwhile, continued to hope that time and chocolates would eventually make the child hers; and to keep his family at a distance, she wrote to Carlo every few days letters which still purported to come from the seaside (she took the trouble to send them to a friend, who posted them back again to Florence), and in which she painted the most charming picture of Guido's happiness.

It was then that Guido wrote his letter to me. Abandoned, as he supposed, by his family—for that they should not take the trouble to come to see him when they were so near was only to be explained on the hypothesis that they really had given him up—he must have looked to me as his last and only hope. And the letter, with its fantastic address, had been nearly a fortnight on its way. A fortnight—it must have seemed hundreds of years; and as the centuries succeeded one another, gradually, no doubt, the poor child became convinced that I too had abandoned him. There was no hope left.

"Here we are," said Carlo.

I looked up and found myself confronted by an enormous monument. In a kind of grotto hollowed in the flanks of a monolith of gray sandstone, Sacred Love, in bronze, was embracing a funeral urn. And in bronze letters riveted into the stone was a long legend to the effect that the inconsolable Ernesto Bondi had raised this monument to the memory of his beloved wife, Anunziata, as a token of his undying love for one whom, snatched from him by a premature death, he hoped very soon to join beneath this stone. The first Signora Bondi had died in 1912. I thought of the old man leashed to his white dog; he must always, I reflected, have been a most uxorious husband.

"They buried him here."

We stood there for a long time in silence. I felt the tears coming into my eyes as I thought of the poor child lying there underground. I thought of those luminous grave eyes, and the curve of that beautiful forehead, the droop of the melancholy mouth, of the expression of delight which illumined his face when he learned of some new idea that pleased him, when he heard a piece of music that he liked. And this beautiful small being was dead; and the spirit that inhabited this form, the amazing spirit, that too had been destroyed almost before it had begun to exist.

And the unhappiness that must have preceded the final act, the child's despair, the conviction of his utter abandonment—those were terrible to think of, terrible.

"I think we had better come away now," I said at last, and touched Carlo on the arm. He was standing there like a blind man, his eyes shut, his face slightly lifted towards the light; from between his closed eyelids the tears welled out, hung for a moment, and trickled down his cheeks. His lips trembled and I could see that he was making an effort to keep them still. "Come away," I repeated.

The face which had been still in its sorrow was suddenly convulsed; he opened his eyes, and through the tears they were bright with a violent anger. "I shall kill her," he said, "I shall kill her. When I think of him throwing himself out, falling through the air…" With his two hands he made a violent gesture, bringing them down from over his head and arresting them with a sudden jerk when they were on the level with his breast. "And then crash." He shuddered. "She's as much responsible as though she had pushed him down herself. I shall kill her." He clenched his teeth.

To be angry is easier than to be sad, less painful. It is comforting to think of revenge. "Don't talk like that," I said. "It's no good. It's stupid. And what would be the point?" He had had those fists before, when grief became too painful and he had tried to escape from it. Anger had been the easiest way of escape. I had had, before this, to persuade him back into the harder path of grief. "It's stupid to talk like that," I repeated, and I led him away through the ghastly labyrinth of tombs, where death seemed more terrible even than it is.

By the time we had left the cemetery, and were walking down from San Miniato towards the Piazzale Michelangelo below, he had become calmer. His anger had subsided again into the sorrow from which it had derived all its strength and its bitterness. In the Piazzale we halted for a moment to look down at the city in the valley below us. It was a day of floating clouds—great shapes, white, golden, and gray; and between them patches of a thin, transparent blue. Its lantern level, almost, with our eyes, the dome of the cathedral revealed itself in all its grandiose lightness, its vastness and aerial strength. On the innumerable brown and rosy roofs of the city the afternoon sunlight lay softly, sumptuously, and the towers were as though varnished and enameled with an old gold. I thought of all the Men who had lived here and left the visible traces of their spirit and conceived extraordinary things. I thought of the dead child.

# COMMENTARY ON MR. FORTUNE

The Reverend Timothy Fortune had been a bank clerk but his heart was set neither on riches nor advancement. When, at middle age, a small sum was left to him by an aunt, he went to a training college, was ordained a deacon and quitted England to become a missionary at St. Fabien, a port on an island of the fictional Raritongan Archipelago in the Pacific. After a time he felt the call to go to Fanua, a small, remote island, to make Christians of its peaceful, childlike natives. Mr. Fortune was a humble man and easygoing. "Even as a young man he had learnt that to jump in first doesn't make the bus start any sooner; and his favorite psalm was the one which begins: My soul truly waitest still upon God." He intended no pressure to convert the islanders. He knew they were a happy people; after he had dwelt among them, he thought, they would come to him and he would teach them "how they might be as happy in another life as they were in this."

Three years Mr. Fortune spent on Fanua; he made not a single convert. At first he thought he had converted a beautiful native boy named Lueli; he loved the boy and was loved by him. But one day he discovered that Lueli had only feigned to be a Christian so as not to offend Mr. Fortune; "in secret, in the reality of secretness," he continued to worship an idol. Mr. Fortune was angry, then puzzled, and finally ashamed of his failure. It was clear that he himself was unworthy and this failure his punishment. His tormenting reflections were interrupted by a terrific earthquake that suddenly struck the island. The hut he occupied collapsed and but for Lueli's efforts, at the risk of his life, Mr. Fortune would have perished. The earthquake had awful consequences for both the boy and the priest. The fire in the hut had destroyed Lueli's idol—which he might have saved had he not thought first of his friend's safety; the hideous shaking of the earth, "the flames, that had burst roaring and devouring from the mountain top" had also destroyed Mr. Fortune's belief in God. He had "departed in clouds of smoke, He had gone up and was lost in space."

Lueli felt the loss of the idol more acutely than Mr. Fortune the evaporation of his faith. The boy was listless and utterly miserable. His friends teased him for having lost his God. Mr. Fortune knew he must find ways to draw Lueli from his despair. "After three years of such familiarity it would not be easy to reconstruct his first fascination as something rich and strange. But it must be done if he were to compete successfully with his rival in Lueli's affections. It must be done because that rival was death." He tried to bring about a change in Lueli's mood by introducing him to games: ping-pong, spellikins, dicing, skittles. He caught a baby flying-fox and reared it for Lueli as a pet. He introduced him, with the aid of a magnifying glass, to the wonderful details of natural history. Nothing worked. Lueli got hit in the nose playing ping-pong, stoutly resisted spellikins, was bored by dicing and was regularly scratched and bitten by the fox. Then one morning Mr. Fortune remembered mathematics. The sequel to this inspiration is recounted in the excerpt below, taken from Sylvia Townsend Warner's gentle satire, *Mr. Fortune's Maggot*. It is a witty, enchanting episode of modern literature.

## 3

*He knew what's what, and that's as high*
*As metaphysic wit can fly.*

—SAMUEL BUTLER (*Hudibras*)

# GEOMETRY IN THE SOUTH PACIFIC

## By Sylvia Townsend Warner

[EXCERPT FROM "MR. FORTUNE'S MAGGOT"]

And then one morning when they had been living in the new hut for about six weeks he [Mr. Fortune] woke up inspired. Why had he wasted so much time displaying his most trivial and uncompelling charms, opposing to the magnetism of death such fripperies and titbits of this world, such gewgaws of civilization as a path serpentining to a parrot-cote (a parrot-cote which hadn't even allured the parrots), or a pocket magnifying glass, while all the time he carried within him the inestimable treasures of intellectual enjoyment? Now he would pipe Lueli a tune worth dancing to, now he would open for him a new world. He would teach him mathematics.

He sprang up from bed, full of enthusiasm. At the thought of all those stretches of white beach he was like a bridegroom. There they were, hard and smooth from the tread of the sea, waiting for that noble consummation of blank surfaces, to show forth a truth; waiting, in this particular instance, to show forth the elements of plane geometry.

At breakfast Mr. Fortune was so glorified and gay that Lueli caught a reflection of his high spirits and began to look more life-like than he had done for weeks. On their way down to the beach they met a party of islanders who were off on a picnic. Mr. Fortune with delight heard Lueli answering their greetings with something like his former sociability, and even plucking up heart enough for a repartee. His delight gave a momentary stagger when Lueli decided to go a-picnicking too. But, after all, it didn't matter a pin. The beach would be as smooth again to-morrow, the air as sweet and nimble; Lueli would be in better trim for learning after a spree, and, now he came to think of it, he himself wouldn't teach any the worse for a little private rubbing-up beforehand.

It must be going on for forty years since he had done any mathematics; for he had gone into the Bank the same year that his father died, leaving Rugby at seventeen because, in the state that things were then in, the Bank was too good an opening to be missed. He had once got a prize—The Poetical Works of Longfellow—for Algebra, and he had scrambled along well enough in other branches of mathematics; but he had not learnt with any particular thrill or realized that thrill there might be until he was in the Bank, and learning a thing of the past.

Then, perhaps because of that never-ending entering and adding up and striking balances, and turning on to the next page to enter, add up and strike balances again, a mental occupation minute, immediate and yet, so to speak, wool-gathering, as he imagined knitting to be, the absolute quality of mathematics began to take on for him an inexpressibly romantic air. "Pure Mathematics." He used to speak of them to his fellow clerks as though he were hinting at some kind of transcendental debauchery of which he had been made free—and indeed there does seem to be a kind of unnatural vice in being so completely pure. After a spell of this holy boasting he would grow a little uneasy; and going to the Free Library he took out mathematical treatises, just to make sure that he could follow step by step as well as soar. For twenty pages perhaps, he read slowly, carefully, dutifully, with pauses for self-examination and working out the examples. Then, just as it was working up and the pauses should have been more scrupulous than ever, a kind of swoon and ecstasy would fall on him, and he read ravening on, sitting up till dawn to finish the book, as though it were a novel. After that his passion was stayed; the book went back to the Library and he was done with mathematics till the next bout. Not much remained with him after these orgies, but something remained: a sensation in the mind, a worshipping acknowledgment of something isolated and unassailable, or a remembered mental joy at the rightness of thoughts coming together to a conclusion, accurate thoughts, thoughts in just intonation, coming together like unaccompanied voices coming to a close.

But often his pleasure flowered from quite simple things that any fool could grasp. For instance he would look out of the bank windows, which had green shades in their lower halves; and rising above the green shades he would see a row of triangles, equilateral, isosceles, acute-angled, right-angled, obtuse-angled. These triangles were a range of dazzling mountain peaks, eternally snowy, eternally untrodden; and he could feel the keen wind which blew from their summits. Yet they were also a row of triangles, equilateral, isosceles, acute-angled, right-angled, obtuse-angled.

This was the sort of thing he designed for Lueli's comfort. Geometry would be much better than algebra, though he had not the same certificate from Longfellow for teaching it. Algebra is always dancing over the pit of the unknown, and he had no wish to direct Lueli's thoughts to that quarter. Geometry would be best to begin with, plain plane geometry, immutably plane. Surely if anything could minister to the mind diseased it would be the steadfast contemplation of a right angle, an existence that no mist of human tears could blur, no blow of fate deflect.

Walking up and down the beach, admiring the surface which to-morrow with so much epiphany and glory was going to reveal the first axioms of Euclid, Mr. Fortune began to think of himself as possessing an universal elixir and charm. A wave of missionary ardour swept him along and he seemed to view, not Lueli only, but all the islanders rejoicing in this new dispensation. There was beach-board enough for all and to spare. The picture grew in his mind's eye, somewhat indebted to Raphael's Cartoon of the School of Athens. Here a group bent over an equation, there they pointed out to each other with admiration that the square on the hypotenuse equalled the sum of the squares on the sides containing the right angle; here was one delighting in a rhomboid and another in conic sections, that enraptured figure had secured the twelfth root of two, while the children might be filling up the foreground with a little long division.

By the morrow he had slept off most of his fervour. Calm, methodical, with a mind prepared for the onset, he guided Lueli down to the beach and with a stick prodded a small hole in it.

"What is this?"

"A hole."

"No, Lueli, it may seem like a hole, but it is a point."

Perhaps he had prodded a little too emphatically. Lueli's mistake was quite natural. Anyhow, there were bound to be a few misunderstandings at the start.

He took out his pocket knife and whittled the end of the stick. Then he tried again.

"What is this?"

"A smaller hole."

"Point," said Mr. Fortune suggestively.

"Yes, I mean a smaller point."

"No, not quite. It is a point, but it is not smaller. Holes may be of different sizes, but no point is larger or smaller than another point."

Lueli looked from the first point to the second. He seemed to be about to speak, but to think better of it. He removed his gaze to the sea.

Meanwhile, Mr. Fortune had moved about, prodding more points. It was rather awkward that he should have to walk on the beach-board, for his footmarks distracted the eye from the demonstration.

"Look, Lueli!"

Lueli turned his gaze inland.

"Where?" said he.

"At all these. Here; and here; and here. But don't tread on them."

Lueli stepped back hastily. When he was well out of the danger-zone he stood looking at Mr. Fortune with great attention and some uneasiness.

"These are all points."

Lueli recoiled a step further. Standing on one leg he furtively inspected the sole of his foot.

"As you see, Lueli, these points are in different places. This one is to the west of that and consequently that one is to the east of this. Here is one to the south. Here are two close together, and there is one quite apart from all the others. Now look at them, remember what I have said, think carefully and tell me what you think."

Inclining his head and screwing up his eyes Lueli inspected the demonstration with an air of painstaking connoisseurship. At length he ventured the opinion that the hole lying apart from the others was perhaps the neatest. But if Mr. Fortune would give him the knife he would whittle the stick even finer.

"Now what did I tell you? Have you forgotten that points cannot be larger or smaller? If they were holes it would be a different matter. But these are points. Will you remember that?"

Lueli nodded. He parted his lips, he was about to ask a question. Mr. Fortune went on hastily.

"Now suppose I were to cover the whole beach with these: what then?" A look of dismay came over Lueli's countenance. Mr. Fortune withdrew the hypothesis.

"I don't intend to. I only ask you to imagine what it would be like if I did." The look of dismay deepened.

"They would all be points," said Mr. Fortune, impressively. "All in different places. And none larger or smaller than another.

"What I have explained to you is summed up in the axiom: a point has position but not magnitude. In other words if a given point were not in a given place it would not be there at all."

Whilst allowing time for this to sink in he began to muse about those other words. Were they quite what he meant? Did they indeed mean anything? Perhaps it would have been better not to try to supplement Euclid. He turned to his pupil. The last words had sunk in at any rate, had been received without scruple and acted upon. Lueli was out of sight.

Compared with his intentions actuality had been a little quelling. It became more quelling as time went on. Lueli did not again remove himself without leave; he soon discovered that Mr. Fortune was extremely in earnest, and was resigned to regular instruction every morning and a good deal of rubbing-in and evocation during the rest of the day. No one ever had a finer capacity for listening than he, or a more docile and obliging temperament. But whereas in the old days these good gifts had flowed from him spontaneously and pleasurably he now seemed to be exhibiting them by rote and in a manner almost desperate, as though he were listening and obliging as a circus animal does its tricks. Humane visitors to circuses often point out with what alacrity the beasts run into the ring to perform their turn. They do not understand that in the choice of two evils most animals would rather flourish round a spacious ring than be shut up in a cage. The activity and the task is a distraction from their unnatural lot, and they tear through paper hoops all the better because so much of their time is spent behind iron bars.

It had been a very different affair when Lueli was learning Bible history and the Church Catechism, *The King of Love my Shepherd is* and *The Old Hundredth*. Then there had been no call for this blatant submission; lessons had been an easy-going conversation, with Lueli keeping his end up as an intelligent pupil should and Mr. Fortune feeling like a cross between wise old Chiron and good Mr. Barlow. Now they were a succession of harangues, and rather strained harangues to boot. Theology, Mr. Fortune found, is a more accommodating subject than mathematics; its technique of exposition allows greater latitude. For instance when you are gravelled for matter there is always the moral to fall back upon. Comparisons too may be drawn, leading cases cited, types and antetypes analysed and anecdotes introduced. Except for Archimedes mathematics is singularly naked of anecdotes.

Not that he thought any the worse of it for this. On the contrary he compared its austere and integral beauty to theology decked out in her flaunting charms and wielding all her bribes and spiritual bonuses; and like Dante at the rebuke of Beatrice he blushed that he should ever have followed aught but the noblest. No, there was nothing lacking in mathematics. The deficiency was in him. He added line to line, precept to precept; he exhausted himself and his pupil by hours of demonstration and exposition; leagues of sand were scarred, and smoothed again by the tide, and scarred afresh: never an answering spark rewarded him. He might as well have made the sands into a rope-walk.

Sometimes he thought that he was taxing Lueli too heavily, and desisted. But if he desisted for pity's sake, pity soon drove him to work again, for if it were bad to see Lueli sighing over the properties of parallel lines, it was worse to see him moping and pining for his god. Teioa's words, uttered so matter-of-factly, haunted his mind. "I expect he will die soon." Mr. Fortune was thinking so too. Lueli grew steadily more lacklustre, his eyes were dull, his voice was flat; he appeared to be retreating behind a film that thickened and toughened and would soon obliterate him.

"If only, if only I could teach him to enjoy an abstract notion! If he could once grasp how it all hangs together, and is everlasting and harmonious, he would be saved. Nothing else can save him, nothing that I or his fellows can offer him. For it must be new to excite him and it must be true to hold him, and what else is there that is both new and true?"

There were women, of course, a race of beings neither new nor true, yet much vaunted by some as a cure for melancholy and a tether for the soul. Mr. Fortune would have cheerfully procured a damsel (not that they were likely to need much of that), dressed her hair, hung the whistle and the Parnell medal round her neck, dowered her with the nineteen counters and the tape measure and settled her in Lueli's bed if he had supposed that this would avail. But he feared that Lueli was past the comfort of women, and in any case that sort of thing is best arranged by the parties concerned.

So he resorted to geometry again, and once more Lueli was hurling himself with frantic docility through the paper hoops. It was really rather astonishing, how dense he could be! Once out of twenty, perhaps, he would make the right answer. Mr. Fortune, too anxious to be lightly elated, would probe a little into his reasons for making it. Either they were the wrong reasons or he had no reasons at all. Mr. Fortune was often horribly tempted to let a mistake pass. He was not impatient: he was far more patient than in the palmiest days of theology—but he found it almost unendurable to be for ever saying with various inflexions of kindness: "No, Lueli. Try again," or: "Well, no, not exactly," or: "I fear you have not quite understood," or: "Let me try to make that clearer." He withstood the temptation. His easy acceptance (though in good faith) of a sham had brought them to this pass, and tenderness over a false currency was not likely to help them out of it. No, he would not be caught that way twice. Similarly he pruned and repressed Lueli's talent for leaking away down side-issues, though this was hard too, for it involved snubbing him almost every time he spoke on his own initiative.

Just as he had been so mistaken about the nature of points, confounding them with holes and agitating himself at the prospect of a beach pitted all over, Lueli contrived to apply the same sort of well-meaning misconceptions to every stage of his progress—if progress be the word to apply to one who is hauled along in a state of semiconsciousness by the scruff of his neck. When the points seemed to be tolerably well-established in his mind Mr. Fortune led him on to lines, and by joining up points he illustrated such simple figures as the square, the triangle and the parallelogram. Lueli perked up, seemed interested, borrowed the stick and began joining up points too. At first he copied Mr. Fortune, glancing up after each stroke to see if it had been properly directed. Then growing rather more confident, and pleased—as who is not?—with the act of drawing on sand, he launched out into a more complicated design.

"This is a man," he said.

Mr. Fortune was compelled to reply coldly:

"A man is not a geometrical figure."

At length Mr. Fortune decided that he had better take in sail. Pure mathematics were obviously beyond Lueli; perhaps applied mathematics would work better. Mr. Fortune, as it happened, had never applied any, but he knew that other people did so, and though he considered it a rather lower line of business he was prepared to try it.

"If I were to ask you to find out the height of that tree, how would you set about it?"

Lueli replied with disconcerting readiness:

"I should climb up to the top and let down a string."

"But suppose you couldn't climb up it?"

"Then I should cut it down."

"That would be very wasteful: and the other might be dangerous. I can show you a better plan than either of those."

The first thing was to select a tree, an upright tree, because in all elementary demonstrations it is best to keep things as clear as possible. He would never have credited the rarity of upright trees had he not been pressed to find one. Coco-palms, of course, were hopeless: they all had a curve or a list. At length he remembered a tree near the bathing-pool, a perfect specimen of everything a tree should be, tall, straight as a die, growing by itself; set apart, as it were, for purposes of demonstration.

He marched Lueli thither, and when he saw him rambling towards the pool he recalled him with a cough.

"Now I will show you how to discover the height of that tree. Attend. You will find it very interesting. The first thing to do is to lie down."

Mr. Fortune lay down on his back and Lueli followed his example.

Many people find that they can think more clearly in a recumbent position. Mr. Fortune found it so too. No sooner was he on his back than he remembered that he had no measuring stick. But the sun was delicious and the grass soft; he might well spare a few minutes in exposing the theory.

"It is all a question of measurements. Now my height is six foot two inches, but for the sake of argument we will assume it to be six foot exactly. The distance from my eye to the base of the tree is so far an unknown quantity. My six feet however are already known to you."

Now Lueli had sat up, and was looking him up and down with an intense and curious scrutiny, as though he were something utterly unfamiliar. This was confusing, it made him lose the thread of his explanation. He felt a little uncertain as to how it should proceed.

Long ago on dark January mornings, when a septic thumb (bestowed on him by a cat which he had rescued from a fierce poodle) obliged him to stay away from the Bank, he had observed young men with woollen comforters and raw-looking wind-bitten hands practising surveying under the snarling elms and whimpering poplars of Finsbury Park. They had tapes and tripods, and the girls in charge of perambulators dawdled on the asphalt paths to watch their proceedings. It was odd how vividly fragments of his old life had been coming back to him during these last few months.

He resumed:

"In order to ascertain the height of the tree I must be in such a position that the top of the tree is exactly in a line with the top of a measuring stick—or any straight object would do, such as an umbrella—which I shall secure in an upright position between my feet. Knowing then that the ratio that the height of the tree bears to the length of the measuring-stick must equal the ratio that the distance from my eye to the base of the tree bears to my height, and knowing (or being able to find out) my height, the length of the measuring stick and the distance from my eye to the base of the tree, I can, therefore, calculate the height of the tree."

"What is an umbrella?"

Again the past flowed back, insurgent and actual. He was at the Oval, and out of an overcharged sky it had begun to rain again. In a moment the insignificant tapestry of lightish faces was exchanged for a noble pattern of domes, blackish, blueish and greenish domes, sprouting like a crop of miraculous and religious mushrooms. The rain fell harder and harder, presently the little white figures were gone from the field and, as with an abnegation of humanity, the green plain, so much smaller for their departure, lay empty and forsaken, ringed round with tier upon tier of blackly glistening umbrellas.

He longed to describe it all to Lueli, it seemed to him at the moment that he could talk with the tongues of angels about umbrellas. But this was a lesson in mathematics: applied mathematics, moreover, a compromise, so that all further compromises must be sternly nipped. Unbending to no red herrings he replied:

"An umbrella, Lueli, when in use resembles the—the shell that would be formed by rotating an arc of curve about its axis of symmetry, attached to a cylinder of small radius whose axis is the same as the axis of symmetry of the generating curve of the shell. When not in use it is properly an elongated cone, but it is more usually helicoidal in form."

Lueli made no answer. He lay down again, this time face downward.

Mr. Fortune continued: "An umbrella, however, is not essential. A stick will do just as well, so find me one, and we will go on to the actual measurement."

Lueli was very slow in finding a stick. He looked for it rather languidly and stupidly, but Mr. Fortune tried to hope that this was because his mind was engaged on what he had just learnt.

Holding the stick between his feet, Mr. Fortune wriggled about on his back trying to get into the proper position. He knew he was making a fool of himself. The young men in Finsbury Park had never wriggled about on their backs. Obviously there must be some more dignified way of getting the top of the stick in line with the top of the tree and his eye, but just then it was not obvious to him. Lueli made it worse by standing about and looking miserably on. When he had placed himself properly he remembered that he had not measured the stick. It measured (he had had the forethought to bring the tape with him) three foot seven, very tiresome: those odd inches would only serve to make it seem harder to his pupil. So he broke it again, drove it into the ground, and wriggled on his stomach till his eye was in the right place, which was a slight improvement in method, at any rate. He then handed the tape to Lueli, and lay strictly motionless, admonishing and directing while Lueli did the measuring of the ground. In the interests of accuracy he did it thrice, each time with a different result. A few minutes before noon the height of the tree was discovered to be fifty-seven foot, nine inches.

Mr. Fortune now had leisure for compassion. He thought Lueli was looking hot and fagged, so he said:

"Why don't you have a bathe? It will freshen you up."

Lueli raised his head and looked at him with a long dubious look, as though he had heard the words but without understanding what they meant. Then he turned his eyes to the tree and looked at that. A sort of shadowy wrinkle, like the blurring on the surface of milk before it boils, crossed his face.

"Don't worry any more about that tree. If you hate all this so much we won't do any more of it, I will never speak of geometry again. Put it all out of your head and go and bathe."

# COMMENTARY ON STATISTICS AS A LITERARY STIMULUS

Y ou would not, perhaps, think statistics a subject likely to inspire the literary imagination; yet offhand I can recall at least half a dozen fables based on the theory of probability, among them Clerk Maxwell's celebrated conjecture about the demon who could reverse the second law of thermodynamics (making heat flow the wrong way) and one or two remarkable anecdotes by Augustus de Morgan. Charles Dickens issued an interesting tribute to theoretical statistics by refusing, one day late in December, to travel by train, on the ground that the average annual quota of railroad accidents in Britain had not been filled and therefore further disasters were obviously imminent.

All of us, I suppose, are a little afraid of statistics. Like Atropos, the sister who cut the thread, they are inexorable; like her too, they are not only impersonal but terribly personal. One dreams of flouting them. A modern Prometheus would not waste his time showing up the gods by stuffing a sacrificial bull with bones; he would flaunt his artfulness and independence by juggling the law of large numbers. Neither heaven nor earth could be straightened out thereafter. That bit of mischief is essentially what the next two fables are about.

*Inflexible Logic* by Russell Maloney is a widely known and admired story built around a famous statistical whimsy. Eddington gave currency to it in one of his lectures but I am far from certain that he made it up. Maloney was a writer of short stories, sketches, profiles, anecdotes, many of which appeared in *The New Yorker* magazine between the years 1934 and 1950. He conducted for several years the magazine's popular department, "Talk of the Town," and claimed to have written for it "something like 2600 perfect anecdotes." He died in New York, September 5, 1948, at the age of thirty-eight.

*The Law* is a fascinating, and, in a way, terrifying story of a sudden, mysterious failure of the "law of averages." To be sure there is no such law, but when it fails the consequences are much worse than if death had taken a holiday. If men cannot be depended on to behave like a herd or like the molecules of a gas the entire social order falls to ruin. Robert Coates (1897–1973) the author of this and other equally entrancing tales, was a writer and art critic. He too contributed frequently to *The New Yorker*. His books include *Wisteria Cottage* (1948), *The Bitter Season* (1946), *The Outlaw Years* (1930), *The Eater of Darkness* (1929).

4

*How often might a man, after he had jumbled a set of letters in a bag, fling them out upon the ground before they would fall into an exact poem, yea, or so much as make a good discourse in prose. And may not a little book be as easily made by chance as this great volume of the world.*

—Archbishop Tillotson

# Inflexible Logic

## By Russell Maloney

When the six chimpanzees came into his life, Mr. Bainbridge was thirty-eight years old. He was a bachelor and lived comfortably in a remote part of Connecticut, in a large old house with a carriage drive, a conservatory, a tennis court, and a well-selected library. His income was derived from impeccably situated real estate in New York City, and he spent it soberly, in a manner which could give offence to nobody. Once a year, late in April, his tennis court was resurfaced, and after that anybody in the neighborhood was welcome to use it; his monthly statement from Brentano's seldom ran below seventy-five dollars; every third year, in November, he turned in his old Cadillac coupé for a new one; he ordered his cigars, which were mild and rather moderately priced, in shipments of one thousand, from a tobacconist in Havana; because of the international situation he had cancelled arrangements to travel abroad, and after due thought had decided to spend his travelling allowance on wines, which seemed likely to get scarcer and more expensive if the war lasted. On the whole, Mr. Bainbridge's life was deliberately, and not too unsuccessfully, modelled after that of an English country gentleman of the late eighteenth century, a gentleman interested in the arts and in the expansion of science, and so sure of himself that he didn't care if some people thought him eccentric.

Mr. Bainbridge had many friends in New York, and he spent several days of the month in the city, staying at his club and looking around. Sometimes he called up a girl and took her out to a theatre and a night club. Sometimes he and a couple of classmates got a little tight and went to a prizefight. Mr. Bainbridge also looked in now and then at some of the conservative art galleries, and liked occasionally to go to a concert. And he liked cocktail parties, too, because of the fine footling conversation and the extraordinary number of pretty girls who had nothing else to do with the rest of their evening. It was at a New York cocktail party, however, that Mr. Bainbridge kept his preliminary appointment with doom. At one of the parties given by Hobie Packard, the stockbroker, he learned about the theory of the six chimpanzees.

It was almost six-forty. The people who had intended to have one drink and go had already gone, and the people who intended to stay were fortifying themselves with slightly dried canapés and talking animatedly. A group of stage and radio people had coagulated in one corner, near Packard's Capehart, and were wrangling about various methods of cheating the Collector of Internal Revenue. In another corner was a group of stockbrokers, talking about the greatest stockbroker of them all, Gauguin. Little Marcia Lupton was sitting with a young man, saying earnestly, "Do you really want to know what my greatest ambition is? I want to be myself," and Mr. Bainbridge smiled gently, thinking of the time Marcia had said that to him. Then he heard the voice of Bernard Weiss, the critic, saying, "Of course he wrote one good novel. It's not surprising. After all, we know that if six chimpanzees were set to work pounding six typewriters at random, they would, in a million years, write all the books in the British Museum."

Mr. Bainbridge drifted over to Weiss and was introduced to Weiss's companion, a Mr. Noble. "What's this about a million chimpanzees, Weiss?" he asked.

"Six chimpanzees," Mr. Weiss said. "It's an old cliché of the mathematicians. I thought everybody was told about it in school. Law of averages, you know, or maybe it's permutation and combination. The six chimps, just pounding away at the typewriter keys, would be bound to copy out all the books ever written by man. There are only so many possible combinations of letters and numerals, and they'd produce all of them—see? Of course they'd also turn out a mountain of gibberish, but they'd work the books in, too. All the books in the British Museum."

Mr. Bainbridge was delighted; this was the sort of talk he liked to hear when he came to New York. "Well, but look here," he said, just to keep up his part in the foolish conversation, "what if one of the chimpanzees finally did duplicate a book, right down to the last period, but left that off? Would that count?"

"I suppose not. Probably the chimpanzee would get around to doing the book again, and put the period in."

"What nonsense!" Mr. Noble cried.

"It may be nonsense, but Sir James Jeans believes it," Mr. Weiss said, huffily. "Jeans or Lancelot Hogben. I know I ran across it quite recently."

Mr. Bainbridge was impressed. He read quite a bit of popular science, and both Jeans and Hogben were in his library. "Is that so?" he murmured, no longer feeling frivolous. "Wonder if it has ever actually been tried? I mean, has anybody ever put six chimpanzees in a room with six typewriters and a lot of paper?"

Mr. Weiss glanced at Mr. Bainbridge's empty cocktail glass and said drily, "Probably not."

Nine weeks later, on a winter evening, Mr. Bainbridge was sitting in his study with his friend James Mallard, an assistant professor of mathematics at

New Haven. He was plainly nervous as he poured himself a drink and said, "Mallard, I've asked you to come here—Brandy? Cigar?—for a particular reason. You remember that I wrote you some time ago, asking your opinion of . . . of a certain mathematical hypothesis or supposition."

"Yes," Professor Mallard said, briskly. "I remember perfectly. About the six chimpanzees and the British Museum. And I told you it was a perfectly sound popularization of a principle known to every schoolboy who had studied the science of probabilities."

"Precisely," Mr. Bainbridge said. "Well, Mallard, I made up my mind. . . . It was not difficult for me, because I have, in spite of that fellow in the White House, been able to give something every year to the Museum of Natural History, and they were naturally glad to oblige me. . . . And after all, the only contribution a layman can make to the progress of science is to assist with the drudgery of experiment. . . . In short, I—"

"I suppose you're trying to tell me that you have procured six chimpanzees and set them to work at typewriters in order to see whether they will eventually write all the books in the British Museum. Is that it?"

"Yes, that's it," Mr. Bainbridge said. "What a mind you have, Mallard. Six fine young males, in perfect condition. I had a—I suppose you'd call it a dormitory—built out in back of the stable. The typewriters are in the conservatory. It's light and airy in there, and I moved most of the plants out. Mr. North, the man who owns the circus, very obligingly let me engage one of his best animal men. Really, it was no trouble at all."

Professor Mallard smiled indulgently. "After all, such a thing is not unheard of," he said. "I seem to remember that a man at some university put his graduate students to work flipping coins, to see if heads and tails came up an equal number of times. Of course they did."

Mr. Bainbridge looked at his friend very queerly. "Then you believe that any such principle of the science of probabilities will stand up under an actual test?"

"Certainly."

"You had better see for yourself." Mr. Bainbridge led Professor Mallard downstairs, along a corridor, through a disused music room, and into a large conservatory. The middle of the floor had been cleared of plants and was occupied by a row of six typewriter tables, each one supporting a hooded machine. At the left of each typewriter was a neat stack of yellow copy paper. Empty wastebaskets were under each table. The chairs were the unpadded, spring-backed kind favored by experienced stenographers. A large bunch of ripe bananas was hanging in one corner, and in another stood a Great Bear water-cooler and a rack of Lily cups. Six piles of typescript, each about a foot high, were ranged along the wall on an improvised shelf. Mr. Bainbridge picked up one of the piles, which he could just conveniently lift, and set it on a table before Professor Mallard. "The output to date of Chimpanzee A, known as Bill," he said simply.

"'"Oliver Twist," by Charles Dickens,'" Professor Mallard read out. He read the first and second pages of the manuscript, then feverishly leafed through to the end. "You mean to tell me," he said, "that this chimpanzee has written—"

"Word for word and comma for comma," said Mr. Bainbridge. "Young, my butler, and I took turns comparing it with the edition I own. Having finished 'Oliver Twist,' Bill is, as you see, starting the sociological works of Vilfredo Pareto, in Italian. At the rate he has been going, it should keep him busy for the rest of the month."

"And all the chimpanzees"—Professor Mallard was pale, and enunciated with difficulty—"they aren't all—"

"Oh, yes, all writing books which I have every reason to believe are in the British Museum. The prose of John Donne, some Anatole France, Conan Doyle, Galen, the collected plays of Somerset Maugham, Marcel Proust, the memoirs of the late Marie of Rumania, and a monograph by a Dr. Wiley on the marsh grasses of Maine and Massachusetts. I can sum it up for you, Mallard, by telling you that since I started this experiment, four weeks and some days ago, none of the chimpanzees has spoiled a single sheet of paper."

Professor Mallard straightened up, passed his handkerchief across his brow, and took a deep breath. "I apologize for my weakness," he said. "It was simply the sudden shock. No, looking at the thing scientifically—and I hope I am at least as capable of that as the next man—there is nothing marvellous about the situation. These chimpanzees, or a succession of similar teams of chimpanzees, would in a million years write all the books in the British Museum. I told you some time ago that I believed that statement. Why should my belief be altered by the fact that they produced some of the books at the very outset? After all, I should not be very much surprised if I tossed a coin a hundred times and it came up heads every time. I know that if I kept at it long enough, the ratio would reduce itself to an exact fifty per cent. Rest assured, these chimpanzees will begin to compose gibberish quite soon. It is bound to happen. Science tells us so. Meanwhile, I advise you to keep this experiment secret. Uninformed people might create a sensation if they knew."

"I will, indeed," Mr. Bainbridge said. "And I'm very grateful for your rational analysis. It reassures me. And now, before you go, you must hear the new Schnabel records that arrived today."

During the succeeding three months, Professor Mallard got into the habit of telephoning Mr. Bainbridge every Friday afternoon at five-thirty, immediately after leaving his seminar room. The Professor would say, "Well?," and Mr. Bainbridge would reply, "They're still at it, Mallard. Haven't spoiled a sheet of paper yet." If Mr. Bainbridge had to go out on Friday afternoon, he would leave a written message with his butler, who would read it to Professor Mallard: "Mr. Bainbridge says we now have Trevelyan's 'Life of Macaulay,' the Confessions of St. Augustine, 'Vanity Fair,' part of Irving's 'Life of George

Washington,' the Book of the Dead, and some speeches delivered in Parliament in opposition to the Corn Laws, sir." Professor Mallard would reply, with a hint of a snarl in his voice, "Tell him to remember what I predicted," and hang up with a clash.

The eleventh Friday that Professor Mallard telephoned, Mr. Bainbridge said, "No change. I have had to store the bulk of the manuscript in the cellar. I would have burned it, except that it probably has some scientific value."

"How dare you talk of scientific value?" The voice from New Haven roared faintly in the receiver. "Scientific value! You—you—chimpanzee!" There were further inarticulate sputterings, and Mr. Bainbridge hung up with a disturbed expression. "I am afraid Mallard is overtaxing himself," he murmured.

Next day, however, he was pleasantly surprised. He was leafing through a manuscript that had been completed the previous day by Chimpanzee D, Corky. It was the complete diary of Samuel Pepys, and Mr. Bainbridge was chuckling over the naughty passages, which were omitted in his own edition, when Professor Mallard was shown into the room. "I have come to apologize for my outrageous conduct on the telephone yesterday," the Professor said.

"Please don't think of it any more. I know you have many things on your mind," Mr. Bainbridge said. "Would you like a drink?"

"A large whiskey, straight, please," Professor Mallard said. "I got rather cold driving down. No change, I presume?"

"No, none. Chimpanzee F, Dinty, is just finishing John Florio's translation of Montaigne's essays, but there is no other news of interest."

Professor Mallard squared his shoulders and tossed off his drink in one astonishing gulp. "I should like to see them at work," he said. "Would I disturb them, do you think?"

"Not at all. As a matter of fact, I usually look in on them around this time of day. Dinty may have finished his Montaigne by now, and it is always interesting to see them start a new work. I would have thought that they would continue on the same sheet of paper, but they don't, you know. Always a fresh sheet, and the title in capitals."

Professor Mallard, without apology, poured another drink and slugged it down. "Lead on," he said.

It was dusk in the conservatory, and the chimpanzees were typing by the light of student lamps clamped to their desks. The keeper lounged in a corner, eating a banana and reading *Billboard*. "You might as well take an hour or so off," Mr. Bainbridge said. The man left.

Professor Mallard, who had not taken off his overcoat, stood with his hands in his pockets, looking at the busy chimpanzees. "I wonder if you know, Bainbridge, that the science of probabilities takes everything into account," he said, in a queer, tight voice. "It is certainly almost beyond the bounds of credibility that these chimpanzees should write books without a single error, but that abnormality may be corrected by—*these!*" He took his hands from his pockets, and each one held a .38 revolver. "Stand back out of harm's way!" he shouted.

"Mallard! Stop it!" The revolvers barked, first the right hand, then the left, then the right. Two chimpanzees fell, and a third reeled into a corner. Mr. Bainbridge seized his friend's arm and wrested one of the weapons from him.

"Now I am armed, too, Mallard, and I advise you to stop!" he cried. Professor Mallard's answer was to draw a bead on Chimpanzee E and shoot him dead. Mr. Bainbridge made a rush, and Professor Mallard fired at him. Mr. Bainbridge, in his quick death agony, tightened his finger on the trigger of his revolver. It went off, and Professor Mallard went down. On his hands and knees he fired at the two chimpanzees which were still unhurt, and then collapsed.

There was nobody to hear his last words. "The human equation . . . always the enemy of science . . ." he panted. "This time . . . vice versa . . . I, a mere mortal . . . savior of science . . . deserve a Nobel . . ."

When the old butler came running into the conservatory to investigate the noises, his eyes were met by a truly appalling sight. The student lamps were shattered, but a newly risen moon shone in through the conservatory windows on the corpses of the two gentlemen, each clutching a smoking revolver. Five of the chimpanzees were dead. The sixth was Chimpanzee F. His right arm disabled, obviously bleeding to death, he was slumped before his typewriter. Painfully, with his left hand, he took from the machine the completed last page of Florio's Montaigne. Groping for a fresh sheet, he inserted it, and typed with one finger, "*Uncle Tom's Cabin,* by Harriet Beecher Stowe. Chapte . . ." Then he, too, was dead.

**5**

*Chaos umpire sits
And by decision more
    embroils the fray
By which he reigns: next
    him high arbiter
Chance governs all.*

—MILTON

*Lo! thy dread empire,
    Chaos! is restor'd.*

—ALEXANDER POPE

*"If the law supposes that," said Mr. Bumble, . . . "the law is a
ass—a idiot."*

—DICKENS (*Oliver Twist*)

*Stand not upon the order of your going,
But go at once.*

—SHAKESPEARE (*Macbeth*)

# THE LAW

## By Robert M. Coates

The first intimation that things were getting out of hand came one early-fall evening in the late nineteen-forties. What happened, simply, was that between seven and nine o'clock on that evening the Triborough Bridge had the heaviest concentration of outbound traffic in its entire history.

This was odd, for it was a weekday evening (to be precise, a Wednesday), and though the weather was agreeably mild and clear, with a moon that was close enough to being full to lure a certain number of motorists out of the city, these facts alone were not enough to explain the phenomenon. No other bridge or main highway was affected, and though the two preceding nights had been equally balmy and moonlit, on both of these the bridge traffic had run close to normal.

The bridge personnel, at any rate, was caught entirely unprepared. A main artery of traffic, like the Triborough, operates under fairly predictable conditions. Motor travel, like most other large-scale human activities, obeys the Law of Averages—that great, ancient rule that states that the actions of people in the mass will always follow consistent patterns—and on the basis of past experience it had always been possible to foretell, almost to the last digit, the number of cars that would cross the bridge at any given hour of the day or night. In this case, though, all rules were broken.

The hours from seven till nearly midnight are normally quiet ones on the bridge. But on that night it was as if all the motorists in the city, or at any rate a staggering proportion of them, had conspired together to upset tradition. Beginning almost exactly at seven o'clock, cars poured onto the bridge in such numbers and with such rapidity that the staff at the toll booths was overwhelmed almost from the start. It was soon apparent that this was no momentary congestion, and as it became more and more obvious that the traffic jam promised to be one of truly monumental proportions, added details of police were rushed to the scene to help handle it.

Cars streamed in from all directions—from the Bronx approach and the Manhattan one, from 125th Street and the East River Drive. (At the peak of the crush, about eight-fifteen, observers on the bridge reported that the drive was a solid line of car headlights as far south as the bend at Eighty-ninth Street, while the congestion crosstown in Manhattan disrupted traffic as far west as Amsterdam Avenue.) And perhaps the most confusing thing about the whole manifestation was that there seemed to be no reason for it.

Now and then, as the harried toll-booth attendants made change for the seemingly endless stream of cars, they would question the occupants, and it soon became clear that the very participants in the monstrous tieup were as ignorant of its cause as anyone else was. A report made by Sergeant Alfonse O'Toole, who commanded the detail in charge of the Bronx approach, is typical. "I kept askin' them," he said, "'Is there night football somewhere that we don't know about? Is it the races you're goin' to?' But the funny thing was half the time they'd be askin' *me*. 'What's the crowd for, Mac?' they would say. And I'd just look at them. There was one guy I mind, in a Ford convertible with a girl in the seat beside him, and when he asked me, I said to him, 'Hell, you're *in* the crowd, ain't you?' I said. 'What brings *you* here?' And the dummy just looked at me. 'Me?' he says. 'I just come out for a drive in the moonlight. But if I'd known there'd be a crowd like this. . .' he says. And then he asks me, 'Is there any place I can turn around and get out of this?'" As the *Herald Tribune* summed things up in its story next morning, it "just looked as if everybody in Manhattan who owned a motorcar had decided to drive out on Long Island that evening."

The incident was unusual enough to make all the front pages next morning, and because of this, many similar events, which might otherwise have gone unnoticed, received attention. The proprietor of the Aramis Theatre, on Eighth Avenue, reported that on several nights in the recent past his auditorium had been practically empty, while on others it had been jammed to suffocation. Luncheon owners noted that increasingly their patrons were developing a habit of making runs on specific items; one day it would be the roast shoulder of veal with pan gravy that was ordered almost exclusively, while the next everyone would be taking the Vienna loaf, and the roast veal went begging. A man who ran a small notions store in Bayside revealed that over a period of four days two hundred and seventy-four successive customers had entered his shop and asked for a spool of pink thread.

These were news items that would ordinarily have gone into the papers as fillers or in the sections reserved for oddities. Now, however, they seemed to have a more serious significance. It was apparent at last that something decidedly strange was happening to people's habits, and it was as unsettling as those occasional moments on excursion boats when the passengers are moved, all at once, to rush to one side or the other of the vessel. It was not till one day in December when, almost incredibly, the Twentieth Century Limited left New York for Chicago with just three passengers aboard that business leaders discovered how disastrous the new trend could be, too.

Until then, the New York Central, for instance, could operate confidently on the assumption that although there might be several thousand men in New York who had business relations in Chicago, on any single day no more—and no less—than some hundreds of them would have occasion to go there. The play producer could be sure that his patronage would sort itself out and that roughly as many persons would want to see the performance on Thursday as there had been on Tuesday or Wednesday. Now they couldn't be sure of anything. The Law of Averages had gone by the board, and if the effect on business promised to be catastrophic, it was also singularly unnerving for the general customer.

The lady starting downtown for a day of shopping, for example, could never be sure whether she would find Macy's department store a seething mob of other shoppers or a wilderness of empty, echoing aisles and unoccupied salesgirls. And the uncertainty produced a strange sort of jitteriness in the individual when faced with any impulse to action. "Shall we do it or shan't we?" people kept asking themselves, knowing that if they did it, it might turn out that thousands of other individuals had decided similarly; knowing, too, that if they *didn't*, they might miss the one glorious chance of all chances to have Jones Beach, say, practically to themselves. Business languished, and a sort of desperate uncertainty rode everyone.

At this juncture, it was inevitable that Congress should be called on for action. In fact, Congress called on itself, and it must be said that it rose nobly to the occasion. A committee was appointed, drawn from both Houses and headed by Senator J. Wing Slooper (R.), of Indiana, and though after considerable investigation the committee was forced reluctantly to conclude that there was no evidence of Communist instigation, the unconscious subversiveness of the people's present conduct was obvious at a glance. The problem was what to do about it. You can't indict a whole nation, particularly on such vague grounds as these were. But, as Senator Slooper boldly pointed out, "You can control it," and in the end a system of reëducation and reform was decided upon, designed to lead people back to—again we quote Senator Slooper—"the basic regularities, the homely averageness of the American way of life."

In the course of the committee's investigations, it had been discovered, to everyone's dismay, that the Law of Averages had never been incorporated into the body of federal jurisprudence, and though the upholders of States' Rights rebelled violently, the oversight was at once corrected, both by Constitutional amendment and by a law—the Hills-Slooper Act—implementing it. According to the Act, people were *required* to be average, and, as the simplest way of assuring it, they were divided alphabetically and their permissible activities catalogued accordingly. Thus, by the plan, a person whose name began with "G," "N," or "U," for example, could attend the theatre only on Tuesdays, and he could go to baseball games only on Thursdays, whereas his visits to a haberdashery were confined to the hours between ten o'clock and noon on Mondays.

The law, of course, had its disadvantages. It had a crippling effect on theatre parties, among other social functions, and the cost of enforcing it was unbelievably heavy. In the end, too, so many amendments had to be added to it—such as the one permitting gentlemen to take their fiancées (if accredited) along with them to various events and functions no matter what letter the said fiancées' names began with—that the courts were frequently at a loss to interpret it when confronted with violations.

In its way, though, the law did serve its purpose, for it did induce—rather mechanically, it is true, but still adequately—a return to that average existence that Senator Slooper desired. All, indeed, would have been well if a year or so later disquieting reports had not begun to seep in from the backwoods. It seemed that there, in what had hitherto been considered to be marginal areas, a strange wave of prosperity was making itself felt. Tennessee mountaineers were buying Packard convertibles, and Sears, Roebuck reported that in the Ozarks their sales of luxury items had gone up nine hundred per cent. In the scrub sections of Vermont, men who formerly had barely been able to scratch a living from their rock-strewn acres were now sending their daughters to Europe and ordering expensive cigars from New York. It appeared that the Law of Diminishing Returns was going haywire, too.

# PART XXIV

# MATHEMATICS AND MUSIC

# COMMENTARY ON
# SIR JAMES JEANS

Sir James Jeans was a mathematical physicist whose writings were much admired by Tallulah Bankhead. I do not know that Miss Bankhead was especially moved by Jeans' contributions to the theory of gases or to the study of the equilibrium of rotating fluid masses, but it is recorded that she described the best known of his works, *The Mysterious Universe,* as a book every girl should read.[1]

Jeans had a productive and varied career which divides more or less into two periods. He was born in 1877 in Ormskirk, Lancashire, to parents in comfortable circumstances. His father was a journalist attached to the press gallery of the House of Commons but with interests a good deal broader than the chicaneries and trivia of daily politics. He published two popular books on science,[2] which reflected not only his admiration for scientific knowledge but his conviction that students of the subject have a duty to follow the example of men like Tyndall, Huxley and Clifford in kindling "a love of science among the masses."[3] His outlook, a strange compound of strict Victorian religious orthodoxy and free thinking, must have been confusing to his son. Jeans was a precocious child, inclined to melancholy. He amused himself by memorizing seven-place logarithms and by dissecting and studying the mechanism of clocks; he was also trained at an early age to read the first leader of the *Times* each morning to his parents. One of his biographers, J. G. Crowther, remarks that "there are cases of the balance of infants' minds having been disturbed by this practice."[4]

At nineteen Jeans entered Trinity College, Cambridge, where he read mathematics and soon gave evidence of exceptional powers. One envies him his undergraduate days in a college whose faculty included J. W. L. Glaisher, W. W. Rouse Ball, Alfred North Whitehead and Edmund T. Whittaker. He finished second in the stiff competitive examination known as the tripos, two places above his classmate G. H. Hardy, who later became the foremost British mathematician of his generation.

Jeans first applied his imagination and formidable mathematical technique to problems concerned with the distribution of energy among the molecules in a gas. Clerk Maxwell and Ludwig Boltzmann had invented theories which treated gas molecules as if they were tiny billiard balls—"rigid and geometrically perfect spheres." While these theories fitted the observed facts fairly well, they produced serious dilemmas connected with the law of the conservation of energy and the second law of thermodynamics. In Jeans' first treatise, *Dynamical Theory of Gases* (1904), he refined the older theories and overcame some of the principal difficulties to which they gave rise.

What is generally regarded as Jeans' masterwork, his *Problems of Cosmogony and Stellar Dynamics* (1917), also has to do with the behavior of gases. This book discusses the cosmogonic problems involving "incompressible masses acted on by their own gravitation."[5] What happens to a mass of liquid "spinning about an axis and isolated in space"? What are the changes of form assumed by masses of gas (like the sun) under rotation, and as the masses shrink? What is the bearing of these matters on the evolution of planets and of our solar system? Questions of this kind had attracted the greatest scientists from Newton and Laplace to Poincaré and Sir George Darwin. Jeans' treatise, while claiming no finality for its conclusions, is acknowledged to be a landmark in the history of astronomy, a contribution to the solution of the underlying mathematical problems which must be ranked a "permanent achievement, come what may in the future development of cosmogony."[6]

Some men turn to religion as they grow old; Jeans turned to a mixture of religion and popular science. At the age of fifty-two, when he was knighted, he could look back upon his research and teaching career with unmixed satisfaction. He had been elected to a fellowship in Trinity in 1901, had served (1905–1910) as professor of applied mathematics at Princeton, as Stokes Lecturer at Cambridge, and as one of the joint secretaries of the Royal Society (to which he had been elected when he was only twenty-eight) from 1919 to 1929. Marriage to a wealthy American girl—it was a very happy union—had given him financial independence; Jeans was not compelled to assume teaching duties which might interfere with his research. By 1928 he had published seven books and seventy-six original papers and had earned a reputation as an outstanding mathematician. At this point, perhaps because he felt, as Milne reports, that his powers as a mathematician were declining, he abandoned pure science for popularization.[7]

The conversion was a tremendous success. His first book in a nontechnical vein was *The Universe Around Us* (1929); it was followed by *The Mysterious Universe* (1930) and by half a dozen similar volumes, the last appearing in 1947. Jeans' style was "transparent, trenchant, and dignified. His scientific narrative flowed like a grand river under complete control."[8] His images were vivid, often breathtaking. Their object was to make the reader goggle at immensities and smallnesses, excellent items for conversational gambits. Because he was primarily a mathematician, Jeans expatiated less on physical ideas than on startling numerical contrasts. The scale of matter, he noted, ranges "from electrons of a fraction of a millionth of a millionth of an inch in diameter, to nebulae whose diameters are measured in hundreds of thousands of millions of miles"; a model of the universe in which the sun was represented "by a speck of dust $\frac{1}{3400}$ of an inch in diameter" would have to extend four million miles in every direction to encompass a few of our island-universe neighbors; "Empty Waterloo Station of everything except six specks of dust, and it is still far more crowded with dust than space is with stars"; the molecules in a pint of water "placed end to end . . . would form a chain capable of encircling the earth over 200 million times"; the energy in a thimble of water would drive a large vessel back and forth across the ocean twenty times; a pinhead heated to a temperature equal to that at the center of the sun would "emit enough heat to kill anyone who ventured within a thousand miles of it"; the sun loses daily by radiation 360,000 million tons of its weight, the earth only ninety pounds. Most of us were brought up on these images and have never shaken off their emotional impact, though it is true that the facts of the atomic age make some of Jeans' best flesh-creepers sound commonplace.

The public took Jeans' excitements to its heart, but scientists and philosophers were able to restrain their enthusiasm. As, more and more, a religious, emotional and mystical note crept into his books, they came under sharp attack. In her biting *Philosophy and the Physicists,* Susan Stebbing let fly at both Jeans and Eddington for their philosophical interpretation of physical theories.[9] "Both of these writers approach their task through an emotional fog; they present their views with an amount of personification and metaphor that reduces them to the level of revivalist preachers." Many of Jeans' "devices," wrote Miss Stebbing, "are used apparently for no other purpose than to reduce the reader to a state of abject terror." Other critics were kinder (and less witty) but raised essentially the same objections. It is hard to escape the feeling that Jeans capitalized increasingly on his charm and virtuosity as a popular expositor, that he yielded to an inner need to write more even as his ideas were petering out.[10]

Jeans died of a heart attack on September 16, 1946. He was reading proofs of his last book, *The Growth of Physical Science,* only a few days before his death.

In this biographical sketch, brief though it is, I have said a good deal more about Jeans' work as a popularizer than as an original investigator. This seemed desirable because of his real contribution, now disparaged, to scientific education. He was a first-rate mathematical physicist and for that he will be remembered. It would be unjust, however, to overlook the impetus he gave to scientific understanding in much broader circles, even though some of his ideas were muddled and grossly misleading; even though, in later years, he came close to being a hack. On the whole, I think more persons got a glimpse of the meaning of science, a taste of its excitement, integrity and beauty from Jeans' vivid primers than were led astray by his misty philosophy.

The selection below is from Jeans' *Science and Music.* This is an excellent account of the science of sound, written in a lucid and straightforward style, without theological dissonances; it is expert in discussions of both the experimental and mathematical sides of the subject. I have chosen excerpts which illustrate the remarkable contribution that mathematicians have made to the analysis of musical structure, from the profound discovery made by Pythagoras through the magnificent labors of Helmholtz and his successors. Jeans deeply loved music all his life; he often played the organ for three or four hours a day, and thought in musical images even in his scientific work.[11] His second wife was a brilliant organist and their mutual interest in music led to the writing of this book.

## ENDNOTES

[1]J. G. Crowther, *British Scientists of the Twentieth Century,* London, 1952, p. 95.

[2]*The Creators of the Age of Steel* (1884); *Lives of the Electricians* (1887).

[3]Crowther, *op. cit.,* p. 96.

[4]Crowther, *op. cit.,* p. 97.

[5]E. A. Milne, *Sir James Jeans, Cambridge,* 1952, p. 110.

[6]E. A. Milne, *op. cit.,* p. 114 *passim.*

[7]Milne, *op. cit.,* p. 73. Milne gives this on the authority of Jeans' second wife, whom he married after the death of the first Lady Jeans in 1934.

[8]Crowther, *op. cit.,* p. 93.

[9]Penguin Books, Harmondsworth, 1944; Chapters 1–3 *passim.*

[10]"*The Mysterious Universe* received fierce criticism when it was published. Rutherford was heard to say that Jeans had told him that 'that fellow Eddington has written a book which has sold 50,000 copies; I will write one that will sell 100,000.' And Rutherford added: 'He did.'" Crowther, *op. cit.,* p. 136.

[11]Crowther, *op. cit.,* p. 93

1

*Discoursed with Mr. Hooke about the nature of sounds, and he did make me understand the nature of musicall sounds made by strings, mighty prettily; and he told me that having come to a certain number of vibrations proper to make any tone, he is able to tell how many strokes a fly makes with her wings (those flies that hum in their flying) by the note that it answers to in musique, during their flying. That I suppose is a little too much refined; but his discourse in general of sound was mighty fine.*

—SAMUEL PEPYS (*Diary,* Aug. 8, 1666)

*Through and through the world is infested with quantity: To talk sense is to talk quantities. It is no use saying the nation is large— How large? It is no use saying that radium is scarce—How scarce? You cannot evade quantity. You may fly to poetry and music, and quantity and number will face you in your rhythms and your octaves.*

—ALFRED NORTH WHITEHEAD

# MATHEMATICS OF MUSIC

## By Sir James Jeans

### TUNING-FORKS AND PURE TONES

We have seen that every sound, and every succession of sounds, can be represented by a curve, and our first problem must obviously be to find the relation between such a curve and the sound or sequence of sounds it represents—in brief, we must learn to interpret a sound-curve.

#### PURE TONES

Let us start by taking an ordinary tuning-fork as our source of sound. We begin with this rather than, let us say, a violin or an organ-pipe, because it gives a perfectly pure musical note, as we shall shortly see. If we strike its prongs on something hard, or draw a violin-bow across them, they are set into vibration. We can see that they are in vibration from their fuzzy outline. Or we can feel that they are in vibration by touching them with our fingers, when we shall experience a trembling or a buzzing sensation. Or, without trusting our senses at all, we may gently touch one prong with a light pith ball suspended from a thread, and shall find that the ball is knocked away with some violence.

When the prongs of the fork vibrate, they communicate their vibrations to the air surrounding them, and this in turn transmits the agitation to our ear-drums, with the result that we hear a sound. We can verify that the air is necessary to the hearing of the sound by standing the vibrating fork inside an air-pump and extracting the air. The fuzzy appearance of the prongs shews that the fork is still in vibration, but we can no longer hear the sound, because the air no longer provides a path by which the vibrations can travel to our ears.

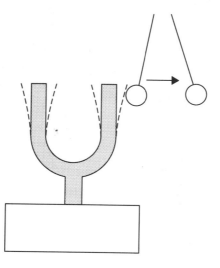

**FIGURE 1**

*The vibrations of a tuning-fork give a fuzzy appearance to the prongs and cause them to repel a light pith ball with some violence.*

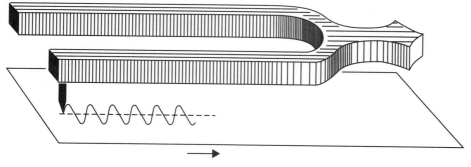

**FIGURE 2**

*The trace of a vibrating fork can be obtained by drawing a piece of paper or smoked glass under it.*

To study these vibrations in detail, we may attach a stiff bristle or a light gramophone needle to the end of one prong of the fork, and while the fork is in vibration, run a piece of smoked glass under it as shewn in Figure 2, taking care that it moves in a perfectly straight line and at a perfectly steady speed. If the fork were not vibrating, the point of the needle would naturally cut a straight furrow through the smoky deposit on the glass; if we held the glass up to the light, it would look like Figure 3. In actual fact, we shall find it looks like Figure 4, which is a copy of an actual photograph; the vibrations have left their record in the smoke, so that the needle has not cut a straight but a wavy furrow.

Each complete wave obviously corresponds to a single to-and-fro motion of the needle point, and so to a complete vibration of the prong of the tuning-fork.

**FIGURE 3**
*The trace of a non-vibrating fork.*

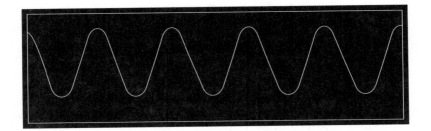

**FIGURE 4**
*The trace of a vibrating fork. The waves are produced by the vibrations of the fork, one complete wave by one complete vibration.*

This wavy curve must clearly be the sound-curve of the sound emitted by the vibrating fork. For if we reverse the motion and compel the needle to follow the furrow, the sideways motions of the needle will set up similar motions in the prong to which it is attached, and these will produce exactly the same sound as was produced when the fork vibrated freely of itself. In fact, the whole process is like that of listening to a gramophone record, except that the tuning-fork, instead of a mica diaphragm, transmits the sound-vibrations to the air.

This simple experiment has disclosed the relation between the musical sound produced by a tuning-fork and its curve, which we now find to consist of a succession of similar waves.

The extreme regularity of these waves is striking; they are all of precisely the same shape, so that their lengths are all exactly the same, and they recur at perfectly regular intervals. Indeed, it is this regularity which distinguishes music from mere noise. So long as a gramophone needle is moving regularly to-and-fro in its groove we hear music; the moment it comes upon an accidental scratch on the record, so that its motion experiences a sudden irregular jerk, we hear mere noise. In such ways as this, we discover that regularity is the essential of a musical sound-curve. Yet the regularity can be overdone, and absolute unending regularity produces mere unpleasing monotony. The

2260 Sir James Jeans

problem of designing a curve which shall give pleasure to the ear is not altogether unlike that of designing a building which shall give pleasure to the eye. A mere collection of random oddments thrown together anyhow is not satisfying; our aesthetic sense calls for a certain amount of regularity, rhythm and balance. Yet these qualities carried to excess produce monotony and lifelessness—the barracks in architecture and the dull flat hum of the tuning-fork in music.

## PERIOD, FREQUENCY AND PITCH

When a tuning-fork is first set into vibration, we hear a fairly loud note, but this gradually weakens in intensity as the vibrations transfer their energy to the surrounding air. Unless the fork was struck very violently in the first instance, we notice that the pitch of this note remains the same throughout; if the fork sounded middle C when it was first struck, it will continue to sound this same note until its sound dies away into silence.[1] On taking a trace of the whole motion, in the manner shewn in Figure 2, we find that the waves slowly decrease in height as the sound diminishes in strength, but they remain always of the same length.

If we measure the speed at which the fork is drawn over the smoked glass in taking this trace, we can easily calculate the amount of time the needle takes to make each wave. This is, of course, the time of a single vibration of the fork, and is only a minute fraction of a second; we call it the "period" of the vibration. The number of complete vibrations which occur in a second is called the "frequency" of the vibration. Actual experiment shews that a tuning-fork which is tuned to middle C of the pianoforte executes 261 vibrations in a second, regardless of whether the sound is loud or soft.

This frequency of 261 is associated with the pitch of middle C not only for the sound of a tuning-fork, but also for all musical sounds, no matter how they are produced. For instance, a siren which runs at such a rate that 261 blasts of air escape in a second will sound middle C. Or we may hold the edge of a card against a rotating toothed wheel; if 261 teeth strike the card every second we again hear middle C. If a steam-saw runs at such a rate that 261 teeth cut into the wood every second, it is again middle C that we hear. The hum of a dynamo is middle C when the current alternates at the rate of 261 cycles a second, and this is true of all electric machinery. There are electric organs on the market in which the sound of a middle C pipe is copied, sometimes very faithfully, by an electric current which is made to alternate at the rate of 261 cycles a second. Again, when a motor-car is running at such a rate that the pistons make 261 strokes a second, a vibration of frequency 261 is set up, and we hear a note of pitch middle C in the noise of the engine.

All this shews that the pitch of a sound depends only on the frequency of the vibration by which it is produced. It does not depend on the nature of the vibration. Thus we may say that it is the frequency of vibration that determines the pitch of a sound. If there is no clearly defined frequency, there is no clearly defined pitch, because the sound is no longer musical.

When a siren or steam-saw or dynamo is increasing its speed, the sound we hear rises in pitch, and conversely. Thus we learn to associate high pitch with high frequency, and vice versa. If we experiment with a series of forks tuned to all the notes in the middle octave of the piano, we shall find the following frequencies:

| c  | 261.0 | f  | 348.4 | a  | 438.9 |
|----|-------|----|-------|----|-------|
| c# | 276.5 | f# | 369.1 | a# | 465.0 |
| d  | 293.0 | g  | 391.1 | b  | 492.7 |
| d# | 310.4 | g# | 414.3 | c' | 522.0 |
| e  | 328.8 |    |       |    |       |

. . . These frequencies might at first sight be thought to be a mere random collection of numbers, but a little study shews that they are not.

We notice at once that the first number 261 is just half of the last number 522. Thus our experiments have shewn that in this particular case the interval of an octave corresponds to a 2 to 1 ratio of frequencies, and other experiments shew that this is universally true—doubling the frequency invariably raises the pitch by an octave. The octave interval is fundamental in the music of all ages and of all countries; we now see its physical significance.

We may further notice that the interval from c to c# represents a rise in frequency of just about 6 per cent., and a little arithmetic will shew that the same is true for every other interval of a semitone. The rise cannot be precisely 6 per cent. for each semitone, since if it were, the rise in the whole octave, consisting of twelve such intervals, would be equal to $1.06 \times 1.06 \times 1.06 \times$ . . . etc., there being twelve factors in all, each equal to 1.06. This is the quantity which the mathematician describes as $(1.06)^{12}$, and it is equal to 2.0122, and not to exactly 2.

In an instrument such as the piano or organ, which is tuned to "equal temperament" the exact interval of 2 is spread equally over the twelve semitone intervals which make the octave. Each step accordingly represents a frequency ratio of 1.05946, since this is the exact twelfth root of 2. . . .

## SIMPLE HARMONIC CURVES

Having learned all we can from the regularity and length of the waves in Figure 4, let us next examine their form. The extreme simplicity of their shape is very noticeable, although it must be said at once that this is not a property of all sound-curves; these particular curves are simple because they are produced by the simplest of all musical instruments—the tuning-fork. Exact measurement shews that the curve has a shape with which the mathematician is very well acquainted. It is called a "sine" curve, or a "simple harmonic" curve, while the motion of the needle which produces it is described as "simple harmonic motion."

These simple harmonic curves and the simple harmonic motion by which they are produced are of fundamental importance in all departments of mechanics and physics, as well as in many other branches of science. They are particularly important in the theory of vibrations, and this makes them of especial interest in the study of music, since musical sound is almost invariably produced by the vibrations of some mechanical structure—a stretched string, a column of air, a drum-skin, or some metallic object such as a cymbal, triangle, tube or bell. For this reason, we shall discuss vibrations in some detail.

## GENERAL THEORY OF VIBRATIONS

Generally speaking, every material structure can find at least one position in which it can remain at rest—otherwise it would be a perpetual motion machine. Such a position is called a "position of equilibrium." When a structure is in such a position, the forces on each particle of it—as for instance the weight of the particle, and the pushes and pulls from neighbouring particles—are exactly balanced. Any slight disturbance, such as a push, pull or knock from outside, will cause the structure to move out of this position of equilibrium to some new position, in which the forces on a particle are no longer evenly balanced; each particle then experiences a "restoring force" which tends to pull it back to the position it originally occupied.

This force starts by dragging the particle back towards its original position of equilibrium. In time it regains this position, but as it is now moving with a certain amount of speed, it overshoots the position and travels a certain distance on the other side before coming to rest. Here it experiences a new force tending to pull it back; again it yields to this force, gets up speed, overshoots the mark, and so on, the motion repeating itself time after time. Clearly the trace of the motion of any particle will be a succession of waves, like those we have already obtained from the tuning-fork in Figure 4 (p. 2259).

Motion of this kind is described by the general term "oscillation." In the special case in which each particle only moves through a very small distance, the motion is called a "vibration." Thus a vibration is a special kind of oscillation, and, as it happens, possesses certain very simple properties which are not possessed by oscillations in general. It is usually true of oscillations that the farther a particle moves from its position of equilibrium, the greater is the restoring force pulling it back. But in a vibration the restoring force is *exactly proportional* to the distance the particle has moved from its position of equilibrium; draw it twice as far from this position, and we double the force pulling it back.

A simple mathematical investigation shews that when this relation holds, the motion of every particle will be of the same kind, whatever the structure to which it belongs. Motion of this kind is defined to be "simple harmonic motion."

**FIGURE 5**

*A position of equilibrium. The weight can rest in equilibrium at* C *but nowhere else. If we pull it*
*aside to* B, *it tends to return to* C.

We have already found a concrete instance of this kind of motion in the
tuning-fork. Another is provided by what is perhaps the simplest mechanical
structure we can imagine—a weight suspended by a fine thread. The position
of equilibrium is one in which the weight lies at a point *C* exactly under the
point of suspension. When the weight is drawn a short distance aside to an
adjacent position *B,* there is no longer equilibrium, and the weight tends to fall
back to *C.* In technical language, a restoring force acts on the weight, tending
to draw it back to its position of equilibrium *C,* and it is a simple problem in
dynamics to find its amount. So long as the displacement of the weight is not
too large, we find that the restoring force is exactly proportional to the extent of
the displacement *BC,* so that the condition for simple harmonic motion is
fulfilled. Indeed, if we take a trace by attaching a needle to the weight and
running a piece of paper horizontally under it, as in Figure 6, we shall find that
this trace is a simple harmonic curve exactly like that made by our tuning-fork.

If we set our suspended weight swinging more violently, and again take a
trace of its motion, we shall again obtain a simple harmonic curve. The waves
will, of course, be greater in size, but their period will be exactly the same as
before. We find that the swinging weight makes just as many swings per
second, no matter what the extent of these swings may be, provided always that
they are small enough to qualify as vibrations. This illustrates the well-known
fact that the period of vibration of a pendulum depends only on its length, and
not on the extent of its swing; it is because of this that our pendulum clocks
keep time.

We found a similar property in the tuning-fork, the period of its vibrations
being the same whether we struck it fairly hard or only very softly. And all true
vibrations possess the same property—the period is independent of the extent

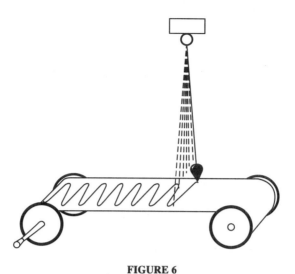

**FIGURE 6**

*Taking the trace of a swinging pendulum. The trace is found to be a simple harmonic curve, exactly similar to that given by a vibrating tuning-fork (Figure 2).*

and energy of the swing. This is a most important fact for the musician. It means that every musical instrument in which the sound is produced by vibrations will "keep time" like a pendulum clock, and so will give a note of the same frequency, and therefore of the same pitch, whether it is played soft or loud. Without this property it may almost be said that music, as we know it, would be impossible. We can hardly imagine an orchestra acquitting itself with credit if every note was out of tune unless it was played with exactly the right degree of force. Crescendos and diminuendos could only be produced by adding and subtracting instruments. As the note of a piano or any percussion instrument decreased in strength it would also change in pitch, and every piece would inevitably begin with a howl and end with a wail.

At the same time, every musician is familiar with cases in which the pitch of an instrument is changed appreciably by playing it softer or louder. The flautist can always pull his instrument a bit out of tune by blowing strong or weak, while the organist knows only too well the dismal wail of flattened notes which is heard when his wind gives out. We shall discuss the theory of such sounds as these later, and shall find that they are not produced by absolutely simple vibrations like those of the tuning fork or pendulum.

## SIMULTANEOUS VIBRATIONS

Many structures are capable of vibrating in more than one way, and so may often be performing several different vibrations at the same time. There is a very general principle in mechanics, which asserts that when any structure whatever is set into vibration—provided only that the displacement of each particle is small—the motion of every particle is either a simple harmonic motion or else is a more complicated motion which results from superposing a number of simple harmonic motions, one for each vibration which is in progress.

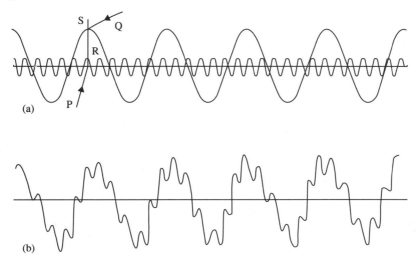

**FIGURE 7**

*The superposition of two vibrations. The two wavy curves in (a) have periods which stand in the ratio of 6¼ to 1. On superposing them we obtain the curve (b), which represents very closely the sound-curve of a tuning-fork which is sounding its clang tone.*

A simple illustration will shew how this can be. Let us suppose that while our tuning-fork is in vibration we hit it on the top of one of the prongs with a hammer. We shall hear a sharp metallic click, which is known as the "clang tone" of the fork. A good musical ear may perhaps recognise that its pitch lies about 2½ octaves above the ordinary note of the fork. Clearly the blow of the hammer has started new vibrations in the fork, of much higher frequency than the original vibration. If we had taken a trace of the motion when the original vibration was acting alone we should have obtained a curve like that shewn in Figure 8. This is reproduced as the long-waved curve in Figure 7(*a*). If we take a trace of the clang tone alone, it will be like the short-waved curve in Figure 7(*a*), this representing a simple harmonic motion having 6¼ times the frequency of the main vibration.

Now suppose we take a trace when the two vibrations are going on together. At the instant of time represented at the point *P*, the particle under consideration is displaced through a distance *PQ* by the main vibration, and through a distance *PR* by the vibration which produces the clang tone. Thus the operation of the two vibrations together displaces it through a distance *PQ* + *PR*, and this is equal to *PS* if we make *QS* equal to *PR*. By adding together displacements in this way all along the curve, we obtain the curve shewn in Figure 7(*b*) as the trace to be expected when both vibrations are in action together. The photograph of an actual trace is shewn in Figure 9.

In addition to the clang tone just mentioned, we may often hear a second clang tone about four octaves higher than the fundamental note of the fork. Indeed, it is difficult to start the fork sounding in such a way that the pure tone of the fork is heard without any admixture of these higher tones. We more usually obtain a mixture of all three tones, but this does not interfere with the utility of the tuning-fork as a source of pure musical tone, since the sounds of higher frequency die away quite rapidly, and the ear soon hears nothing but the fundamental tone of the fork.

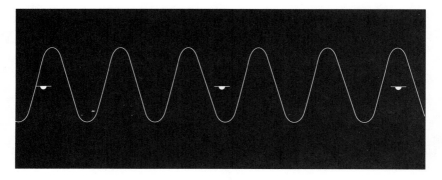

**FIGURE 8**
*The sound-curve of the simple tone from a tuning-fork. The note is of frequency 256 (middle C), and the dots indicate intervals of ¹⁄₁₀₀ second.*

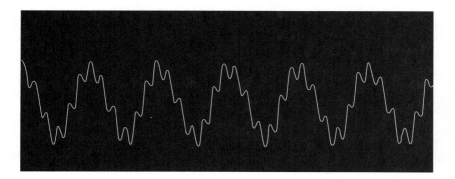

**FIGURE 9**
*The sound-curve of the note from a tuning-fork when the clang tone is sounding. The clang tone superposes small waves onto the longer waves, shewn in Figure 8 above, which represent the main tone of the fork.*

A second example of simultaneous vibrations can be made to teach us something new. If we return to our weight suspended by a string and knock it sideways, it will swing from side to side pendulum-wise through some such path as *AB* in Figure 5 (p. 2263), and its motion, as we have already seen, will be simple harmonic motion. Suppose, however, that when the weight is at *B*, we give it another slight knock in the direction at right angles to *AB*, i.e., *through* the paper of our page in Figure 5. This sets up a new vibration in a direction at right angles to *AB,* and the motion in this direction also must be simple harmonic motion. As we have seen that the period of a pendulum depends only on its length, the new motion will have the same period as the original motion. The whole motion is accordingly obtained from the superposition of two simple harmonic motions whose periods are equal.

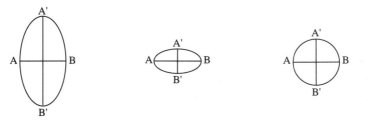

**FIGURE 10**          **FIGURE 11**          **FIGURE 12**
*Three different types of motion which can be executed by the bob of a conical pendulum.*

If we watch the weight from a point directly above it, we shall see it moving in a curved path round its central position *C.* If the second knock was violent, its path will be an elongated ellipse such as *AA'BB'* in Figure 10. If the knock was gentle, its path will be an ellipse elongated in the other direction such as *AA'BB'* in Figure 11. But if the knock was of precisely the same strength as that which originally set the pendulum in motion along *AB,* then the weight will move in the circle *AA'BB'* in Figure 12, forming the arrangement which is generally described as a conical pendulum. It must move with the same speed at each point of its journey, for it is moving in a perfectly level path, so that there is no reason why it should move faster at any one point than at any other.

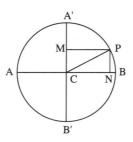

**FIGURE 13**
*A geometrical interpretation of simple harmonic motion. As the point P moves steadily around the circle, the point N moves backwards and forwards along AB, and its motion is simple harmonic motion.*

Thus we learn that each of the motions illustrated in Figures 10, 11 and 12 can be regarded as the superposition of two simple harmonic motions of equal periods. The last of the three is by far the most interesting, because it shews us that a simple circular motion performed at uniform speed can be regarded as made up of two simple harmonic motions in directions at right angles to one another. To put this more definitely, let us imagine that the point $P$ in Figure 13 moves round the circle $AA'BB'$ with uniform speed, like the hand of a clock. Wherever $P$ is, let us draw perpendiculars $PN$, $PM$ on to the lines $AB$, $A'B'$. Then, as $P$ moves steadily round the circle, $N$ moves backwards and forwards along $AB$, while $M$ moves backwards and forwards along $A'B'$. We have learnt that the motion of each of these points will be simple harmonic motion.

This gives us a simple geometrical explanation of simple harmonic motion—as $P$ moves steadily round in a circle, the point $N$ moves in simple harmonic motion. It is easy to see from this definition that the motion of the piston in the cylinder of a locomotive or a motor-car must be approximately simple harmonic motion.

Or we may look at the problem from the other end, and see that as the point $N$ moves to-and-fro in simple harmonic motion along $AB$, the point $P$ moves steadily round the circle $AA'BB'$. This circle is called the "circle of reference" of the simple harmonic motion. Its diameter $AB$ is called the "extent" of the motion, while its radius $CA$ or $CB$ is called the "amplitude" of the motion.

## ENERGY

The amplitude of a vibration gives an indication of its energy, for it is a general law that the energy of a vibration is proportional to the square of the amplitude. For instance, a vibration which has twice the amplitude of another has four times the energy of the other; in other words, the vibrating structure to which it belongs has four times as much capacity for doing work stored up within itself, and it must get rid of this in some way or other before it can come to rest. The energy stored up in a musical instrument is usually expended in setting the air around it into vibration; indeed it is only through its steady outpouring of energy into the surrounding air that we hear the instrument at all.

It follows that if we want to maintain a vibration at the same level of energy we must continually supply energy to it—as we do with an organpipe or a violin-string. If energy is not supplied the vibration will die away —as with a piano-string or a bell or a cymbal. The amplitude of the vibration then slowly decreases, and the circle of reference shrinks in size.

When a structure is performing several vibrations at the same time, energy does not usually pass from one vibration to another. The vibrations are independent, each possessing its own private store of energy which it preserves intact, except for what it may pass on to other outside structures—as for instance, the air around it. Thus the energy of a number of simultaneous vibrations may be thought of as the sum of the energies of the separate vibrations.

## SIMULTANEOUS SOUNDS

When a tuning-fork is sounding, every particle of its substance moves in simple harmonic motion, and those particles which form its surface transmit their motion to the surrounding air. The final result is that every particle of air which is at all near to the tuning-fork is set into motion and moves with a simple harmonic motion, which will naturally have the same period as the tuning-fork. This period is still preserved when the vibration is passed on to the ear-drum of a listener—that is why the note heard by the ear has the same pitch as the fork.

A more complicated situation arises when two tuning-forks are standing side by side. Each then imposes a simple harmonic motion on to the particle of air, so that this has a motion which is obtained by superposing the two motions.

We must study motions of this kind in some detail, because they are of great importance in the practical problems of music. We begin with the simplest problem of all—the superposition of two motions which have the same period. The resulting motion is that which would be forced on a particle of air by the simultaneous vibrations of two forks of the same pitch standing side by side.

## SUPERPOSING VIBRATIONS OF THE SAME PERIOD

The two simple harmonic motions can be represented by two simple harmonic curves, such as those which pass through $X$ and $Y$ in Figure 14. These particular curves have been drawn with their amplitudes in the ratio of 5 to 2, so that $YN = \frac{2}{5}XN$, and the same relation holds all along the curves. At the instant of time represented at the point $N$, the first harmonic motion produces a displacement through a distance $XN$, while the second produces a displacement through a distance $YN$ which is $\frac{2}{5}$ times $XN$. Thus the combined effect of the two motions is a displacement through a distance equal to $1\frac{2}{5}$ times $XN$. This is represented by $ZN$ in Figure 14.

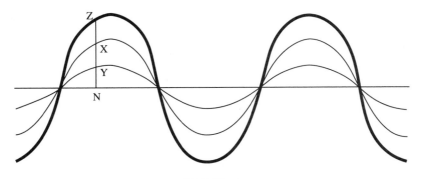

**FIGURE 14**

*The superpostion of two simple harmonic motions of equal period. Here the vibratory motions (represented by the thin curves) are "in the same phase"—crest over crest and trough over trough. The vibrations now reinforce one another, and their resultant (represented by the thick curve) has an amplitude which is equal to the sum of the amplitudes of the two constituents.*

The thick curve through $Z$ is drawn so that its distance above or below the central line is everywhere exactly $1\frac{2}{5}$ times that of the thin curve through $X$. This curve must then represent the motion of which we are in search. It is simply the thin curve through $X$ magnified $1\frac{2}{5}$ times vertically, while its horizontal dimensions remain unchanged. Thus the new motion is a simple harmonic motion having an amplitude equal to the sum of the amplitudes of the constituent motions, and the same period as both.

The foregoing instance is only a very special case of the general problem, for the thin curves in Figure 14 are drawn in a very special way. The crests of the waves of the two curves occur at the same instants as also the troughs; in the diagram, crest lies directly over crest and trough over trough. Vibrations in which this relation holds are said to be "in the same phase."

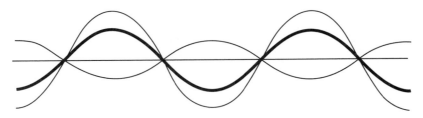

**FIGURE 15**

*The superposition of two simple harmonic motions of equal period. Here the vibratory motions (represented by the thin curves) are "in opposite phase"—crest over trough and trough over crest. The constituent vibrations now pull in opposite directions, and so partially neutralise one another, the amplitude of their resultant (represented by the thick curve) being equal to the difference of the amplitudes of the two constituents.*

The curves might equally well have been drawn as in Figure 15, the crests of one set of waves occurring at the same instants as the troughs of the other set. Vibrations in which this relation holds are said to be "in opposite phase." Crest lies over trough and vice versa, so that the two constituents produce displacements in opposite directions. The resultant motion is again that shewn in the thick curve, but its amplitude is no longer $(1 + \frac{2}{5})$ times the amplitude of the larger constituent, but only $(1 - \frac{2}{5})$ times.

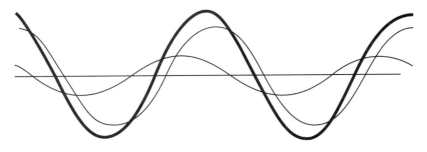

**FIGURE 16**

*The superposition of two simple harmonic motions of equal period. Here there is no simple phase relation between the two constituent vibratory motions (represented by the thin curves), but their resultant is still a simple harmonic motion (represented by the thick curve).*

We must not, however, expect as a matter of course that two motions which occur simultaneously will be either in the same, or in opposite, phase. Such simplicity is unusual, and it is far more likely that the crests of one set of waves will be neither over the crests nor over the troughs of the other set, but somewhere in between, as shewn in Figure 16. If we add together the displacements represented by the two thin curves here, using the method illustrated in Figure 14 (i.e., making $ZN = XN + YN$, and so on), we shall find that the resultant motion is represented by the thick curve shewn in the figure. We may judge by eye that this is yet another simple harmonic curve, as in actual fact it is, but we can only prove this by a new method of attack on the problem, to which we now turn.

We have seen that any simple harmonic motion can be derived from the steady motion of a point round a circle. For instance, as the point $P$ moves round the circle in Figure 13, the point $N$ moves backwards and forwards along the line $AB$ in simple harmonic motion. The two simple harmonic motions which we now want to superpose can of course be derived from the motions of two points, each moving steadily round a circle of its own. Let the two points be $P$ and $Q$ in Figure 17, so that the points $N$, $O$ immediately beneath them execute the simple harmonic motions with which we are concerned.

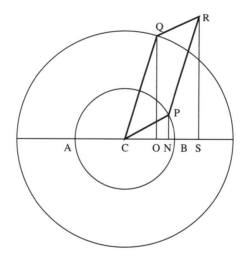

**FIGURE 17**

*The superposition of two simple harmonic motions. As* P *and* Q *move round their respective circles,* N *and* O *execute simple harmonic motions. The resultant motion is that executed by* S, *because* CO + CN = CS.

At the instant to which Figure 17 refers, the motion of $P$ has produced a displacement $CN$, while that of $Q$ has produced a displacement $CO$, so that the total displacement, being the sum of the two, is equal to $CO + CN$.

To represent this in Figure 17, we start from $Q$, and draw the line $QR$ in a direction parallel to $CP$ and of length equal to $CP$. Then, because $QR$ and $CP$ are parallel and equal, the length $OS$ which lies directly under $PR$ must be exactly equal to the length $CN$ which lies directly under $CP$. Hence the sum we need, namely $CO + CN$, must be equal to $CO + OS$, and so to $CS$.

Thus as $P$ and $Q$ move round their respective circles, the points $N$ and $O$ execute the two constituent simple harmonic motions, and the point $S$ executes the motion which results from their superposition.

We are at present supposing the two simple harmonic motions performed by $N$ and $O$ to be of the same frequency, so that the radii $CP$ and $CQ$ rotate at exactly the same rate and the angle $PCQ$ remains always the same. Indeed, we can visualise the whole motion by imagining that we cut the parallelogram $CPRQ$ out of cardboard, and then make it rotate round $C$ at the same rate as $P$ and $Q$. We see that $R$ will move in a circle at uniform speed, so that $S$ will move backwards and forwards along $AB$ in simple harmonic motion. This shews that when two simple harmonic motions have the same frequency, the result of superposing them is a third simple harmonic motion of the same frequency as both. In terms of music, the simultaneous sounding of two pure tones of the same pitch produces a pure tone which is still of the same pitch. . . .

## THE VIBRATIONS OF STRINGS AND HARMONICS

We began our study of sound-curves by examining the curve produced by a tuning-fork. A tuning-fork was chosen, because it emits a perfectly pure tone. But, as every musician knows, its sound is not only perfectly pure, but is also perfectly uninteresting to a musical ear—just because it is so pure.

The artistic eye does not find pleasure in the simple figures of the geometer—the straight line, the triangle or the circle—but rather in a subtle blend of these in which the separate ingredients can hardly be distinguished. In the same way, the painter finds but little interest in the pure colours of his paint-box; his real interest lies in creating subtle, rich or delicate blends of these. It is the same in music; our ears do not find pleasure in the simple tones we have so far been studying but in intricate blends of these. The various musical instruments provide us with ready-made blends, which we can combine still further at our discretion.

In the present chapter we shall consider the sounds which are emitted by stretched strings—such as, for instance, are employed in the piano, violin, harp, zither and guitar—and we shall find how to interpret these as blends of the pure tones we have already had under consideration.

## EXPERIMENTS WITH THE MONOCHORD

Our source of sound will no longer be a tuning-fork but an instrument which was known to the ancient Greek mathematicians, Pythagoras in particular, and is still to be found in every acoustical laboratory—the monochord.

Its essentials are shewn in Figure 18. A wire, with one end $A$ fastened rigidly to a solid framework of wood, passes over a fixed bridge $B$ and a movable bridge $C$, after which it passes over a freely turning wheel $D$, its other end supporting a weight $W$. This weight of course keeps the wire in a state of tension, and we can make the tension as large or small as we please by altering

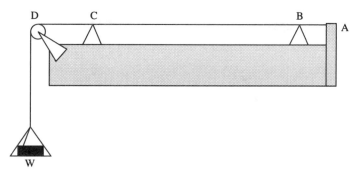

**FIGURE 18**

*The monochord. The string is kept in a state of tension by the suspended weight W, while "bridges" like those of a violin limit the vibration to a range BC. The instrument is arranged so that both range and tension are under control.*

the weight. Only the piece $BC$ of the string is set into vibration, and as the bridge $C$ can be moved backwards and forwards, this can be made of any length we please. It can be set in vibration in a variety of ways—by striking it, as in the piano; by stroking it with a bow, as in the violin; by plucking it, as in the harp; possibly even by blowing over it as in the Aeolian harp, or as the wind makes the telegraph wires whistle on a cold windy day.

On setting the string vibrating in any of these ways, we hear a musical note of definite pitch. While this is still sounding, let us press with our hand on the weight $W$. We shall find that the note rises in pitch, and the harder we press on the weight, the greater the rise will be. The pressure of our hand has of course increased the tension in the string, so that we learn that increasing the tension of a string raises the pitch of the note it emits. This is the way in which the violinist and piano-tuner tune their strings and wires; when one of these is too low in pitch, they screw up the tuning-key.

A series of experiments will disclose the exact relation between the pitch and the tension of a string. Suppose that the string originally sounds c′ (middle C), the tension being 10 lb. To raise the note an octave, to c″, we shall find we must increase the tension to 40 lb.; to raise it yet another octave to c‴, we need a total tension of 160 lb., and so on. In each case a fourfold increase in the tension is needed to double the frequency of the note sounded, and we shall find that this is always the case. It is a general law that the frequency is proportional to the square root of the tension.

We can also experiment on the effect of changing the length of our string, repeating experiments such as were performed by Pythagoras some 2500 years ago. Sliding the bridge $C$ in Figure 18 to the right shortens the effective length $BC$ of the string, but leaves the tension the same—that necessary to support the weight $W$. When we shorten the string, we find that the pitch of the sound rises. If we halve its length, the pitch rises exactly an octave, shewing that the period of vibration has also been halved. By experimenting with the bridge $C$ in all sorts of positions, we discover the general law that the period is exactly proportional to the length of the string, so that the frequency of vibration varies inversely as the length of the string. This law is exemplified in all stringed instruments. In the violin, the same string is made to give out different notes by altering its effective length by touching it with the finger. In the pianoforte different notes are obtained from wires of different lengths.

We may experiment in the same way on the effect of changing the thickness or the material of our wire.

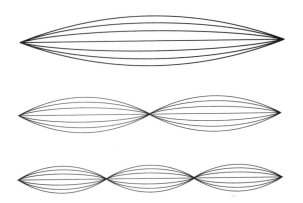

**FIGURES 19, 20 and 21**
*Characteristic vibrations of a stretched string. The string vibrates in one, two and three equal parts respectively, and emits its fundamental tone, the octave and the twelfth of this in so doing.*

## MERSENNE'S LAWS

The knowledge gained from all these experiments can be summed up in the following laws, which were first formulated by the French mathematician Mersenne (*Harmonie Universelle,* 1636):

I. When a string and its tension remain unaltered, but the length is varied, the period of vibration is proportional to the length. (The law of Pythagoras.)

II. When a string and its length remain unaltered, but the tension is varied, the frequency of vibration is proportional to the square root of the tension.

III. For different strings of the same length and tension, the period of vibration is proportional to the square root of the weight of the string.

The operation of all these laws is illustrated in the ordinary pianoforte. The piano-maker could obtain any range of frequencies he wanted by using strings of different lengths but similar structure, the material and tension being the same in all. But the 7¼ octaves range of the modern pianoforte contains notes whose frequencies range from 27 to 4096. If the piano-maker relied on the law of Pythagoras alone, his longest string would have to be more than 150 times the length of his shortest, so that either the former would be inconveniently long, or the latter inconveniently short. He accordingly avails himself of the two other laws of Mersenne. He avoids undue length of his bass strings by increasing their weight—usually by twisting thinner copper wire spirally round them. He avoids inconvenient shortness of his treble strings by increasing their tension. This had to be done with caution in the old wooden-frame piano, since the combined tension of more than 200 stretched strings imposed a great strain on a wooden structure. The modern steel frame can, however, support a total tension of about 30 tons with safety, so that piano-wires can now be screwed up to tensions which were formerly quite impracticable. . . .

## HARMONIC ANALYSIS

Several times already we have superposed two simple harmonic curves, and studied the new curves resulting from the superposition. The essence of the process of superposition has already been illustrated in Figure 7(*a*) on p. 2265, and Figure 14 on p. 2269. In each of these cases the number of superposed curves is only two; when a greater number of such curves is superposed, the resultant curve may be of a highly complicated form.

There is a branch of mathematics known as "harmonic analysis" which deals with the converse problem of sorting out the resultant curve into its constituents. Superposing a number of curves is as simple as mixing chemicals in a test-tube; anyone can do it. But to take the final mixture and discover what ingredients have gone into its composition may require great skill.

Fortunately the problem is easier for the mathematician than for the analytical chemist. There is a very simple technique for analysing any curve, no matter how complicated it may be, into its constituent simple harmonic curves. It is based on a mathematical theorem known as Fourier's theorem, after its discoverer, the famous French mathematician J. B. J. Fourier (1768–1830).

The theorem tells us that every curve, no matter what its nature may be, or in what way it was originally obtained, can be exactly reproduced by superposing a sufficient number of simple harmonic curves—in brief, every curve can be built up by piling up waves.

The theorem further tells us that we need only use waves of certain specified lengths. If, for instance, the original curve repeats itself regularly at intervals of one foot, we need only employ curves which repeat themselves regularly 1, 2, 3, 4, etc. times every foot—i.e., waves of lengths 12, 6, 4, 3, etc. inches. This is almost obvious, for waves of other lengths, such as 18 or 5 inches, would prevent the composite curve repeating regularly every foot. If the original curve does not repeat regularly, we treat its whole length as the first half-period[2] of a curve which does repeat, and obtain the theorem in its more usual form. It tells us that the original curve can be built up out of simple harmonic constituents such that the first has one complete half-wave within the range of the original curve, the second has two complete half-waves, the third has three, and so on; constituents which contain fractional parts of half-waves need not be employed at all. There is a fairly simple rule for calculating the amplitudes of the various constituents, but this lies beyond the scope of the present book.

We obtain a first glimpse into the way of using this theorem if we suppose our original curve to be the curve assumed by a stretched string at any instant of its vibration. Figures 19, 20 and 21 on p. 2274 shew groups of simple harmonic curves which contain one, two and three complete half-waves respectively within the range of the string. Let us imagine this series of diagrams extended indefinitely so as to exhibit further simple harmonic curves containing 4, 5, 6, 7 and all other numbers of complete half-waves. Then the series of curves obtained in this way is precisely the series of constituent curves required by the theorem. We take one curve out of each diagram, and superpose them all; the theorem tells us that by a suitable choice of these curves, the final resultant curve can be made to agree with any curve we happen to have before us. Or, to state it the other way round, any curve we please can be analysed into constituent curves, one of which will be taken from Figure 19, one from Figure 20, one from Figure 21, and so on.

This is not, of course, the only way in which a curve can be decomposed into a number of other curves. Indeed, the number of ways is infinite, just as there is an infinite number of ways in which a piece of paper can be torn into smaller pieces. But the way just mentioned is unique in one respect, and this makes it of the utmost importance in the theory of music. For when we decompose the curve of a vibrating string into simple harmonic curves in this particular way, we are in effect decomposing the motion of the string into its separate free vibrations, and these represent the constituent tones in the note sounded by the vibration. As the vibratory motion proceeds, each of these free vibrations persists without any change of strength, apart from the gradual dying away already explained. If, on the other hand, we had decomposed the vibration in any other way, the strength of the constituent vibrations would be continually changing—probably hundreds of times a second—and so would have no reference to the musical quality of the sound produced by the main vibration.

So general a theory as this may well seem confused and highly complicated, but a single detailed illustration will bring it into sharp focus and shew its importance.

## STRING PLUCKED AT ITS MIDDLE POINT

Let us displace the middle point of a stretched string *AB* to *C,* so that the string forms a flat triangle *ACB* as in Figure 22. The shape of the string *ACB* may still be regarded as a curve, although a somewhat unusual one, and our theorem tells us that this "curve" can be obtained from the superposition of a number of simple harmonic curves. In actual fact, Figure 23 shews how the curve *ACB* can be resolved into its constituent curves; if we superpose all the curves shewn in this latter figure, we shall find we have restored our original broken line *ACB,* except for a difference in scale; the vertical scale in Figure 23 has been made ten times the horizontal in order that the fluctuations of the higher harmonics may be the more clearly seen.

Suppose we now let go of the point *C,* and allow the natural motion of the string to proceed. We may imagine each of the curves shewn in Figure 23 to decrease and increase rhythmically in its own proper period, and the superposition of the curves at any instant will give us the shape of the string at that instant. These curves correspond to the various harmonics that are sounded on plucking a string at its middle point.

We notice that the second, fourth and sixth harmonics are absent. This is not a general property of harmonics, but is peculiar to the special case we have chosen. We have plucked the string in such a way that its two halves are bound to move in similar fashion, and as a consequence the second, fourth and sixth harmonics, which necessarily imply dissimilarity in the two halves, cannot possibly appear. If we had plucked it anywhere else than at its middle point, some at least of these harmonics would have been present.

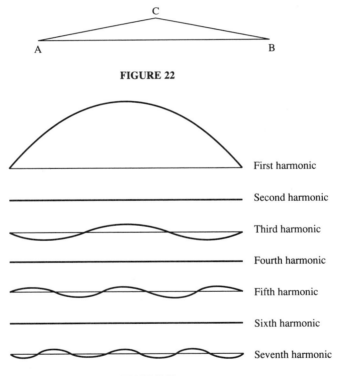

**FIGURE 22**

First harmonic

Second harmonic

Third harmonic

Fourth harmonic

Fifth harmonic

Sixth harmonic

Seventh harmonic

**FIGURE 23**

*The string is displaced to form the triangle ACB. This "curve" can be analysed into the simple harmonic curves shewn in Figure 20. On superposing these we restore the "curve" ACB of Figure 19. (The vertical scales in Figure 23 are all magnified ten-fold.)*

## ANALYSIS OF A SOUND-CURVE

Let us next apply Fourier's theorem to a piece of a sound-curve. The theorem tells us that any sound-curve whatever can be reproduced by the superposition of suitably chosen simple harmonic waves. Consequently any sound, no matter how complex—whether the voice of a singer or a motor-bus changing gear—

can be analysed into pure tones and reproduced exactly by a battery of tuning-forks, or other sources of pure tone. Professor Dayton Miller has built up groups of organ-pipes, which produce the various vowels when sounded in unison; other groups say *papa* and *mama*.

The sound-curve of a musical sound is periodic; it recurs at perfectly regular intervals. Indeed, we have seen that this is the quality which distinguishes music from noise. Fourier's theorem tells us that such a sound-curve can be made up by the superposition of simple harmonic curves such that 1, 2, 3, or some other integral number of complete waves occur within each period of the original curve. If, for instance, the sound-curve has a frequency of 100, it can be reproduced by the superposition of simple harmonic curves of frequencies 100, 200, 300, etc.

Each of these curves represents a pure tone, whence we see that any musical sound of frequency 100 is made up of pure tones having respectively 1, 2, 3, etc. times the frequency of the original sound. These tones are called the "natural harmonics" of the note in question.

## NATURAL HARMONICS AND RESONANCE

Vibrations are often set up in a vibrating structure by a force or disturbance which continually varies in strength; such a force may be periodic in the sense that the variations repeat themselves at regular intervals. Fourier's theorem now tells us that a variable force of this kind can be resolved into a number of constituent forces each of which varies in a simple harmonic manner, and that the frequencies of these forces will be 1, 2, 3 . . . times that of the total force. For instance, if the force repeats itself 100 times a second, the simple harmonic constituents of the force will repeat themselves 100, 200, 300, etc., times a second.

If the structure has free vibrations of frequencies 100, 200, 300, etc., these will be set vibrating strongly by resonance, while any vibrations of other frequencies that the structure may possess will not be set going in any appreciable strength. In other words, a disturbing force only excites by resonance the "natural harmonics" of a tone of the same period as itself.

This is a result of great importance to music in general. Amongst other things, it explains why the stretched string has such outstanding musical qualities; the reason is simply that its free vibrations coincide exactly in frequency with the natural harmonics of its fundamental tone, so that when the fundamental tone is set going, the harmonics are set going as well. . . .

# HEARING

We have now considered the generation of sound and its transmission through the air to the ear; we must finally consider its reception by the ear, and transmission to the brain.

When the air is being traversed by sound-waves, we have seen that the pressure at every point changes rhythmically, being now above and now below the average steady pressure of the atmosphere—just as, when ripples pass over the surface of a pond, the height of water in the pond changes rhythmically at every point, being now above and now below the average steady height when the water is at rest. The same is of course true of the small layer of air which lies in contact with the ear-drum, and it is changes of pressure in this layer which cause the sensation of hearing. The greater the changes of pressure, the more intense the sound, for we have seen that the energy of a sound-wave is proportional to the square of the range through which the pressure varies.

The pressure changes with which we are most familiar are those shewn on our barometers—half an inch of mercury, for instance. The pressure changes which enter into the propagation of sound are far smaller; indeed they are so much smaller that a new unit is needed for measuring them—the "bar." For exact scientific purposes, this is defined as a pressure of a dyne per square centimetre, but for our present purpose it is enough to know that a bar is very approximately a millionth part of the whole pressure of the atmosphere. When we change the height of our ears above the earth's surface by about a third of an inch, the pressure on our ear-drums changes by a bar; when we hear a fairly loud musical sound, the pressure on our ear-drums again changes about a bar.

## THE THRESHOLD OF HEARING

Suppose that we gradually walk away from a spot where a musical note is being continuously sounded. The amount of energy received by our ears gradually diminishes, and we might perhaps expect that the intensity of the sound heard by our brains would diminish in the same proportion. We shall, however, find that this is not so; the sound diminishes for a time, and then quite suddenly becomes inaudible. This shews that the loudness of the sound we hear is not proportional to the energy which falls on our ears; if the energy is below a certain amount we hear nothing at all. The smallest intensity of sound which we can hear is said to be at "the threshold of hearing."

We obtain direct evidence that such a threshold exists if we strike a tuning-fork and let its vibrations gradually die away. A point is soon reached at which we hear nothing. Yet the fork is still vibrating, and emitting sound, as can be proved by pressing its handle against any large hard surface, such as a table-top. This, acting as a sound-board, amplifies the sound so much that we can hear it again. Without this amplification the sound lay below the threshold of hearing; the amplification has raised it above the threshold.

In possessing a threshold of this kind, hearing is exactly in line with all the other senses; with each our brains are conscious of nothing at all until the stimulus reaches a certain "threshold" degree of intensity. The threshold of seeing, for instance, is of special importance in astronomy; our eyes see stars down to a certain limit of faintness, roughly about 6. 5 magnitudes, and beyond this see nothing at all. Just as a sound-board may raise the sound of a tuning-fork above the threshold of hearing, so a telescope raises the light of a faint star above the threshold of seeing.

We naturally enquire what is the smallest amount of energy that must fall on our ears in order to make an impression on our brains? In other words, how much energy do our ears receive at their threshold of hearing?

The answer depends enormously on the pitch of the sound we are trying to hear. Somewhere in the top octave of the pianoforte there is a pitch at which the sensitivity of the ear is a maximum, and here a very small amount of sound energy can make itself heard, but when we pass to tones of either higher or lower pitch, the ear is less sensitive, so that more energy is needed to produce the same impression of hearing. Beyond these tones we come to others of very high and very low pitch, which we cannot hear at all unless a large amount of energy falls on our ears, and finally, still beyond these, tones which no amount of energy can make us hear, because they lie beyond the limits of hearing.

The following table contains results which have been obtained by Fletcher and Munson.[3] The first two columns give the pitch and frequency of the tone under discussion, the next column gives the pressure variation at which the tone first becomes audible, while the last column gives the amount of energy needed at this pitch in terms of that needed at $f^{iv}$, at which the energy required is least:

| Tone | Frequency | Pressure variation at which note is first heard | Energy required in terms of minimum |
|---|---|---|---|
| CCCC (32-ft. pipe of organ; close to lower limit of hearing) | 16 | 100 bars | 1,500,000,000,000 |
| AAA (bottom note of piano) | 27 | 1 bar | 150,000,000 |
| CCC (lowest C on piano) | 32 | ⅖ bar | 25,000,000 |
| CC | 64 | ¹⁄₄₀ bar | 100,000 |
| C | 128 | ¹⁄₂₀₀ bar | 3,800 |
| c′ (middle C) | 256 | ¹⁄₁₀₀₀ bar | 150 |
| c″ | 512 | ¹⁄₂₅₀₀ bar | 25 |
| c‴ | 1,024 | ¹⁄₅₀₀₀ bar | 6 |
| c$^{iv}$ | 2,048 | ¹⁄₁₀₀₀₀ bar | 1.5 |
| f$^{iv}$ (maximum sensitivity) | 2,734 | ¹⁄₁₂₅₀₀ bar | 1.0 |
| c$^{v}$ (top of piano) | 4,096 | ¹⁄₁₀₀₀₀ bar | 1.5 |
| c$^{vi}$ | 8,192 | ¹⁄₂₀₀₀ bar | 38 |
| c$^{vii}$ | 16,384 | ¹⁄₁₀₀ bar | 15,000 |
| Close to upper limit of hearing | 20,000 | 500 bars | 38,000,000,000,000 |

We see that the ear can respond to a very small variation of pressure when the tone is of suitable pitch. Throughout the top octave of the piano, less than a ten-thousand-millionth part of an atmosphere suffices; as already mentioned, this is produced by an air-displacement of less than a ten-thousand-millionth part of an inch, which again is only about a hundredth part of the diameter of a molecule.

We also notice the immense range of figures in the last column. Our ears are acutely sensitive to sound within the top two octaves of the piano, and quite deaf, at least by comparison, to tones which are far below or above this range; to make a pure tone of pitch CCCC audible needs a million million times more energy than is needed for one seven octaves higher.

The structure of an ordinary organ provides visual confirmation of this. The pipe of pitch CCCC is a huge 32-foot monster, with a foot opening which absorbs an enormous amount of wind, and yet it hardly sounds louder than a tiny metal pipe perhaps three inches long taken from the treble. A child can blow the latter pipe quite easily from its mouth, but the whole force of a man's lungs will not make the 32-foot pipe sound audibly. . . .

## THE SCALE OF SOUND INTENSITY

The change in the intensity of a sound which results from a tenfold increase in the energy causing this sound is called a "bel." The word has nothing to do with beauty or charm, but is merely three-quarters of the surname of Graham Bell, the inventor of the telephone.

We have already thought of this tenfold increase as produced by ten equal steps of approximately 25 per cent each. More exactly, each of these must represent an increase by a factor of $\sqrt[10]{(10)}$, of which the value is 1.2589. Each of these steps of a tenth of a bel is known as a "decibel"; as we have seen, it represents just about the smallest change in sound intensity which our ears notice under ordinary conditions.

The intensity at the threshold of hearing is usually taken as zero point, so that, if we take the smallest amount of energy we can hear as unit:

|      |                                              |             |
|-----:|----------------------------------------------|-------------|
| 1    | unit of energy gives a sound intensity of    | 0 decibels  |
| 1.26 | unit of energy gives a sound intensity of    | 1 decibel   |
| 1.58 | unit of energy gives a sound intensity of    | 2 decibels  |
| 2    | unit of energy gives a sound intensity of    | 3 decibels  |
| 4    | unit of energy gives a sound intensity of    | 6 decibels  |
| 8    | unit of energy gives a sound intensity of    | 9 decibels  |
| 10   | unit of energy gives a sound intensity of    | 10 decibels |
| 100  | unit of energy gives a sound intensity of    | 20 decibels |
| 1000 | unit of energy gives a sound intensity of    | 30 decibels |

## THE SCALE OF LOUDNESS

The scale of sound intensity had its zero fixed at the threshold of hearing, but as the position of this depends enormously on the pitch of the sound under discussion, this scale is only useful in comparing the relative loudnesses of two sounds of the same pitch. It is of no use for the comparison of two sounds of different pitches. For this latter purpose we must introduce a new scale, the scale of loudness.

The zero point of this scale is taken to be the loudness, as heard by the average normal hearer, of a sound-wave in air, which has a frequency of 1000 and a pressure range of $\frac{1}{5000}$ bar—or, more precisely, 0.0002 dyne—at the ear of the listener. This, as we have already seen, is just about the threshold of hearing for a sound of this particular frequency.

The unit on this scale is called a "phon." So long as we limit ourselves to sounds of frequency 1000, the phon is taken to be the same thing as the decibel, both as regards its amount and its zero point. Thus if a sound of frequency 1000 has an intensity of $x$ decibels on the scale of sound intensity, it has a loudness of $x$ phons on the new scale of loudness. But the phon and decibel diverge when the frequency of the sound is different from 1000. Two sounds of different pitch are said to have the same number of phons of loudness when they sound equally loud to the ear. Thus we say that a sound has a loudness of $x$ phons when it sounds as loud to the ear as a sound of frequency 1000 and an intensity of $x$ decibels. Such a sound lies at $x$ decibels above the threshold of hearing for a sound of frequency 1000, not above that for a sound of its own pitch.[4]

## THE THRESHOLD OF PAIN

We have already considered what is the smallest amount of sound we can hear; we consider next what is the largest amount. This is not a meaningless problem. For, if we continually supply more and more energy to a source of sound—as for instance by beating a gong harder and harder—the sound will get louder and louder and, in time, we shall find it becoming too loud for pleasure. At first it is merely disagreeable, but from being disagreeable it soon passes to being uncomfortable. Finally the vibrations set up in our ear-drums and inner ear may become so violent as to give us acute pain, and possibly injure our ears.

If we note the number of bels our ears can endure without discomfort, we shall find that this again, like the position of the threshold of hearing, depends on the pitch of the sound. At the bass end of the pianoforte it is about six bels; it has risen to eleven bels by middle C; it rises further to twelve bels in the top octave of the pianoforte, after which it probably falls rapidly.

The intensity of sound at the threshold of hearing, and also the range above the threshold which we can endure without undue discomfort, both vary greatly with the pitch of the sound, but their sum, which fixes a sort of threshold of pain, varies much less. Throughout the greater part of the range used in music, the intensity at this threshold is given by a pressure variation of about 600 bars, except that it falls to about 200 bars in the region of maximum sensitivity.

We can represent this in a diagram as in Figure 24, and the shaded area which is the area of hearing can be divided up further by curves of equivalent loudness as shewn in Figure 25. Both the limits of the area of hearing and the curves of equal loudness have been determined by Fletcher and Munson.

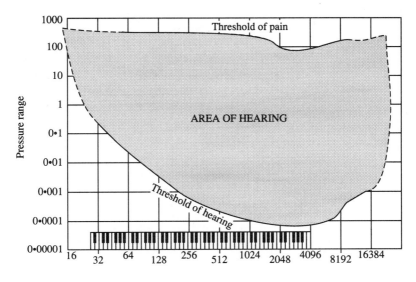

**FIGURE 24**

*The limits of the area of hearing, as determined by Fletcher and Munson. Each point in this diagram represents a sound of a certain specified frequency (as shewn on the scale at the bottom) and of a certain specified intensity (as shewn by the scale on the left). If the point lies within the shaded area, the sound can be heard with comfort. If the point lies above the shaded area, the hearing of the sound is painful. If the point lies below the shaded area, the sound lies below the threshold of hearing, and so cannot be heard at all.*

We see at a glance how the ear is both most sensitive to faint sound, and also least tolerant to excessive sound, in the range of the upper half of the piano. To be heard at a moderate comfortable loudness of say 50 or 60 phons, treble music needs but little energy, while bass music needs a great deal. This is confirmed by exact measurements of the energy employed in playing various instruments. The following table gives the results of experiments made at the Bell Telephone Laboratories:

| *Origin of Sound* | *Energy Watts* |
|---|---|
| Orchestra of seventy-five performers, at loudest . . . . . . | 70 |
| Bass drum at loudest . . . . . . . . . . . . . . . . . . . . . . . . . . . | 25 |
| Pipe organ at loudest . . . . . . . . . . . . . . . . . . . . . . . . . . | 13 |
| Trombone at loudest . . . . . . . . . . . . . . . . . . . . . . . . . . . | 6 |
| Piano at loudest . . . . . . . . . . . . . . . . . . . . . . . . . . . . . . . | 0.4 |
| Trumpet at loudest . . . . . . . . . . . . . . . . . . . . . . . . . . . . | 0.3 |
| Orchestra of seventy-five performers, at average . . . . . . | 0.09 |
| Piccolo at loudest . . . . . . . . . . . . . . . . . . . . . . . . . . . . . | 0.08 |
| Clarinet at loudest . . . . . . . . . . . . . . . . . . . . . . . . . . . . | 0.05 |
| Human voice { Bass singing *ff* . . . . . . . . . . . . . . . . . . . | 0.03 |
| Human voice { Alto singing *pp* . . . . . . . . . . . . . . . . . . . | 0.001 |
| Human voice { Average speaking voice . . . . . . . . . . . . | 0.000024 |
| Violin at softest used in a concert . . . . . . . . . . . . . . . . . | 0.0000038 |

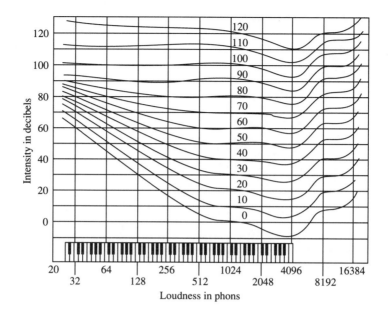

**FIGURE 25**

*The loudness of sounds which lie within the area of hearing, as determined by Fletcher and Munson. As in Figure 24, each point of the diagram represents a sound of specified frequency (as shewn on the scale at the bottom) and of specified intensity in decibels (as shewn on the scale on the left), the zero point being the faintest sound of frequency 1000 which can be heard at all. The loudness of the sound in phons is the number written on the curved line which passes through the point; thus these curves are curves of equal loudness.*

We may notice in passing how very small is the energy of even a loud sound. A fair-sized pipe organ may need a 10,000-watt motor to blow it; of this energy only 13 watts reappears as sound, while the other 9987 watts is wasted in friction and heat. A strong man soon tires of playing a piano at its loudest, his energy output being perhaps 200 watts; of this only 0.4 watts goes into sound. A thousand basses singing *fortissimo* only give out enough energy to keep one 30-watt lamp alight; if they turned dynamos with equal vigour, 6000 such lamps could be kept alight.

The first and last entries in the preceding table represent the extreme range of sounds heard in a concert room, and we notice that the former is more than eighteen million times the latter. Yet this range, large though it is, is only one of 7¼ bels, and so is not much more than half of the range of 12 bels which the ear can tolerate in treble sounds.

For a person well away from the instruments, we may perhaps estimate the violin at its softest as being about 1 bel above the threshold of hearing for the note it is playing, so that the full orchestra is about 8.3 bels, or 83 decibels. This may be compared with the intensities of various other sounds, as shewn in the following table:

| | |
|---|---|
| Threshold of hearing | 0 decibels |
| Gentle rustle of leaves | 10 decibels |
| Quiet London garden | 20 decibels |
| Whisper at 4 feet | 20 decibels |
| Quiet suburban street, London | 30 decibels |
| Quietest time at night, Central New York | 40 decibels |
| Conversation at 12 feet | 50 decibels |
| Busy traffic, London | 60 decibels |
| Busy traffic, New York | 68 decibels |
| Very heavy traffic, New York | 82 decibels |
| Lion roaring at 18 feet | 88 decibels |
| Subway station with express passing, New York | 95 decibels |
| Boiler factory | 98 decibels |
| Steel plate hammered by four men, 2 feet away | 112 decibels |

Owing to the different thresholds of hearing, the sounds in the above tables are not strictly comparable, unless they happen to be of the same pitch. The following table shews the differences of subjective loudness for a few common sounds:

| | |
|---|---|
| Threshold of hearing | 0 phons |
| Ticking of a watch at 3 feet | 20 phons |
| Sounds in a quiet residential street | 40 phons |
| Quiet conversation | 60 phons |
| Sounds in a busy main street | 75 phons |
| Sounds in a tube train | 90 phons |
| Sounds in a busy machine shop | 100 phons |
| Proximity of aeroplane engine | 120 phons |

Experiments shew that a faint sound will not be heard at all through a louder sound of the same pitch, if the difference in intensity is more than about 1.2 bels, but the difference in loudness may be greater if the sounds are of very different pitch. Conversation at 12 feet should just be heard against busy traffic in London, because the difference in intensity only amounts to 1.0 bels; it will not, however, be heard against busy traffic in New York, because the difference here is 1.8 bels. In the same way a roaring lion would only just be heard in a boiler factory, although he might hope to attract considerable attention in a New York subway station. . . .

## ENDNOTES

[1]If the fork was struck very violently in the first instance, there may be a very slight sharpening of pitch as the vibrations become of more usual intensity.

[2]It might seem simpler to treat the original curve as a whole period of a repeating curve, but there are mathematical reasons against this.

[3]Many other investigators have worked at the problem, their results generally agreeing fairly closely, although not always exactly, with those stated in the table. Investigations, by Andrade and Parker (1937), yield results which are in very close agreement with those of Fletcher and Munson.

[4]This defines the British standard phon. The Americans use the same phon as the British, but frequently describe it as a decibel. The Germans use a different zero point, 0.0003 dyne in place of 0.0002.

# PART XXV

# MATHEMATICS AS A CULTURE CLUE

# COMMENTARY ON
# OSWALD SPENGLER

O swald Spengler (1880–1936) was a sickly German high-school teacher of apocalyptic inclination who at the age of thirty-one retired from his post to write an immense and sensational book on the philosophy of history. The book which brought fame to this obscure scholar was *The Decline of the West*; it was conceived in 1911 and completed in 1917. Exempt from military service because of a weak heart and defective eyesight, Spengler had ample time during the war years to elaborate his theme; but throughout he was harassed by poverty and other adversities, and only a sense of mission based on the conviction that he had discovered a great truth about "history and the philosophy of destiny" sustained him.[1] In 1918 a reluctant Viennese publisher was persuaded to take the book. It appeared in a small edition but in a few weeks began to sell. In Germany, where its gloomy tone suited the post-war mood, the book provoked vehement controversy; abroad "it won the admiration of the half-educated and the scorn of the judicious."[2] Now in another post-war period, the issues raised by Spengler are again at the focus of attention. His theory has found few adherents, yet it has impelled many thoughtful men to sober reflection. "It is easy," as one critic has observed, "to criticize Spengler, but not so easy to get rid of him."[3]

Spengler's main thesis is that the patterns of history are cyclical, not linear. Man does not improve. He experiences the inexorable biological progression of "birth, growth, maturity and decay"; he accommodates himself to circumstance, changing his ways in order to survive; but his basic attitudes remain unchanged. The same principle applies to the several cultures man has produced over the centuries. Like living organisms cultures flourish, decline and die. They do not progress; they merely recur. Their course is as predestined as the course of their creators. In the history of every culture there is discernible a "master pattern"[4]—"a characteristic cast of the human spirit working itself out." This master pattern shapes each of the activities which compose the culture. While the master patterns differ, and thus distinguish one culture from another, they pass inevitably through the same "morphological" stages. Spengler has epitomized his ambitious program in these words:

"I hope to show that without exception all great creations and forms in religion, art, politics, social life, economy and science appear, fulfill themselves and die down *contemporaneously* in all the Cultures; that the inner structure of one corresponds strictly with that of all the others; that there is not a single phenomenon of deep physiognomic importance in the record of one for which we could not find a counterpart in the record of every other; and that this counterpart is to be found under a characteristic form and in a perfectly definite chronological position."[5]

Even in two massive volumes Spengler was unable to persuade men of insight and dispassionate judgment that he had fulfilled his promise. Both his arguments and his presentation were vulnerable. Scholars hacked at his blunders, scientists at his pseudo-scientific reasoning, philosophers at his conclusions, literary critics at his swollen, unlovely style. It was pointed out that the cyclical view of history was a "hoary commonplace"; that Spengler had borrowed his main ideas from his betters; that he was antirational, pompously prophetic, crude and melodramatic. None of these charges were altogether baseless; some, indeed, were painfully true. Yet *The Decline of the West* contains elements of great originality, flashes of extraordinary insight. Spengler exaggerated but he also brilliantly illuminated corners of history which less passionate philosophers had overlooked; his expression was pretentious but it was powerful. Above all, as H. Stuart Hughes said in his recent excellent study: ". . . *The Decline of the West* offers the nearest thing we have to a key to our times. It formulates more comprehensively than any other single book the modern *malaise* that so many feel and so few can express."[6]

Spengler was convinced that mathematics is no exception to his principle of cultural parallelism. There are no eternal verities even in this most abstract, seemingly disembodied intellectual activity. *"There is not, and cannot be, number as such.* There are several number-worlds as there are several Cultures."[7] Mathematics, like art or religion or politics, expresses man's basic attitudes, his conception of himself; like the other elements in a culture it exemplifies "the way in which a soul seeks to actualitize itself in the picture of its outer world."[8] The first selection below is taken from the chapter "Meaning of Numbers," one of the most remarkable discussions in *The Decline of the West*. It is unnecessary to agree with Spengler's thesis to be stimulated by this performance. No one else has made even a comparable attempt to cast a synoptic eye over the evolving concept of number. A good deal of what Spengler has to say on this subject strikes one as far-fetched and misty. But he was a capable mathematician; his ideas cannot be dismissed as hollow; and I think you will find this a disturbing and exciting essay.

The second selection is less disturbing but no less exciting. It attacks the question "Do mathematical truths reside in the external world, or are they man-made inventions?" There are very few sensible discussions of this problem. Leslie White's approach is that of a cultural anthropologist. What he has to say is balanced and persuasive. It is worth comparing both with Spengler's vehement opinions and with the moderate, lucidly reasonable views of Richard von Mises (see pp. 1695–1724). Dr. White (1900–1975) was chairman of the department of anthropology at the University of Michigan. He also taught at the University of Chicago, Yale, Columbia and Yenching University, Peiping,

China. He had extensive experience as a field investigator among the Pueblo Indians. His best known book *The Science of Culture,* was published in 1949.

## ENDNOTES

[1]The words in quotation marks are Spengler's and are taken from the preface to the English translation of his book, by Charles Francis Atkinson: *The Decline of the West: Form and Actuality,* New York, 1926, p. XIV. For biographical and other details I have drawn on H. Stuart Hughes, *Oswald Spengler, A Critical Estimate,* New York, 1952.

[2]Hughes, *op. cit.*, p. 1.

[3]The *Times Literary Supplement,* October 3, 1952, p, 637.

[4]This is A. L. Kroeber's expression (*Configurations of Culture Growth,* Berkeley and Los Angeles, 1944, p. 826) quoted by Hughes, *op. cit.*, p. 10.

[5]Spengler, *op. cit.*, p. 112.

[6]Hughes, *op. cit.*, p. 165.

[7]Spengler, *op. cit.*, p. 59.

[8]*Ibid.,* p. 56.

*In the study of ideas, it is necessary to remember that insistence on hard-headed clarity issues from sentimental feeling, as it were a mist, cloaking the perplexities of fact. Insistence on clarity at all costs is based on sheer superstition as to the mode in which human intelligence functions. Our reasonings grasp at straws for premises and float on gossamers for deductions.*

—ALFRED NORTH WHITEHEAD (*Adventures in Ideas*)

# MEANING OF NUMBERS

## By Oswald Spengler

In order to exemplify the way in which a soul seeks to actualize itself in the picture of its outer world—to show, that is, in how far Culture in the "become" state can express or portray an idea of human existence—I have chosen *number*, the primary element on which all mathematics rests. I have done so because mathematics, accessible in its full depth only to the very few, holds a quite peculiar position amongst the creations of the mind. It is a science of the most rigorous kind, like logic but more comprehensive and very much fuller; it is a true art, along with sculpture and music, as needing the guidance of inspiration and as developing under great conventions of form; it is, lastly, a metaphysic of the highest rank, as Plato and above all Leibniz show us. Every philosophy has hitherto grown up in conjunction with a mathematic *belonging* to it. Number is the symbol of causal necessity. Like the conception of God, it contains the ultimate meaning of the world-as-nature. The existence of numbers may therefore be called a mystery, and the religious thought of every Culture has felt their impress.

Just as all becoming possesses the original property of *direction* (irreversibility), all things-become possess the property of *extension*. But these two words seem unsatisfactory in that only an artificial distinction can be made between them. The real secret of all things-become, which are *ipso facto* things extended (spatially and materially), is embodied in mathematical number as contrasted with chronological number. Mathematical number contains in its very essence the notion of a *mechanical demarcation*, number being in that respect akin to *word*, which, in the very fact of its comprising and denoting, fences off world-impressions. The deepest depths, it is true, are here both incomprehensible and inexpressible. But the actual number with which the mathematician works, the figure, formula, sign, diagram, in short the *number-sign which he thinks, speaks or writes exactly,* is (like the exactly-used word)

2293

from the first a symbol of these depths, something imaginable, communicable, comprehensible to the inner and the outer eye, which can be accepted as representing the demarcation. The origin of numbers resembles that of the myth. Primitive man elevates indefinable nature-impressions (the "alien," in our terminology) into deities, *numina,* at the same time capturing and impounding them by a *name* which limits them. So also numbers are something that marks off and captures nature-impressions, and it is by means of names and numbers that the human understanding obtains power over the world. In the last analysis, the number-language of a mathematic and the grammar of a tongue are structurally alike. Logic is always a kind of mathematic and vice versa. Consequently, in all acts of the intellect germane to mathematical number—measuring, counting, drawing, weighing, arranging and dividing[1]— men strive to delimit the extended in words as well, i.e., to set it forth in the form of proofs, conclusions, theorems and systems; and it is only through acts of this kind (which may be more or less unintentioned) that waking man begins to be able to use numbers, normatively, to specify objects and properties, relations and differentiae, unities and pluralities—briefly, that structure of the world-picture which he feels as necessary and unshakable, calls "Nature" and "cognizes." *Nature is the numerable,* while History, on the other hand, is the aggregate of that which has no relation to mathematics—hence the mathematical certainty of the laws of Nature, the astounding rightness of Galileo's saying that Nature is "written in mathematical language," and the fact, emphasized by Kant, that exact natural science reaches just as far as the possibilities of applied mathematics allow it to reach. In number, then, as the *sign of completed demarcation,* lies the *essence* of everything actual, which is cognized, is delimited, and has become all at once—as Pythagoras and certain others have been able to see with complete inward certitude by a mighty and truly religious intuition. Nevertheless, mathematics—meaning thereby the capacity to think practically in figures—must not be confused with the far narrower scientific mathematics, that is, the *theory* of numbers as developed in lecture and treatise. The mathematical vision and thought that a Culture possesses within itself is as inadequately represented by its written mathematic as its philosophical vision and thought by its philosophical treatises. Number springs from a source that has also quite other outlets. Thus at the beginning of every Culture we find an archaic style, which might fairly have been called geometrical in other cases as well as the Early Hellenic. There is a common factor which is expressly mathematical in this early Classical style of the 10th Century B.C., in the temple style of the Egyptian Fourth Dynasty with its absolutism of straight line and right angle, in the Early Christian sarcophagus-relief, and in Romanesque construction and ornament. Here every line, every deliberately non-imitative figure of man and beast, reveals a mystic number-thought in direct connexion with the mystery of death (the hardset).

Gothic cathedrals and Doric temples are *mathematics in stone*. Doubtless Pythagoras was the first in the Classical Culture to conceive number scientifically as the principle of a world-order of comprehensible things—as *standard* and as *magnitude*—but even before him it had found expression, as a noble arraying of sensuous-material units, in the strict canon of the statue and the Doric order of columns. The great arts are, one and all, modes of interpretation by means of limits based on number (consider, for example, the problem of space-representation in oil painting). A high mathematical endowment may, without any mathematical science whatsoever, come to fruition and full self-knowledge in *technical* spheres.

In the presence of so powerful a number-sense as that evidenced, even in the Old Kingdom,[2] in the dimensioning of pyramid temples and in the technique of building, water-control and public administration (not to mention the calendar), no one surely would maintain that the valueless arithmetic of Ahmes belonging to the New Empire represents the level of Egyptian mathematics. The Australian natives, who rank intellectually as thorough primitives, possess a mathematical instinct (or, what comes to the same thing, a power of thinking in numbers which is not yet communicable by signs or words) that as regards the interpretation of pure space is far superior to that of the Greeks. Their discovery of the boomerang can only be attributed to their having a sure feeling for numbers of a class that we should refer to the higher geometry. *Accordingly*— we shall justify the adverb later—they possess an extraordinarily complicated ceremonial and, for expressing degrees of affinity, such fine shades of language as not even the higher Cultures themselves can show.

There is analogy, again, between the Euclidean mathematic and the absence, in the Greek of the mature Periclean age, of any feeling either for ceremonial public life or for loneliness, while the Baroque, differing sharply from the Classical, presents us with a mathematic of spatial analysis, a court of Versailles and a state system resting on dynastic relations.

It is the style of a Soul that comes out in the world of numbers, and the world of numbers includes something more than the science thereof.

From this there follows a fact of decisive importance which has hitherto been hidden from the mathematicians themselves.

*There is not, and cannot be, number as such.* There are several number-worlds as there are several Cultures. We find an Indian, an Arabian, a Classical, a Western type of mathematical thought and, corresponding with each, a type of number—each type fundamentally peculiar and unique, an expression of a specific world-feeling, a symbol having a specific validity which is even capable of scientific definition, a principle of ordering the Become which reflects the central essence of one and only one soul, viz., the soul of that particular Culture. Consequently, there are more mathematics than one. For indubitably the inner structure of the Euclidean geometry is something quite different from that of the Cartesian, the analysis of Archimedes is something other than the analysis of Gauss, and not merely in matters of form, intuition and method but above all in essence, in the intrinsic and obligatory meaning of number which they respectively develop and set forth. This number, the horizon within which it has been able to make phenomena self-explanatory, and therefore the whole of the "nature" or world-extended that is confined in the given limits and amenable to its particular sort of mathematic, are not common to all mankind, but specific in each case to one definite sort of mankind.

The style of any mathematic which comes into being, then, depends wholly on the Culture in which it is rooted, the sort of mankind it is that ponders it. The soul can bring its inherent possibilities to scientific development, can manage them practically, can attain the highest levels in its treatment of them—but is quite impotent to alter them. The idea of the Euclidean geometry is actualized in the earliest forms of Classical ornament, and that of the Infinitesimal Calculus in the earliest forms of Gothic architecture, centuries before the first learned mathematicians of the respective Cultures were born.

A deep inward experience, the genuine *awakening of the* ego, which turns the child into the higher man and initiates him into community of his Culture, marks the beginning of number-sense as it does that of language-sense. It is only after this that objects come to exist for the waking consciousness as things limitable and distinguishable as to number and kind; only after this that properties, concepts, causal necessity, system in the world-around, *a form of the world,* and *world laws* (for that which is set and settled is *ipso facto* bounded, hardened, number-governed) are susceptible of exact definition. And therewith comes too a sudden, almost metaphysical, feeling of anxiety and awe regarding the deeper meaning of measuring and counting, drawing and form.

Now, Kant has classified the sum of human knowledge according to syntheses *a priori* (necessary and universally valid) and *a posteriori* (experiential and variable from case to case) and in the former class has included mathematical knowledge. Thereby, doubtless, he was enabled to reduce a strong inward feeling to abstract form. But, quite apart from the fact (amply evidenced in modern mathematics and mechanics) that there is no such sharp distinction between the two as is originally and unconditionally implied in the principle, the *a priori* itself, though certainly one of the most inspired conceptions of philosophy, is a notion that seems to involve enormous difficulties. With it Kant postulates—without attempting to prove what is quite incapable of proof—both *unalterableness of form* in all intellectual activity and *identity of form for all men* in the same. And, in consequence, a factor of incalculable importance is—thanks to the intellectual prepossessions of his period, not to mention his own—simply ignored. This factor is the *varying degree* of this alleged "universal validity." There are doubtless certain characters of very wide-ranging validity which are (seemingly at any rate) independent of the Culture and century to which the cognizing individual may belong, but along with these there is a quite particular necessity of form which underlies all his thought as axiomatic and to which he is subject by virtue of belonging to his own Culture and no other. Here, then, we have two very different kinds of *a priori* thought-content, and the definition of a frontier between them, or even the demonstration that such exists, is a problem that lies beyond all possibilities of knowing and will never be solved. So far, no one has dared to assume that the supposed constant structure of the intellect is an illusion and that the history spread out before us contains more than one *style of knowing*. But we must not forget that unanimity about things that have not yet become problems may just as well imply universal error as universal truth. True, there has always been a certain sense of doubt and obscurity—so much so, that the correct guess might have been made from that non-agreement of the philosophers which every glance at the history of philosophy shows us. But that this non-agreement is not due to imperfections of the human intellect or present gaps in a perfectible knowledge, in a word, is not due to defect, but to destiny and historical necessity—this is a *discovery*. Conclusions on the deep and final things are to be reached not by predicating constants but by studying differentiae and developing the *organic logic* of differences. The *comparative morphology of knowledge forms* is a domain which Western thought has still to attack.

If mathematics were a mere science like astronomy or mineralogy, it would be possible to define their object. This man is not and never has been able to do. We West-Europeans may put our own scientific notion of number to perform the same tasks as those with which the mathematicians of Athens and Baghdad busied themselves, but the fact remains that the theme, the intention and the methods of the like-named science in Athens and in Baghdad were quite different from those of our own. *There is no mathematic but only mathematics.* What we call "the history of mathematics"—implying merely the progressive actualizing of a single invariable ideal—is in fact, below the deceptive surface of history, a complex of self-contained and independent developments, an ever-repeated process of bringing to birth new form-worlds and appropriating, transforming and sloughing alien form-worlds, a purely organic story of blossoming, ripening, wilting and dying within the set period. The student must not let himself be deceived. The mathematic of the Classical soul sprouted almost out of nothingness, the historically-constituted Western soul, already possessing the Classical science (not inwardly, but outwardly as a thing learnt), had to win its own by apparently altering and perfecting, but in reality destroying the essentially alien Euclidean system. In the first case, the agent was Pythagoras, in the second Descartes. In both cases the act is, at bottom, the same.

The relationship between the form-language of a mathematic and that of the cognate major arts,[3] is in this way put beyond doubt. The temperament of the thinker and that of the artist differ widely indeed, but the expression-methods of the waking consciousness are inwardly the same for each. The sense of form of the sculptor, the painter, the composer is essentially mathematical in its nature. The same inspired ordering of an infinite world which manifested itself in the geometrical analysis and projective geometry of the 17th Century, could vivify, energize, and suffuse contemporary music with the harmony that it developed out of the art of thoroughbass, (which is the geometry of the sound-world) and contemporary painting with the principle of perspective (the felt geometry of the space-world that only the West knows). This inspired ordering is that which Goethe called *"The Idea, of which the form is immediately apprehended in the domain of intuition,* whereas pure science does not apprehend but observes and dissects." The Mathematic goes beyond observation and dissection, and in its highest moments finds the way by vision, not abstraction. To Goethe again we owe the profound saying: "the mathematician is only complete in so far as he feels within himself the *beauty* of the true." Here we feel how nearly the secret of number is related to the secret of artistic creation. And so the born mathematician takes his place by the side of the great masters of the fugue, the chisel and the brush; he and they alike strive, and must strive, to actualize the grand order of all things by clothing it in symbol and so to communicate it to the plain fellow-man who hears that order within himself but cannot effectively possess it; the domain of number, like the domains of tone, line and colour, becomes an image of the world-form. For this reason the word "creative" means more in the mathematical sphere than it does in the pure sciences—Newton, Gauss, and Riemann were artist-natures, and we know with what suddenness their great conceptions came upon them.[4] "A mathematician," said old Weierstrass, "who is not at the same time a bit of a poet will never be a full mathematician."

The mathematic, then, is an art. As such it has its styles and style-periods. It is not, as the layman and the philosopher (who is in this matter a layman too) imagine, substantially unalterable, but subject like every art to unnoticed changes from epoch to epoch. The development of the great arts ought never to be treated without an (assuredly not unprofitable) side-glance at contemporary mathematics. In the very deep relation between changes of musical theory and the analysis of the infinite, the details have never yet been investigated, although æsthetics might have learned a great deal more from these than from all so-called "psychology." Still more revealing would be a history of musical instruments written, not (as it always is) from the technical standpoint of tone-production, but as a study of the deep spiritual bases of the tone-colours and tone-effects aimed at. For it was the wish, intensified to the point of a longing, to fill a spatial infinity with sound which produced—in contrast to the Classical lyre and reed (lyra, kithara; aulos, syrinx) and the Arabian lute—the two great families of keyboard instruments (organ, pianoforte, etc.) and bow instruments, and that as early as the Gothic time. The development of both these families belongs spiritually (and possibly also in point of technical origin) to the Celtic-Germanic North lying between Ireland, the Weser and the Seine. The organ and clavichord belong certainly to England, the bow instruments reached their definite forms in Upper Italy between 1480 and 1530, while it was principally in Germany that the organ was developed into the *space-command-ing* giant that we know, an instrument the like of which does not exist in all musical history. The free organ-playing of Bach and his time was nothing if it was not analysis—analysis of a strange and vast tone-world. And, similarly, it is in conformity with the Western number-thinking, and in opposition to the Classical, that our string and wind instruments have been developed not singly but in great groups (strings, woodwind, brass), ordered within themselves according to the compass of the four human voices; the history of the modern orchestra, with all its discoveries of new and modification of old instruments, is in reality the self-contained history of one tone-world—a world, moreover, that is quite capable of being expressed in the forms of the higher analysis.

When, about 540 B.C., the circle of the Pythagoreans arrived at the idea that *number is the essence of all things,* it was not "a step in the development of mathematics" that was made, but a wholly new mathematic that was born. Long heralded by metaphysical problem-posings and artistic form-tendencies, now it came forth from the depths of the Classical soul as a formulated theory, a mathematic born in one act at one great historical moment—just as the mathematic of the Egyptians had been, and the algebra-astronomy of the Babylonian Culture with its ecliptic co-ordinate system—and new—for these older mathematics had long been extinguished and the Egyptian was never written down. Fulfilled by the 2nd century A.D., the Classical mathematic vanished in its turn (for though it seemingly exists even to-day, it is only as a convenience of notation that it does so), and gave place to the Arabian. From what we know of the Alexandrian mathematic, it is a necessary presumption that there was a great movement within the Middle East, of which the centre of gravity must have lain in the Persian-Babylonian schools (such as Edessa, Gundisapora and Ctesiphon) and of which only details found their way into the regions of Classical speech. In spite of their Greek names, the Alexandrian mathematicians—Zenodorus who dealt with figures of equal perimeter, Serenus who worked on the properties of a harmonic pencil in space, Hypsicles who introduced the Chaldean circle-division, Diophantus above all—were all without doubt Aramæans, and their works only a small part of a literature which was written principally in Syriac. This mathematic found its completion in the investigations of the Arabian-Islamic thinkers, and after these there was again a long interval. And then a perfectly new mathematic was born, the Western, *our own,* which in our infatuation we regard as "Mathematics," as the culmination and the implicit purpose of two thousand years' evolution, though in reality its centuries are (strictly) numbered and to-day almost spent.

The most valuable thing in the Classical mathematic is its proposition that number is the essence of all things *perceptible to the senses.* Defining number as a measure, it contains the whole world-feeling of a soul passionately devoted to the "here" and the "now." Measurement in this sense means the measurement of something near and corporeal. Consider the content of the Classical art-work, say the free-standing statue of a naked man; here every essential and important element of Being, its whole rhythm, is exhaustively rendered by surfaces, dimensions and the sensuous relations of the parts. The Pythagorean notion of the harmony of numbers, although it was probably deduced from music—a music, be it noted, that knew not polyphony or harmony, and formed its instruments to render single plump, almost fleshy, tones—seems to be the very mould for a sculpture that has this ideal. The worked stone is only a something in so far as it has considered limits and measured form; what it *is* is what it *has become* under the sculptor's chisel. Apart from this it is a *chaos,* something not yet actualized, in fact for the time being a null. The same feeling transferred to the grander stage produces, as an opposite to the state of chaos, that of *cosmos,* which for the Classical soul implies a cleared-up situation of the external world, a harmonic order which includes each separate thing as a well-defined, comprehensible and present entity. The sum of such things constitutes neither more nor less than the whole world, and the interspaces between them, which for us are filled with the impressive symbol of the Universe of Space, are for the nonent (τὸ μὴ ὄν).

Extension means, for Classical mankind body, and for us space, and it is as a function of space that, to us, things "appear." And, looking backward from this standpoint, we may perhaps see into the deepest concept of the Classical metaphysics, Anaximander's ἄπειρον—a word that is quite untranslatable into any Western tongue. It is that which possesses no "number" in the Pythagorean sense of the word, no measurable dimensions or definable limits, and therefore no being; the measureless, the negation of form, the statue not yet carved out of the block; the ἀρχή optically boundless and formless, which only becomes a something (namely, the world) after being split up by the senses. It is the underlying form *a priori* of Classical cognition, bodiliness as such, which is replaced exactly in the Kantian world-picture by that Space out of which Kant maintained that all things could be "thought forth."

We can now understand what it is that divides one mathematic from another, and in particular the Classical from the Western. The whole world-feeling of the matured Classical world led it to see mathematics only as the theory of relations of magnitude, dimension and form between bodies. When, from out of this feeling, Pythagoras evolved and expressed the decisive formula, number had come, for him, to be an *optical* symbol—not a measure of form generally, an abstract relation, but a frontier-post of the domain of the Become, or rather of that part of it which the senses were able to split up and pass under review. By the whole Classical world without exception numbers are conceived as units of measure, as magnitude, lengths, or surfaces, and for it no other sort of extension is imaginable. The whole Classical mathematic is at bottom *Stereometry* (solid geometry). To Euclid, who rounded off its system in the third century, the triangle is of deep necessity the bounding surface of a body, never a system of three intersecting straight lines or a group of three points in three-dimensional space. He defines a line as "length without breadth" (μῆκος ἀ πλατές). In our mouths such a definition would be pitiful—in the Classical mathematic it was brilliant.

The Western number, too, is not, as Kant and even Helmholtz thought, something proceeding out of Time as an *a priori* form of conception, but is something specifically spatial, in that it is an order (or ordering) of like units. Actual time (as we shall see more and more clearly in the sequel) has not the slightest relation with mathematical things. Numbers belong exclusively to the domain of extension. But there are precisely as many possibilities—and therefore necessities—of ordered presentation of the extended as there are Cultures. Classical number is a thought-process dealing not with spatial relations but with visibly limitable and tangible units, and it follows naturally and necessarily that the Classical knows only the "natural" (positive and whole) numbers, which on the contrary play in our Western mathematics a quite undistinguished part in the midst of complex, hypercomplex, non-Archimedean and other number-systems.

On this account, the idea of irrational numbers—the unending decimal fractions of our notation—was unrealizable within the Greek spirit. Euclid says—and he ought to have been better understood—that incommensurable lines are *"not related to one another like numbers."* In fact, it is the idea of irrational number that, once achieved, separates the notion of number from that of magnitude, for the magnitude of such a number ($\pi$, for example) can never be defined or exactly represented by any straight line. Moreover, it follows from this that in considering the relation, say, between diagonal and side in a square the Greek would be brought up suddenly against a quite other sort of number, which was fundamentally alien to the Classical soul, and was consequently feared as a secret of its proper existence too dangerous to be unveiled. There is a singular and significant late-Greek legend, according to which the man who first published the hidden mystery of the irrational perished by shipwreck, "for the unspeakable and the formless must be left hidden for ever."[5]

The fear that underlies this legend is the selfsame notion that prevented even the ripest Greeks from extending their tiny city-states so as to organize the country-side politically, from laying out their streets to end in prospects and their alleys to give vistas, that made them recoil time and again from the Babylonian astronomy with its penetration of endless starry space,[6] and refuse to venture out of the Mediterranean along sea paths long before dared by the Phoenicians and the Egyptians. It is the deep metaphysical fear that the sense-comprehensible and present in which the Classical existence had entrenched itself would collapse and precipitate its cosmos (largely created and sustained by art) into unknown primitive abysses. And to understand this fear is to understand the final significance of Classical number—that is, *measure in contrast to the immeasurable*—and to grasp the high ethical significance of its limitation. Goethe too, as a nature-student, felt it—hence his almost terrified aversion to mathematics, which as we can now see was really an involuntary reaction against the *non-Classical* mathematic, the Infinitesimal Calculus which underlay the natural philosophy of his time.

Religious feeling in Classical man focused itself ever more and more intensely upon physical present, *localized* cults which alone expressed a college of Euclidean deities. Abstractions, *dogmas* floating homeless in the space of thought, were ever alien to it. A cult of this kind has as much in common with a Roman Catholic dogma as the statue has with the cathedral organ. There is no doubt that something of cult was comprised in the Euclidean mathematic— consider, for instance, the secret doctrines of the Pythagoreans and the Theorems of regular polyhedrons with their esoteric significance in the circle of Plato. Just so, there is a deep relation between Descartes' analysis of the infinite and contemporary dogmatic theology as it progressed from the final decisions of the Reformation and the Counter-Reformation to entirely desensualized deism. Descartes and Pascal were mathematicians and Jansenists, Leibniz a mathematician and pietist. Voltaire, Lagrange and D'Alembert were contemporaries. Now, the Classical soul felt the principle of the irrational, which overturned the statuesquely-ordered array of whole numbers and the complete and self-sufficing world-order for which these stood, as an impiety against the Divine itself. In Plato's "Timæus" this feeling is unmistakable. For the transformation of a series of discrete numbers into a continuum challenged not merely the Classical notion of number but the Classical world-idea itself, and so it is understandable that even *negative* numbers, which to us offer no conceptual difficulty, were impossible in the Classical mathematic, let alone *zero as a number,* that refined creation of a wonderful abstractive power which, for the Indian soul that conceived it as base for a positional numeration, was nothing more nor less than the key to the meaning of existence. *Negative magnitudes* have no existence. The expression $(-2) \times (-3) = +6$ is neither something perceivable nor a representation of magnitude. The series of magnitudes ends with $+1$, and in graphic representation of negative numbers $(+3 \ +2 \ +1 \ 0 \ -1 \ -2 \ -3)$ we have suddenly, from zero onwards, *positive* symbols of something negative; they *mean* something, but they no longer *are.* But the fulfillment of this act did not lie within the direction of Classical number-thinking.

Every product of the waking consciousness of the Classical world, then, is elevated to the rank of actuality by way of sculptural definition. That which cannot be drawn is not "number." Archytas and Eudoxus use the terms surface- and volume-numbers to mean what we call second and third powers, and it is easy to understand that the notion of higher integral powers did not exist for them, for a fourth power would predicate at once, for the mind based on the plastic feeling, an extension in four dimensions, and four *material* dimensions into the bargain, "which is absurd." Expressions like $\epsilon^{ix}$ which we constantly use, or even the fractional index (e.g., $5^{1/2}$) which is employed in the Western mathematics as early as Oresme (14th Century), would have been to them utter nonsense. Euclid calls the factors of a product its sides πλευραί and fractions (finite of course) were treated as whole-number relationships between two lines. Clearly, out of this no conception of zero as a number could possibly come, for from the point of view of a draughtsman it is meaningless. We, having minds differently constituted, must not argue from our habits to theirs and treat their mathematic as a "first stage" in the development of "Mathematics." Within and for the purpose of the world that Classical man evolved for himself, the Classical mathematic was a complete thing—it is merely not so *for us*. Babylonian and Indian mathematics had long contained, as essential elements of *their* number-worlds, things which the Classical number-feeling regarded as nonsense—and not from ignorance either, since many a Greek thinker was acquainted with them. It must be repeated, "Mathematics" is an illusion. A mathematical, and, generally, a scientific way of thinking is right, convincing, a "necessity of thought," when it completely expresses the life-feeling proper to it. Otherwise it is either impossible, futile and senseless, or else, as we in the arrogance of our historical soul like to say, "primitive." The modern mathematic, though "true" only for the Western spirit, is undeniably a master—work of that spirit; and yet to Plato it would have seemed a ridiculous and painful aberration from the path leading to the "true"—to wit, the Classical-mathematic. And so with ourselves. Plainly, we have almost no notion of the multitude of great ideas belonging to other Cultures that we have suffered to lapse because *our* thought with its limitations has not permitted us to assimilate them, or (which comes to the same thing) has led us to reject them as false, superfluous, and nonsensical.

The Greek mathematic, as a science of perceivable magnitudes, deliberately confines itself to facts of the comprehensibly present, and limits its researches and their validity to the near and the small. As compared with this impeccable consistency, the position of the Western mathematic is seen to be, practically, somewhat illogical, though it is only since the discovery of Non-Euclidean Geometry that the fact has been really recognized. Numbers are images of the perfectly desensualized understanding, of pure thought, and contain their abstract validity within themselves. Their exact application to the actuality of conscious experience is therefore a problem in itself—a problem which is always being posed anew and never solved—and the congruence of mathematical system with empirical observation is at present anything but self-evident. Although the lay idea—as found in Schopenhauer—is that mathematics rest upon the direct evidences of the senses, Euclidean geometry, superficially identical though it is with the popular geometry of all ages, is only in agreement with the phenomenal world approximately and within very narrow limits—in fact, the limits of a drawing-board. Extend these limits, and what becomes, for instance, of Euclidean parallels? They meet at the line of the horizon—a simple fact upon which all our art-perspective is grounded.

Now, it is unpardonable that Kant, a Western thinker, should have evaded the mathematic of distance, and appealed to a set of figure-examples that their mere pettiness excludes from treatment by the specifically Western infinitesimal methods. But Euclid, as a thinker of the Classical age, was entirely consistent with its spirit when he refrained from proving the phenomenal truth of his axioms by referring to, say, the triangle formed by an observer and two infinitely distant fixed stars. For these can neither be drawn nor "intuitively apprehended" and his feeling was precisely the feeling which shrank from the irrationals, which did not dare to give nothingness a value as zero (i.e., a number) and even in the contemplation of cosmic relations shut its eyes to the Infinite and held to its symbol of Proportion.

Aristarchus of Samos, who in 288–277 belonged to a circle of astronomers at Alexandria that doubtless had relations with Chaldaeo-Persian schools, projected the elements of a heliocentric world-system.[7] Rediscovered by Copernicus, it was to shake the metaphysical passions of the West to their foundations—witness Giordano Bruno[8]—to become the fulfillment of mighty premonitions, and to justify that Faustian, Gothic world-feeling which had already professed its faith in infinity through the forms of its cathedrals. But the world of Aristarchus received his work with entire indifference and in a brief space of time it was forgotten—designedly, we may surmise. His few followers were nearly all natives of Asia Minor, his most prominent supporter Seleucus (about 150) being from the Persian Seleucia on Tigris. In fact, the Aristarchian system had no spiritual appeal to the Classical Culture and might indeed have become dangerous to it. And yet it was differentiated from the Copernican (a point always missed) by something which made it perfectly comfortable to the Classical world-feeling, viz., the assumption that the cosmos is *contained* in a materially finite and optically appreciable *hollow sphere,* in the middle of which the planetary system, arranged as such on Copernican lines, moved. In the Classical astronomy, the earth and the heavenly bodies are consistently regarded as entities of two different kinds, however variously their movements in detail might be interpreted. Equally, the opposite idea that the earth is *only a star among stars*[9] is not inconsistent in itself with either the Ptolemaic or the Copernican systems and in fact was pioneered by Nicolaus Cusanus and Leonardo da Vinci. But by this device of a celestial sphere the principle of infinity which would have endangered the sensuous-Classical notion of bounds was smothered. One would have supposed that the infinity-conception was inevitably implied by the system of Aristarchus—long before his time, the Babylonian thinkers had reached it. But no such thought emerges. On the contrary, in the famous treatise on the grains of sand[10] Archimedes [see selection, *The Sand Reckoner*, p. 411. ED.] proves that the filling of this stereometric body (for that is what Aristarchus's Cosmos is, after all) with atoms of sand leads to very high, but *not* to infinite, figure-results. This proposition, quoted though it may be, time and again, as being a first step towards the Integral Calculus, amounts to a denial (implicit indeed in the very title) of everything that we mean by the word analysis. Whereas in our physics, the constantly-surging hypotheses of a material (i.e., directly cogniz-able) æther, break themselves one after the other against our refusal to ack-nowledge material limitations of any kind, Eudoxus, Apollonius and Archi-medes, certainly the keenest and boldest of the Classical mathematicians, completely worked out, in the main with rule and compass, a *purely optical* analysis of things-become on the basis of sculptural-Classical bounds. They used deeply-thought-out (and for us hardly understandable) methods of integra-tion, but these possess only a superficial resemblance even to Leibniz's definite-integral method. They employed geometrical loci and co-ordinates, but these are always specified lengths and units of measurement and never, as in Fermat and above all in Descartes, unspecified spatial relations, values of points in terms of their positions in space. With these methods also should be classed the exhaustion-method of Archimedes,[11] given by him in his recently discovered letter to Eratosthenes on such subjects as the quadrature of the parabola section by means of inscribed rectangles (instead of through similar polygons). But the very subtlety and extreme complication of his methods, which are grounded in certain of Plato's geometrical ideas, make us realize, in spite of superficial analogies, what an enormous difference separates him from Pascal. Apart altogether from the idea of Riemann's integral, what sharper contrast could there be to these ideas than the so-called quadratures of to-day? The name itself is now no more than an unfortunate survival, the "surface" is indicated by a bounding function, and the *drawing,* as such, has vanished. Nowhere else did the two mathematical minds approach each other more closely than in this instance, and nowhere is it more evident that the gulf between the two souls thus expressing themselves is impassable.

In the cubic style of their early architecture the Egyptians, so to say, con-cealed pure numbers, fearful of stumbling upon their secret, and for the Hel-lenes too they were the key to the meaning of the become, the stiffened, the mortal. The stone statue and the scientific system deny life. Mathematical number, the formal principle of an extension-world of which the phenomenal existence is only the derivative and servant of waking human consciousness, bears the hall-mark of causal necessity and so is linked with *death* as chrono-logical number is with becoming, with *life,* with the necessity of destiny. This connexion of strict mathematical form with the *end* of organic being, with the phenomenon of its organic remainder the corpse, we shall see more and more clearly to be the origin of all great art. We have already noticed the develop-ment of early ornament on funerary equipments and receptacles. *Numbers are symbols of the mortal.* Stiff forms are the negation of life, formulæ and laws spread rigidity over the face of nature, numbers make dead—and the "Mothers" of Faust II sit enthroned, majestic and withdrawn, in

> The realms of Image unconfined.
> . . . Formation, transformation,
> Eternal play of the eternal mind
> With semblances of all things in creation
> For ever and for ever sweeping round.

Goethe draws very near to Plato in this divination of one of the final secrets. For his unapproachable Mothers are Plato's Ideas—the possibilities of a spirituality, the unborn forms to be realized as active and purposed Culture, as art, thought, polity and religion, in a world ordered and determined by that spirituality. And so the number-thought and the world-idea of a Culture are related, and by this relation, the former is elevated above mere knowledge and experience and becomes a view of the universe, there being consequently as many mathematics—as many number-worlds—as there are higher Cultures. Only so can we understand, as something *necessary,* the fact that the greatest mathematical thinkers, the creative artists of the realm of numbers, have been brought to the decisive mathematical discoveries of their several Cultures by a deep religious intuition.

Classical, Apollinian number we must regard as the creation of Pythagoras—*who founded a religion.* It was an instinct that guided Nicolaus Cusanus, the great Bishop of Brixen (about 1450), from the idea of the unendingness of God in nature to the elements of the Infinitesimal Calculus. Leibniz himself, who two centuries later definitely settled the methods and notation of the Calculus, was led by purely metaphysical speculations about the divine principle and its relation to infinite extent to conceive and develop the notion of an *analysis situs*—probably the most inspired of all interpretations of pure and emancipated space—the possibilities of which were to be developed later by Grassmann in his *Ausdehnungslehre* and above all by Riemann, their real creator, in his symbolism of two-sided planes representative of the nature of equations. And Kepler and Newton, strictly religious natures both, were and remained convinced, like Plato, that it was precisely through the medium of number that they had been able to apprehend intuitively the essence of the divine world-order.

The Classical arithmetic, we are always told, was first liberated from its sense-bondage, widened and extended by Diophantus, who did not indeed create algebra (the science of undefined magnitudes) but brought it to expression within the framework of the Classical mathematic that we know —and so suddenly that we have to assume that there was a pre-existent stock of ideas which he worked out. But this amounts, not to an enrichment of, but a complete victory over, the Classical world-feeling, and the mere fact should have suf-ficed in itself to show that, inwardly, Diophantus does not belong to the Classical Culture at all. What is active in him is a new number-feeling, or let us say a new limit-feeling with respect to the actual and become, and no longer that Hellenic feeling of sensuously-present limits which had produced the Euclidean geometry, the nude statue and the coin. Details of the formation of this new mathematic we do not know—Diophantus stands so completely by himself in the history of so-called late-Classical mathematics that an Indian influence has been presumed. But here also the influence must really have been that of those early-Arabian schools whose studies (apart from the dogmatic) have hitherto been so imperfectly investigated. In Diophantus, unconscious though he may be of his own essential antagonism to the Classical foundations on which he attempted to build, there emerges from under the surface of Euclidean *intention* the new limit-*feeling* which I designate the "Magian." He did not widen the idea of number as magnitude, but (unwittingly) eliminated it. No Greek could have stated anything about an *undefined* number *a* or an *undenominated* number 3—which are neither magnitudes nor lines—whereas the new limit-feeling sensibly expressed by numbers of this sort at least under-lay, if it did not constitute, Diophantine treatment; and the letter-notation which we employ to clothe our own (again transvalued) algebra was first introduced by Vieta in 1591, an unmistakable, if unintended, protest against the classiciz-ing tendency of Renaissance mathematics.

Diophantus lived about 250 A.D., *that is, in the third century of that Arabian Culture* whose organic history, till now smothered under the surface-forms of the Roman Empire and the "Middle Ages," comprises everything that happened after the beginning of our era in the region that was later to be Islam's. It was precisely in the time of Diophantus that the last shadow of the Attic statuary art paled before the new space-sense of cupola, mosaic and sarcophagus-relief that we have in the Early-Christian-Syrian style. In that time there was once more *archaic* art and strictly geometrical ornament; and at that time too Diocletian completed the transformation of the now merely sham Empire into a Caliphate. The four centuries that separate Euclid and Diophantus, separate also Plato and Plotinus—the last and conclusive thinker, the Kant, of a fulfilled Culture and the first schoolman, the Duns Scotus, of a Culture just awakened.

It is here that we are made aware for the first time of the existence of those higher individualities whose coming, growth and decay constitute the *real substance of history* underlying the myriad colours and changes of the surface. The Classical spirituality, which reached its final phase in the cold intelligence of the Romans and of which the whole Classical Culture with all its works, thoughts, deeds and ruins forms the "body," had been born about 1100 B.C. in the country about the Ægean Sea. The Arabian Culture, which, under cover of the Classical Civilization, had been germinating in the East since Augustus, came wholly out of the region between Armenia and Southern Arabia,

Alexandria and Ctesiphon, and we have to consider as expressions of this new soul almost the whole "late-Classical" art of the Empire, all the young ardent religions of the East—Mandæanism, Manichæism, Christianity, Neo-Platonism, and in Rome itself, as well as the Imperial Fora, that Pantheon which is the *first of all mosques*.

That Alexandria and Antioch still wrote in Greek and imagined that they were thinking in Greek is a fact of no more importance than the facts that Latin was the scientific language of the West right up to the time of Kant and that Charlemagne "renewed" the Roman Empire.

In Diophantus, number has ceased to be the measure and essence of *plastic things*. In the Ravennate mosaics man has ceased to be a *body*. Unnoticed, Greek designations have lost their original connotations. We have left the realm of Attic καλοκἀγαθία the Stoic ἀταραξία and γαλήνη. Diophantus does not yet know zero and negative numbers, it is true, but he has *ceased* to know Pythagorean numbers. And this Arabian indeterminateness of number is, in its turn, something quite different from the controlled variability of the later Western mathematics, the variability of the *function*.

The Magian mathematic—we can see the outline, though we are ignorant of the details—advanced through Diophantus (who is obviously not a starting-point) boldly and logically to a culmination in the Abbassid period (9th century) that we can appreciate in Al-Khwarizmi and Alsidzshi. And as Euclidean geometry is to Attic statuary (the same expression-form in a different medium) and the analysis of space to polyphonic music, so this algebra is to the Magian art with its mosaic, its arabesque (which the Sassanid Empire and later Byzantium produced with an ever-increasing profusion and luxury of tangible-intangible organic motives) and its Constantinian high-relief in which uncertain deep-darks divide the freely-handled figures of the foreground. As algebra is to Classical arithmetic and Western analysis, so is the cupola-church to the Doric temple and the Gothic cathedral. It is not as though Diophantus were one of the great mathematicians. On the contrary, much of what we have been accustomed to associate with his name is not his work alone. His accidental importance lies in the fact that, so far as our knowledge goes, he was the first mathematician in whom the new number-feeling is unmistakably present. In comparison with the masters who *conclude* the development of a mathematic—with Apollonius and Archimedes, with Gauss, Cauchy, Riemann—Diophantus has, in his form-language especially, something *primitive*. This something, which till now we have been pleased to refer to "late-Classical" decadence, we shall presently learn to understand and value, just as we are revising our ideas as to the despised "late-Classical" art and beginning to see in it the tentative expression of the nascent Early Arabian Culture. Similarly archaic, primitive, and groping was the mathematic of Nicolas Oresme, Bishop of Liseux (1323–1382),[12] who was the first Western who used co-ordinates so to say elastically[13] and, more important still, to employ fractional powers—

both of which presuppose a number-feeling, obscure it may be but quite unmistakable, which is completely non-Classical and *also* non-Arabic. But if, further, we think of Diophantus together with the early-Christian sarcophagi of the Roman collections, and of Oresme together with the Gothic wall-statuary of the German cathedrals, we see that the mathematicians as well as the artists have something in common, which is, that they stand in their respective Cultures at *the same* (viz., the primitive) level of abstract understanding. In the world and age of Diophantus the stereometric sense of bounds, which had long ago reached in Archimedes the last stages of refinement and elegance proper to the megalopolitan intelligence, had passed away. Throughout that world men were unclear, longing, mystic, and no longer bright and free in the Attic way; they were men rooted in the earth of a young country-side, not megalopolitans like Euclid and D'Alembert. They no longer understood the deep and complicated forms of the Classical thought, and their own were confused and new, far as yet from urban clarity and tidiness. Their Culture was in the *Gothic* condition, as all Cultures have been in their youth—as even the Classical was in the early Doric period which is known to us now only by its Dipylon pottery. Only in Baghdad and in the 9th and 10th Centuries were the young ideas of the age of Diophantus carried through to completion by ripe masters of the calibre of Plato and Gauss.

The decisive act of Descartes, whose geometry appeared in 1637, consisted not in the introduction of a new method or idea in the domain of traditional geometry (as we are so frequently told), but in the definitive conception of *a new number-idea,* which conception was expressed in the emancipation of geometry from servitude to optically-realizable constructions and to measured and measurable lines generally. With that, the analysis of the infinite became a fact. The rigid, so-called Cartesian, system of co-ordinates—a semi-Euclidean method of ideally representing measurable magnitudes—had long been known (witness Oresme) and regarded as of high importance, and when we get to the bottom of Descartes' thought we find that what he did was not to round off the system but to overcome it. Its last historic representative was Descartes' contemporary Fermat.

In place of the sensuous element of concrete lines and planes—the specific character of the Classical feeling of bounds—there emerged the abstract, spatial, un-Classical element of the *point* which from then on was regarded as a group of co-ordered pure numbers. The idea of magnitude and of perceivable dimension derived from Classical texts and Arabian traditions was destroyed and replaced by that of variable relation-values between positions in space. It is not in general realized that this amounted to the *supersession of geometry,* which thenceforward enjoyed only a fictitious existence behind a facade of Classical tradition. The word "geometry" has an inextensible Apollinian meaning, and from the time of Descartes what is called the "new geometry" is made up in part of synthetic work upon the *position* of *points* in a space which is no longer necessarily three-dimensional (a "manifold of points"), and in part of analysis, in which numbers are defined through point-positions in space. And this replacement of lengths by positions carries with it a purely spatial, and no longer a material, conception of extension.

The clearest example of this destruction of the inherited optical-finite geometry seems to me to be the conversion of angular functions—which in the Indian mathematic had been numbers (in a sense of the word that is hardly accessible to our minds)—into *periodic* functions, and their passage thence into an infinite number-realm, in which they become series and not the smallest trace remains of the Euclidean figure. In all parts of that realm the circle-number $\pi$ like the Napierian base $\epsilon$, generates relations of all sorts which obliterate all the old distinctions of geometry, trigonometry and algebra, which are neither arithmetical nor geometrical in their nature, and in which no one any longer dreams of actually drawing circles or working out powers.

At the moment exactly corresponding to that at which (c. 540) the Classical Soul in the person of Pythagoras discovered its own proper Apollinian number, the measurable magnitude, the Western soul in the persons of Descartes and his generation (Pascal, Fermat, Desargues) discovered a notion of number that was the child of a passionate *Faustian* tendency towards the infinite. Number as *pure magnitude* inherent in the material presentness of things is paralleled by numbers as *pure relation*[14] and if we may characterize the Classical "world," the cosmos, as being based on a deep need of visible limits and composed accordingly as a sum of material things, so we may say that our world-picture is an actualizing of an infinite space in which things visible appear very nearly as realities of a lower order, limited in the presence of the illimitable. The symbol of the West is an idea of which no other Culture gives even a hint, the idea of *Function*. The function is anything rather than an expansion of, it is complete emancipation from, any pre-existent idea of number. With the function, not only the Euclidean geometry (and with it the common human geometry of children and laymen, based on everyday experience) but also the Archimedean arithmetic, ceased to have any value for the really *significant* mathematic of Western Europe. Henceforward, this consisted solely in abstract analysis. For classical man geometry and arithmetic were self-contained and complete sciences of the highest rank, both phenomenal and both concerned with magnitudes that could be drawn or numbered. For us, on the contrary, those things are only practical auxiliaries of daily life. Addition and multiplication, the two Classical methods of reckoning magnitudes, have, like their sister geometrical-drawing, utterly vanished in the infinity of functional processes. Even the power, which in the beginning denotes numerically a set of multiplications (products of equal magnitudes), is, through the exponential idea (logarithm) and its employment in complex, negative and fractional forms, dissociated from all connexion with magnitude and transferred to a transcendent relational world which the Greeks, knowing only the two positive whole-number powers that represent areas and volumes, were unable to approach. Think, for instance, of expressions like $\epsilon^{-x}, \sqrt[\pi]{x}, \alpha^{1/i}$.

Every one of the significant creations which succeded one another so rapidly from the Renaissance onward—imaginary and complex numbers, introduced by Cardanus as early as 1550; infinite series, established theoretically by Newton's great discovery of the binomial theorem in 1666; the differential geometry, the definite integral of Leibniz; the aggregate as a new number-unit, hinted at even by Descartes; new processes like those of general integrals; the expansion of functions into series and even into infinite series of other functions—is a victory over the popular and sensuous number-feeling in us, a victory which the new mathematic had to win in order to make the new world-feeling actual.

In all history, so far, there is no second example of one Culture paying to another Culture long extinguished such reverence and submission in matters of science as ours has paid to the Classical. It was very long before we found courage to think our proper thought. But though the wish to emulate the Classical was constantly present, every step of the attempt took us in reality further away from the imagined ideal. The history of Western knowledge is thus one of *progressive emancipation* from Classical thought, an emancipation never willed but enforced in the depths of the unconscious. *And so the development of the new mathematic consists of a long, secret and finally victorious battle against the notion of magnitude.*

One result of this Classicizing tendency has been to prevent us from finding the new notation proper to our Western number as such. The present-day sign-language of mathematics perverts its real content. It is principally owing to that tendency that the belief in numbers as magnitudes still rules to-day even amongst mathematicians, for is it not the base of all our written notation?

But it is not the separate signs (e.g., $x$, $\pi$, $\varsigma$) serving to express the functions *but the function itself as unit,* as element, the variable relation no longer capable of being optically defined, that constitutes the new number; and this new number should have demanded a new notation built up with entire disregard of Classical influences. Consider the difference between two equations (if the same word can be used of two such dissimilar things) such as $3^x + 4^x = 5^x$ and $x^n + y^n = z^n$ (the equation of Fermat's theorem). The first consists of several Classical numbers—i.e., magnitudes—but the second is *one number* of a different sort, veiled by being written down according to Euclidean-Archimedean tradition in the identical form of the first. In the first case, the sign = establishes a rigid connexion between definite and tangible magnitudes, but in the second it states that within a domain of variable images there exists a relation such that from certain alterations certain other alterations necessarily follow. The first equation has as its aim the specification by measurement of a concrete magnitude, viz., a "result," while the second has, in general, no result but is simply the picture and sign of a relation which for $n > 2$ (this is the famous Fermat problem[15]) *can probably be shown to* exclude integers. A Greek mathematician would have found it quite impossible to understand the purport of an operation like this, which was not meant to be "worked out."

As applied to the letters in Fermat's equation, the notion of the unknown is completely misleading. In the first equation $x$ is a magnitude, defined and measurable, which it is our business to compute. In the second, the word "defined" has no meaning at all for $x$, $y$, $z$, $n$, and consequently we do not attempt to compute their "values." Hence they are not numbers at all in the plastic sense but signs representing a connexion that is destitute of the hall-marks of magnitude, shape and unique meaning, an infinity of possible positions of like character, an ensemble unified and so attaining existence as a *number*. The whole equation, though written in our unfortunate notation as a plurality of terms, is actually *one single* number, $x$, $y$, $z$ being no more numbers than $+$ and $=$ are.

In fact, directly the essentially anti-Hellenic idea of the irrationals is introduced, the foundations of the idea of number as concrete and definite collapse. Thenceforward, the series of such numbers is no longer a visible row of increasing, discrete, numbers capable of plastic embodiment but a unidimensional *continuum* in which each "cut" (in Dedekind's sense) represents a number. Such a number is already difficult to reconcile with Classical number, for the Classical mathematic knows only one number between 1 and 3, whereas for the Western the totality of such numbers is an infinite aggregate. But when we introduce further the imaginary ($\sqrt{-1}$ or $i$) and finally the complex numbers (general form $a + bi$), the linear continuum is broadened into the highly transcendent form of a number-body, i.e., the content of an aggregate of homogeneous elements in which a "cut" now stands for a number-surface containing an infinite aggregate of numbers of a lower "potency" (for instance, all the real numbers), and there remains not a trace of number in the Classical and popular sense. These number-surfaces, which since Cauchy and Riemann have played an important part in the theory of functions, are *pure thought pictures*. Even positive irrational number (e.g., $\sqrt{2}$) could be conceived in a sort of negative fashion by Classical minds; they had, in fact, enough idea of it to ban it as ἄρρητος and ἄλογος. But expressions of the form $x + yi$ lie beyond every possibility of comprehension by Classical thought, whereas it is on the extension of the mathematical laws over the whole region of the complex numbers, within which these laws remain operative, that we have built up the function theory which has at last exhibited the Western mathematic in all purity and unity. Not until that point was reached could this mathematic be unreservedly brought to bear in the parallel sphere of our *dynamic* Western physics; for the Classical mathematic was fitted precisely to its own stereometric world of individual objects and to *static* mechanics as developed from Leucippus to Archimedes.

The brilliant period of the Baroque mathematic—the counterpart of the Ionian—lies substantially in the 18th Century and extends from the decisive discoveries of Newton and Leibniz through Euler, Lagrange, Laplace and D'Alembert to Gauss. Once this immense creation found wings, its rise was miraculous. Men hardly dared believe their senses. The age of refined scepticism witnessed the emergence of one seemingly impossible truth after another. Regarding the theory of the differential coefficient, D'Alembert had to say: "Go forward, and faith will come to you." Logic itself seemed to raise objections and to prove foundations fallacious. But the goal was reached.

This century was a very carnival of abstract and immaterial thinking, in which the great masters of analysis and, with them, Bach, Gluck, Haydn and Mozart—a small group of rare and deep intellects—revelled in the most refined discoveries and speculations, from which Goethe and Kant remained aloof; and in point of content it is exactly paralleled by the ripest century of the Ionic, the century of Eudoxus and Archytas (440–350) and, we may add, of Phidias, Polycletus, Alcamenes and the Acropolis buildings—in which the form-world of Classical mathematic and sculpture displayed the whole fullness of its possibilities, and so ended.

And now for the first time it is possible to comprehend in full the elemental opposition of the Classical and the Western souls. In the whole panorama of history, innumerable and intense as historical relations are, we find no two things so fundamentally alien to one another as these. And it is because extremes meet—because it may be there is some deep common origin behind their divergence—that we find in the Western Faustian soul this yearning effort towards the Apollinian ideal, the only alien ideal which we have loved and, for its power of intensely living in the pure sensuous present, have envied.

To return to mathematics. In the Classical world the starting-point of every formative act was, as we have seen, the ordering of the "become," in so far as this was present, visible, measurable and numerable. The Western, Gothic, form-feeling on the contrary is that of an unrestrained, strong-willed far-ranging soul, and its chosen badge is pure, imperceptible, unlimited space. But we must not be led into regarding such symbols as unconditional. On the contrary, they are strictly conditional, though apt to be taken as having identical essence and validity. Our universe of infinite space, whose existence, for us, goes without saying, simply does not exist for Classical man. It is not even capable of being presented to him. On the other hand, the Hellenic cosmos, which is (as we might have discovered long ago) entirely foreign to our way of thinking, was for the Hellene something self-evident. The fact is that the infinite space of our physics is a form of very numerous and extremely complicated elements tacitly assumed, which have come into being only as the copy and expression of *our* soul, and are actual, necessary and natural only for *our* type of waking life. The simple notions are always the most difficult. They are simple, in that they comprise a vast deal that not only is incapable of being exhibited in words but does not even need to be stated, because *for men of the particular group* it is anchored in the intuition; and they are difficult because for all alien men their real content is *ipso facto* quite inaccessible. Such a notion, at once simple and difficult, is our specifically Western meaning of the word "space." The whole of our mathematic from Descartes onward is devoted to the theoretical interpretation of this great and wholly religious symbol. The aim of all our physics since Galileo is identical; but in the Classical mathematics and physics the content of this word is simply *not known.*

Here, too, Classical names, inherited from the literature of Greece and retained in use, have veiled the realities. Geometry means the art of measuring, arithmetic the art of numbering. The mathematic of the West has long ceased to have anything to do with both these forms of defining, but it has not managed to find new names for its own elements—for the word "analysis" is hopelessly inadequate.

The beginning and end of the Classical mathematic is consideration of the properties of individual bodies and their boundary-surfaces; thus indirectly taking in conic sections and higher curves. *We,* on the other hand, at bottom know only the abstract space-element of the point, which can neither be seen, nor measured, nor yet named, but represents simply a centre of reference. The straight line, for the Greeks a measurable edge, is for us an infinite continuum of points. Leibniz illustrates his infinitesimal principle by presenting the straight line as one limiting case and the point as the other limiting case of a circle having infinitely great or infinitely little radius. But for the Greek the circle is a *plane* and the problem that interested him was that of bringing it into a commensurable condition. Thus the *squaring of the circle became for the Classical intellect the supreme problem of the finite.* The deepest problem of world-form seemed to it to be to alter surfaces bounded by curved lines, without change of magnitude, into rectangles and so to render them measurable. For us, on the other hand, it has become the usual, and not specifically significant, practice to represent the number $\pi$ by algebraic means, regardless of any geometrical image.

The Classical mathematician knows only what he sees and grasps. Where definite and defining visibility—the domain of his thought—ceases, his science comes to an end. The Western mathematician, as soon as he has quite shaken off the trammels of Classical prejudice, goes off into a wholly abstract region of infinitely numerous "manifolds" of $n$ (no longer 3) dimensions, in which his so-called geometry always can and generally must do without every commonplace aid. When Classical man turns to artistic expressions of his form-feeling, he tries with marble and bronze to give the dancing or the wrestling human form that pose and attitude in which surfaces and contours have all attainable proportion and meaning. But the true artist of the West shuts his eyes and loses himself in the realm of bodiless music, in which harmony and polyphony bring him to images of utter "beyondness" that transcend all possibilities of visual definition. One need only think of the meanings of the word "figure" as used respectively by the Greek sculptor and the Northern contrapuntist, and the opposition of the two worlds, the two mathematics, is immediately presented. The Greek mathematicians ever use the word σῶμα for their entities, just as the Greek lawyers used it for persons as distinct from things (σώματα καὶ πράγματα: *personæ et res*).

Classical number, integral and corporeal, therefore inevitably seeks to relate itself with the birth of bodily man, the σῶμα. The number 1 is hardly yet conceived of as actual number but rather as ἀρχή, the prime stuff of the number-series, the origin of all true numbers and therefore all magnitudes, measures and materiality (Dinglichkeit). In the group of the Pythagoreans (the date does not matter) its figured-sign was also the symbol of the mother-womb, the origin of all life. The digit 2, the first *true* number, which doubles the 1, was therefore correlated with the male principle and given the sign of the phallus. And, finally, 3, the "holy number" of the Pythagoreans, denoted the act of union between man and woman, the act of propagation—the erotic suggestion in adding and multiplying (the only two processes of increasing, of *propagating,* magnitude useful to Classical man) is easily seen—and its sign was the combination of the two first. Now, all this throws quite a new light upon the legends previously alluded to, concerning the sacrilege of disclosing the irrational. The irrational—in our language the employment of unending decimal fractions—implied the destruction of an organic and corporeal and reproductive order that the gods had laid down. There is no doubt that the Pythagorean reforms of the Classical religion were themselves based upon the immemorial Demeter-cult. Demeter, Gæa, is akin to Mother Earth. There is a deep relation between the honour paid to her and this exalted conception of the numbers.

Thus, inevitably, the Classical became by degrees the Culture of the *small.* The Apollinian soul had tried to tie down the meaning of things-become by means of the principle of *visible limits*; its taboo was focused upon the immediately-present and proximate alien. What was far away, invisible, was *ipso facto* "not there." The Greek and the Roman alike sacrificed to the gods of the place in which he happened to stay or reside; all other deities were outside the range of vision. Just as the Greek tongue—again and again we shall note the mighty symbolism of such language-phenomena—possessed *no word for space,* so the Greek himself was destitute of our feeling of landscape, horizons, outlooks, distances, clouds, and of the idea of the far-spread fatherland embracing the great nation. *Home,* for Classical man, is what he can see from the citadel of his native town and no more. All that lay beyond the visual range of this political atom was alien, and hostile to boot; beyond that narrow range, fear set in at once, and hence the appalling bitterness with which these petty towns strove to destroy one another. The Polis is the smallest of all conceivable state-forms, and its policy is frankly short-range, therein differing in the extreme from our own cabinet-diplomacy which is the policy of the unlimited. Similarly, the Classical temple, which can be taken in in one glance, is the smallest of all first-rate architectural forms. Classical geometry from Archytas to Euclid —like the school geometry of to-day which is still dominated by it—concerned itself with small, manageable figures and bodies, and therefore remained unaware of the difficulties that arise in establishing figures of astronomical dimensions, which in many cases are not amenable to Euclidean geometry.[16] Otherwise the subtle Attic spirit would almost surely have arrived at some notion of the problems of non-Euclidean geometry, for its criticism of the well-known "parallel" axiom,[17] the doubtfulness of which soon aroused opposition yet could not in any way be elucidated, brought it very close indeed to the decisive discovery. The Classical mind as unquestioningly devoted and limited itself to the study of the small and the near as ours has to that of the infinite and ultra-visual. All the mathematical ideas that the West found for itself or borrowed from others were automatically subjected to the form-language of the Infinitesimal—and that long before the actual Differential Calculus was discovered. Arabian algebra, Indian trigonometry, Classical mechanics were incorporated as a matter of course in analysis. Even the most "self-evident" propositions of elementary arithmetic such as $2 \times 2 = 4$ become, when considered analytically, problems, and the solution of these problems was only made possible by deductions from the Theory of Aggregates, and is in many points still unaccomplished. Plato and his age would have looked upon this sort of thing not only as a hallucination but also as evidence of an utterly nonmathematical mind. In a certain measure, geometry may be treated algebraically and algebra geometrically, that is, the eye may be switched off or it may be allowed to govern. We take the first alternative, the Greeks the second. Archimedes, in his beautiful management of spirals, touches upon certain general facts that are also fundamentals in Leibniz's method of the definite integral; but his processes, for all their superficial appearance of modernity, are subordinated to stereometric principles; in like case, an Indian mathematician would naturally have found some trigonometrical formulation.[18]

From this fundamental opposition of Classical and Western numbers there arises an equally radical difference in the relationship of element to element in each of these number-worlds. The nexus of *magnitudes* is called *proportion*, that of *relations* is comprised in the notion of *function*. The significance of these two words is not confined to mathematics proper; they are of high importance also in the allied arts of sculpture and music. Quite apart from the rôle of proportion in ordering the parts of the *individual* statue, the typically Classical art-forms of the statue, the relief, and the fresco, admit *enlargements* *and reductions of scale—words that in music have no meaning at all—*as we see in the art of the gems, in which the subjects are essentially reductions from life-sized originals. In the domain of Function, on the contrary, it is the idea of *transformation of groups* that is of decisive importance, and the musician will readily agree that similar ideas play an essential part in modern composition-theory. I need only allude to one of the most elegant orchestral forms of the 18th Century, the *Tema con Variazioni.*

All proportion assumes the constancy, all transformation the variability of the constituents. Compare, for instance, the congruence theorems of Euclid, the proof of which depends in fact on the assumed ratio 1: 1, with the modern deduction of the same by means of angular functions.

The Alpha and Omega of the Classical mathematic is *construction* (which in the broad sense includes elementary arithmetic), that is, the production of a single visually-present figure. The chisel, in this second sculptural art, is the compass. On the other hand, in function-research, where the object is not a result of the magnitude sort but a discussion of general formal possibilities, the way of working is best described as a sort of composition-procedure closely analogous to the musical; and in fact, a great number of the ideas met with in the theory of music (key, phrasing, chromatics, for instance) can be directly employed in physics, and it is at least arguable that many relations would be clarified by so doing.

Every *construction* affirms, and every *operation* denies appearances, in that the one works out that which is optically given and the other dissolves it. And so we meet with yet another contrast between the two kinds of mathematic; the Classical mathematic of small things deals with the concrete *individual instance* and produces a once-for-all construction, while the mathematic of the infinite handles whole *classes* of formal possibilities, *groups* of functions, operations, equations, curves, and does so with an eye, not to any result they may have, but to their course. And so for the last two centuries—though present-day mathematicians hardly realize the fact—there has been growing up *the idea of a general morphology of mathematical operations,* which we are justified in regarding as the real meaning of modern mathematics as a whole. All this, as we shall perceive more and more clearly, is one of the manifestations of a general tendency inherent in the Western intellect, proper to the Faustian spirit and Culture and found in no other. The great majority of the problems which occupy our mathematic, and are regarded as "our" problems in the same sense as the squaring of the circle was the Greeks',—e.g., the investigation of convergence in infinite series (Cauchy) and the transformation of elliptic and algebraic integrals into multiply-periodic functions (Abel, Gauss)—would probably have seemed to the Ancients, who strove for simple and definite quantitative results, to be an exhibition of rather abstruse virtuosity. And so indeed the popular mind regards them even to-day. There is nothing less "popular" than the modern mathematic, and it too contains its symbolism of the infinitely far, of *distance. All* the great works of the West, from the "Divina Commedia" to "Parsifal," are unpopular, whereas everything Classical from Homer to the Altar of Pergamum was popular in the highest degree.

Thus, finally, the whole content of Western number-thought centres itself upon the historic *limit-problem* of the Faustian mathematic, the key which opens the way to the Infinite, that *Faustian infinite* which is so different from the infinity of Arabian and Indian world-ideas. Whatever the guise—infinite series, curves or functions—in which number appears in the particular case, the *essence* of it is *the theory of the limit.* This limit is the absolute opposite of the limit which (without being so called) figures in the Classical problem of the quadrature of the circle. Right into the 18th Century, Euclidean popular prepossessions obscured the real meaning of the differential principle. The idea of infinitely small quantities lay, so to say, ready to hand, and however skilfully they were handled, there was bound to remain a trace of the Classical constancy, the *semblance of magnitude,* about them, though Euclid would never have known them or admitted them as such. Thus, zero is a constant, a whole number in the linear continuum between $+1$ and $-1$; and it was a great hindrance to Euler in his analytical researches that, like many after him, he treated the differentials as zero. Only in the 19th Century was this relic of Classical number-feeling finally removed and the Infinitesimal Calculus made logically secure by Cauchy's definitive elucidation of the *limit-idea;* only the intellectual step from the "infinitely small quantity" to the "lower limit of *every possible* finite magnitude" brought out the conception of a variable number which oscillates beneath any assignable number that is not zero. A number of this sort has ceased to possess any character of magnitude whatever: the limit, as thus finally presented by theory, is no longer that which is approximated to, but *the approximation, the process, the operation itself. It is not a state, but a relation.* And so in this decisive problem of our mathematic, we are suddenly made to see how *historical* is the constitution of the Western soul.

The liberation of geometry from the visual, and of algebra from the notion of magnitude, and the union of both, beyond all elementary limitations of drawing and counting, in the great structure of function-theory—this was the grand course of Western number-thought. The constant number of the Classical mathematic was dissolved into the variable. Geometry *became* analytical and dissolved all concrete forms, replacing the mathematical bodies from which the rigid geometrical values had been obtained, by abstract spatial relations which in the end ceased to have any application at all to sense-present phenomena. It began by substituting for Euclid's optical figures geometrical loci referred to a co-ordinate system of arbitrarily chosen "origin," and reducing the postulated objectiveness of existence of the geometrical object to the one condition that during the operation (which itself was one of equating and not of measurement) the selected co-ordinate system should not be changed. But these co-ordinates immediately came to be regarded as values pure and simple, serving not so much to determine as to represent and replace the position of points as space-elements. Number, the boundary of things-become, was represented, not as before pictorially by a figure, but symbolically by an equation. "Geometry" altered its meaning; the co-ordinate system as a picturing disappeared and the point became an entirely abstract number-group. In architecture, we find this inward transformation of Renaissance into Baroque through the innovations of Michael Angelo and Vignola. Visually pure lines became, in palace and church facades as in mathematics, ineffectual. In place of the clear co-ordinates that we have in Romano-Florentine colonnading and storeying, the "infinitesimal" appears in the graceful flow of elements, the scrollwork, the cartouches. The constructive dissolves in the wealth of the decorative—in mathematical language, the functional. Columns and pilasters, assembled in groups and clusters, break up the façades, gather and disperse again restlessly. The flat surfaces of wall, roof, storey melt into a wealth of stucco work and ornaments, vanish and break into a play of light and shade. The light itself, as it is made to play upon the form-world of mature Baroque—viz., the period from Bernini (1650) to the Rococo of Dresden, Vienna and Paris—has become an essentially musical element. The Dresden Zwinger[19] is a *sinfonia*. Along with 18th Century mathematics, 18th Century architecture develops into a form-world of *musical* characters.

This mathematics of ours was bound in due course to reach the point at which not merely the limits of artificial geometrical form but the limits of the visual itself were felt by theory and by the soul alike as limits indeed, as obstacles to the unreserved expression of inward possibilities—in other words, the point at which the ideal of transcendent extension came into fundamental conflict with the limitations of immediate perception. The Classical soul, with the entire abdication of Platonic and Stoic ἀταραξία, submitted to the sensuous and (as the erotic under-meaning of the Pythagorean numbers shows) it rather *felt* than *emitted* its great symbols. Of transcending the corporeal here-and-now it was quite incapable. But whereas number, as conceived by a Pythagorean, exhibited the essence of individual and discrete *data* in "Nature" Descartes and his successors looked upon number as *something to be conquered,* to be *wrung out,* an abstract relation royally indifferent to all phenomenal support and capable of holding its own against "Nature" on all occasions. The will-to-power (to use Nietzsche's great formula) that from the earliest Gothic of the Eddas, the Cathedrals and Crusades, and even from the old conquering Goths and Vikings, has distinguished the attitude of the Northern soul to its world, appears also in the sense-transcending energy, the *dynamic* of Western number. In the Apollinian mathematic the intellect is the servant of the eye, in the Faustian its master. Mathematical, "absolute" space, we see then, is utterly un-Classical, and from the first, although mathematicians with their reverence for the Hellenic tradition did not dare to observe the fact, it was something different from the indefinite spaciousness of daily experience and customary painting, the *a priori* space of Kant which seemed so unambiguous and sure a concept. It is a pure abstract, an ideal and unfulfillable postulate of a soul which is ever less and less satisfied with sensuous means of expression and in the end passionately brushes them aside. *The inner eye has awakened.*

And then, for the first time, those who thought deeply were obliged to see that the Euclidean geometry, which is the *true and only* geometry of the simple of all ages, is when regarded from the higher standpoint nothing but a *hypothesis,* the general validity of which, since Gauss, we know it to be quite impossible to prove in the face of other and perfectly nonperceptual geometries. The critical proposition of this geometry, Euclid's axiom of parallels, is an *assertion,* for which we are quite at liberty to substitute another assertion. We may assert, in fact, that through a given point, no parallels, or two, or many parallels may be drawn to a given straight line, and all these assumptions lead to completely irreproachable geometries of three dimensions, which can be employed in physics and even in astronomy, and are in some cases preferable to the Euclidean.

Even the simple axiom that extension is boundless (boundlessness, since Riemann and the theory of curved space, is to be distinguished from endlessness) at once contradicts the essential character of all immediate perception, in that the latter depends upon the existence of light-resistances and *ipso facto* has material bounds. But abstract principles of boundary can be imagined which transcend, in an entirely new sense, the possibilities of optical definition. For the deep thinker, there exists even in the Cartesian geometry the tendency to get beyond the three dimensions of *experiential* space, regarded as an unnecessary restriction on the symbolism of number. And although it was not till about 1800 that the notion of *multi-dimensional* space (it is a pity that no better word was found) provided analysis with broader foundations, the real first step was taken at the moment when powers—that is, really, logarithms—were released from their original relation with sensually realizable surfaces and solids and, through the employment of irrational and complex exponents, brought within the realm of function as perfectly general relation-values. It will be admitted by everyone who understands anything of mathematical reasoning that directly we passed from the notion of $a^3$ as a natural maximum to that of $a^n$, the unconditional necessity of three-dimensional space was done away with.

Once the space-element or point had lost its last persistent relic of visualness and, instead of being represented to the eye as a cut in coordinate lines, was defined as a group of three independent numbers, there was no longer any inherent objection to replacing the number 3 by the general number $n$. The notion of dimension was radically changed. It was no longer a matter of treating the properties of a point metrically with reference to its position in a visible system, but of representing the entirely abstract properties of a number-group by means of any dimensions that we please. The number-group—consisting of *n* independent ordered elements—is an *image* of the point and it is *called* a point. Similarly, an equation logically arrived therefrom is *called* a plane and is the image of a plane. And the aggregate of all points of *n* dimensions is *called* an *n*-dimensional space.[20] In these transcendent space-worlds, which are remote from every sort of sensualism, lie the relations which it is the business of analysis to investigate and which are found to be consistently in agreement with the data of experimental physics. This space of higher degree is a symbol which is through-and-through the peculiar property of the Western mind. That mind alone has attempted, and successfully too, to capture the "become" and the extended in *these* forms, to conjure and bind—to "know"—the alien by *this kind* of appropriation or taboo. Not until such spheres of number-thought are reached, and not for any men but the few who have reached them, do such imaginings as systems of hypercomplex numbers (e.g., the quaternions of the calculus of vectors) and apparently quite meaningless symbols like $\infty^n$ acquire the character of something actual. And here if anywhere it must be understood that actuality is not only sensual actuality. The spiritual is in no wise limited to perception-forms for the actualizing of its idea.

From this grand intuition of symbolic space-worlds came the last and conclusive creation of Western mathematic—the expansion and subtilizing of the function theory in that of *groups*. Groups are aggregates or sets of homogeneous mathematical images—e.g., the totality of all differential equations of a certain type—which in structure and ordering are analogous to the Dedekind number-bodies. Here are worlds, we feel, of perfectly new numbers, which are nevertheless not utterly sense-transcendent for the *inner* eye of the adept; and the problem now is to discover in those vast abstract form-systems certain elements which, relatively to a particular group of operations (viz., of transformations of the system), remain unaffected thereby, that is, possess invariance. In mathematical language, the problem, as stated generally by Klein, is—given an *n*-dimensional manifold ("space") and a group of transformations, it is required to examine the forms belonging to the manifold in respect of such properties as are not altered by transformation of the group.

And with this culmination our Western mathematic, having exhausted every inward possibility and fulfilled its destiny as the *copy and purest expression of the idea of the Faustian soul*, closes its development in the same way as the mathematic of the Classical Culture concluded in the third century. Both those sciences (the only ones of which the organic structure can even to-day be examined historically) arose out of a wholly new idea of number, in the one case Pythagoras's, in the other Descartes'. Both, expanding in all beauty, reached their maturity one hundred years later; and both, after flourishing for three centuries, completed the structure of their ideas at the same moment as the Cultures to which they respectively belonged passed over into the phase of megalopolitan Civilization. The deep significance of this interdependence will be made clear in due course. It is enough for the moment that for us the time of the *great* mathematicians is past. Our tasks to-day are those of preserving, rounding off, refining, selection—in place of big dynamic creation, the same clever detail-work which characterized the Alexandrian mathematic of late Hellenism.

A historical paradigm will make this clearer.

| *Classical* | *Western* |
|---|---|
| 1. *Conception of a new number* | |
| About 540 B.C. | About 1630 A.D. |
| Number as magnitude | Number as relation (Descartes, |
| (Pythagoreans) | Pascal, Fermat). (Newton, |
| | Leibniz, 1670) |
| (About 470, sculpture prevails | (About 1670, music prevails over |
| over fresco painting) | oil painting) |
| 2. *Zenith of systematic development* | |
| 450–350 | 1750–1800 |
| Plato, Archytas, Eudoxus | Euler, Lagrange, Laplace |
| (Phidias, Praxiteles) | (Gluck, Haydn, Mozart) |
| 3. *Inward completion and* | |
| *conclusion of the figure-world* | |
| 300–250 | After 1800 |
| Euclid, Apollonius, Archimedes | Gauss, Cauchy, Riemann |
| (Lysippus, Leochares) | (Beethoven) |

## ENDNOTES

[1]Also "thinking in money."

[2]Dynasties I–VIII, or, effectively, I–VI. The Pyramid period coincides with Dynasties IV–VI. Cheops, Chephren and Mycerinus belong to the IV dynasty, under which also great water-control works were carried out between Abydos and the Fayum.—*Tr.*

[3]As also those of law and of money

[4]Poincaré, in his *Science et Méthode* (Ch. III), searchingly analyses the "becoming" of one of his own mathematical discoveries. Each decisive stage in it bears "*les mêmes caractères de brièveté, de oudaineté et de certitudé absolue*" and in most cases this "*certitude*" was such that he merely registered the discovery and put off its working-out to any convenient season.—*Tr.*

[5]One may be permitted to add that according to legend, both Hippasus who took to himself public credit for the discovery of a sphere of twelve pentagons, viz., the regular dodecahedron (regarded by the Pythagoreans as the quintessence—or æther—of a world of real tetrahedrons, octahedrons, icosahedrons and cubes), and Archytas the eighth successor of the Founder are reputed to have been drowned at sea. The pentagon from which this dodecahedron is derived, itself involves incommensurable numbers. The "pentagram" was the recognition badge of Pythagoreans and the ἄλογον (incommensurable) their special secret. It would be noted, too, that Pythagoreanism was popular till its initiates were found to be dealing in these alarming and subversive doctrines, and then they were suppressed and lynched—a persecution which suggests more than one deep analogy with certain heresy-suppressions of Western history. The English student may be referred to G. J. Allman, *Greek Geometry from Thales to Euclid* (Cambridge, 1889), and to his articles "Pythagoras," "Philolaus" and "Archytas" in the *Ency. Brit.*, XI Edition.—*Tr.*

[6]Horace's words (Odes I xi): "Tu ne quæsieris, scire nefas, quem mihi quem tibi finem di dederint, Leuconoë, *nec Babylonios temptaris numeros . . . carpe diem, quam minimum credula postero.*"—*Tr.*

[7]In the only writing of his that survives, indeed, Aristarchus maintains the geocentric view; it may be presumed therefore that it was only temporarily that he let himself be captivated by a hypothesis of the Chaldaean learning.

[8]Giordano Bruno (born 1548, burned for heresy 1600). His whole life might be expressed as a crusade on behalf of God and the Copernican universe against a degenerated orthodoxy and an Aristotelian world-idea long coagulated in death.—*Tr.*

[9]F. Strunz, *Gesch. d. Naturwiss. im Mittelalter* (1910), p. 90.

[10]In the "Psammites," or "Arenarius," Archimedes framed a numerical notation which was to be capable of expressing the number of grains of sand *in a sphere of the size of our universe.—Tr.*

[11]This, for which the ground had been prepared by Eudoxus, was employed for calculating the volume of pyramids and cones: "the means whereby the Greeks were able *to evade* the forbidden notion of infinity" (Heiberg, *Naturwiss. u. Math. i. Klass. Alter.* [1912], p. 27).

[12]Oresme was, equally, prelate, church reformer, scholar, scientist and economist—the very type of the philosopher-leader.—*Tr.*

[13]Oresme in his *Latitudines Formarum* used ordinate and abscissa, not indeed to specify numerically, but certainly to describe, change, i.e., fundamentally, to express functions.—*Tr.*

[14]Similarly, coinage and double-entry book-keeping play analogous parts in the money-thinking of the Classical and the Western Cultures respectively.

[15]That is, "it is impossible to part a cube into two cubes, a biquadrate into two biquadrates, and generally any power above the square into two powers having the same exponent." Fermat claimed to possess a proof of the proposition, but this has not been preserved, and no general proof has hitherto been obtained.—*Tr.*

[16]A beginning is now being made with the application of non-Euclidean geometries to astronomy. The hypothesis of curved space, closed but without limits, filled by the system of fixed stars on a radius of about 470,000,000 earth-distances, would lead to the hypothesis of a counter-image of the sun which to us appears as a star of medium brilliancy.

[17]That only one parallel to a given straight line is possible through a given point—a proposition that is incapable of proof.

[18]It is impossible to say, with certainty, how much of the Indian mathematics that we possess is old, i.e., before Buddha.

[19]Built for August II, in 1711, as barbican or fore-building for a projected palace. —*Tr.*

[20]From the standpoint of the theory of "aggregates" (or "sets of points"), a well-ordered set of points, irrespective of the dimension figure, is called a corpus; and thus an aggregate of $n - 1$ dimensions is considered, *relatively* to one of $n$ dimensions, as a surface. Thus the limit (wall, edge) of an "aggregate" represents an aggregate of lower "potentiality."

**2**     *And diff'ring judgements serve but to declare*
*That truth lies somewhere, if we knew but where.*

—William Cowper

# The Locus of Mathematical Reality: An Anthropological Footnote

### By Leslie A. White

"He's [the Red King's] dreaming now," said Tweedledee: "and what do you think he's dreaming about?"

Alice said, "Nobody can guess that."

"Why, about *you*!" Tweedledee exclaimed, clapping his hands triumphantly. "And if he left off dreaming about you, where do you suppose you'd be?"

"Where I am now, of course," said Alice.

"Not you!" Tweedledee retorted contemptuously. "You'd be nowhere. Why, you're only a sort of thing in his dream!"

"If that there King was to wake," added Tweedledum, "you'd go out—bang!—just like a candle."

"I shouldn't!" Alice exclaimed indignantly. "Besides, if *I'm* only a sort of thing in his dream, what are *you,* I should like to know?"

"Ditto," said Tweedledum.

"Ditto, ditto!" cried Tweedledee.

He shouted this so loud that Alice couldn't help saying "Hush! You'll be waking him, I'm afraid, if you make so much noise."

"Well, it's no use *your* talking about waking him," said Tweedledum, "when you're only one of the things in his dream. You know very well you're not real."

"I *am* real!" said Alice, and began to cry.

"You won't make yourself a bit realler by crying," Tweedledee remarked: "there's nothing to cry about."

"If I wasn't real," Alice said—half laughing through her tears, it all seemed so ridiculous—"I shouldn't be able to cry."

"I hope you don't suppose those are *real* tears?" Tweedledum interrupted in a tone of great contempt.

—Through the Looking Glass

Do mathematical truths reside in the external world, there to be discovered by man, or are they man-made inventions? Does mathematical reality have an existence and a validity independent of the human species or is it merely a function of the human nervous system?

Opinion has been and still is divided on this question. Mrs. Mary Somerville (1780–1872), an Englishwoman who knew or corresponded with such men as Sir John Herschel, Laplace, Gay Lussac, W. Whewell, John Stuart Mill, Baron von Humboldt, Faraday, Cuvier, and De Candolle, and who was herself a scholar of distinction,[1] expressed a view widely held when she said:[2]

"Nothing has afforded me so convincing a proof of the unity of the Deity as these purely mental conceptions of numerical and mathematical science which have been by slow degrees vouchsafed to man, and are still granted in these latter times by the Differential Calculus, now superseded by the Higher Algebra, all of which must have existed in that sublimely omniscient Mind from eternity."

Lest it be thought that Mrs. Somerville was more theological than scientific in her outlook, let it be noted that she was denounced, by name and in public from the pulpit by Dean Cockburn of York Cathedral for her support of science.[3]

In America, Edward Everett (1794–1865), a distinguished scholar (the first American to win a doctorate at Göttingen), reflected the enlightened view of his day when he declared:[4]

"In the pure mathematics we contemplate absolute truths which existed in the divine mind before the morning stars sang together, and which will continue to exist there when the last of their radiant host shall have fallen from heaven."

In our own day, a prominent British mathematician, G. H. Hardy, has expressed the same view with, however, more technicality than rhetorical flourish:[5]

"I believe that mathematical reality lies outside us, and that our function is to discover or *observe* it, and that the theorems which we prove, and which we describe grandiloquently as our 'creations' are simply our notes of our observations."[6]

Taking the opposite view we find the distinguished physicist, P. W. Bridgman, asserting that "it is the merest truism, evident at once to unsophisticated observation, that mathematics is a human invention."[7] Edward Kasner and James Newman state that "we have overcome the notion that mathematical truths have an existence independent and apart from our own minds. It is even strange to us that such a notion could ever have existed."[8]

From a psychological and anthropological point of view, this latter conception is the only one that is scientifically sound and valid. There is no more reason to believe that mathematical realities have an existence independent of the human mind than to believe that mythological realities can have their being apart from man. The square root of minus one is real. So were Wotan and

Osiris. So are the gods and spirits that primitive peoples believe in today. The question at issue, however, is not, Are these things real?, but Where is the locus of their reality? It is a mistake to identify reality with the external world only. Nothing is more real than an hallucination.

Our concern here, however, is not to establish one view of mathematical reality as sound, the other illusory. What we propose to do is to present the phenomenon of mathematical behavior in such a way as to make clear, on the one hand, why the belief in the independent existence of mathematical truths has seemed so plausible and convincing for so many centuries, and, on the other, to show that all of mathematics is nothing more than a particular kind of primate behavior.

Many persons would unhesitatingly subscribe to the proposition that "mathematical reality must lie either within us, or outside us." Are these not the only possibilities? As Descartes once reasoned in discussing the existence of God, "it is impossible we can have the idea or representation of anything whatever, unless there be somewhere, *either in us or out of us,* an original which comprises, in reality . . ."[9] (emphasis ours). Yet, irresistible though this reasoning may appear to be, it is, in our present problem, fallacious or at least treacherously misleading. The following propositions, though apparently precisely opposed to each other, are equally valid; one is as true as the other: 1. "Mathematical truths have an existence and a validity independent of the human mind," and 2. "Mathematical truths have no existence or validity apart from the human mind." Actually, these propositions, phrased as they are, are misleading because the term "the human mind" is used in two different senses. In the first statement, "the human mind" refers to the individual organism; in the second, to the human species. Thus both propositions can be, and actually are, true. Mathematical truths exist in the cultural tradition into which the individual is born, and so enter his mind from the outside. But apart from cultural tradition, mathematical concepts have neither existence nor meaning, and of course, cultural tradition has no existence apart from the human species. Mathematical realities thus have an existence independent of the individual mind, but are wholly dependent upon the mind of the species. Or, to put the matter in anthropological terminology: mathematics in its entirety, its "truths" and its "realities," is a part of human *culture,* nothing more. Every individual is born into a culture which already existed and which is independent of him. Culture traits have an existence outside of the individual mind and independent of it. The individual obtains his culture by learning the customs, beliefs, techniques of his group. But culture itself has, and can have, no existence apart from the human species. Mathematics, therefore—like language, institutions, tools, the arts, etc.—is the cumulative product of ages of endeavor of the human species.

The great French savant Émile Durkheim (1858-1917) was one of the first to make this clear. He discussed it in the early pages of *The Elementary Forms of the Religious Life.*[10] And in *The Rules of Sociological Method*[11] especially he set forth the nature of *culture*[12] and its relationship to the human mind. Others, too, have of course discussed the relationship to the human mind. Others, too, have of course discussed the relationship between man and culture,[13] but Durkheim's formulations are especially appropriate for our present discussion and we shall call upon him to speak for us from time to time.

*Culture* is the anthropologist's technical term for the mode of life of any people, no matter how primitive or advanced. It is the generic term of which *civilization* is a specific term. The mode of life, or culture, of the human species is distinguished from that of all other species by the use of symbols. Man is the only living being that can freely and arbitrarily impose value or meaning upon any thing, which is what we mean by "using symbols." The most important and characteristic form of symbol behavior is articulate speech. All cultures, all of civilization, have come into being, have grown and developed, as a consequence of the symbolic faculty, unique in the human species.[14]

Every culture of the present day, no matter how simple or primitive, is a product of great antiquity. The language, tools, customs, beliefs, forms of art, etc., of any people are things which have been handed down from generation to generation, from age to age, changing and growing as they went, but always keeping unbroken the connection with the past. Every people lives not merely in a habitat of mountains or plains, of lakes, woods, and starry heavens, but in a setting of beliefs, customs, dwellings, tools, and rituals as well. Every individual is born into a man-made world of *culture* as well as the world of nature. But it is the culture rather than the natural habitat that determines man's thought, feelings, and behavior. To be sure, the natural environment may favor one type of activity or render a certain mode of life impossible. But whatever man does, as individual or as society, is determined by the culture into which he, or they, are born.[15] Culture is a great organization of stimuli that flows down through the ages, shaping and directing the behavior of each generation of human organisms as it goes. Human behavior is response to these cultural stimuli which seize upon each organism at birth—indeed, from the moment of conception, and even before this—and hold it in their embrace until death—and beyond, through mortuary customs and beliefs in a land of the dead.

The language a people speaks is the response to the linguistic stimuli which impinge upon the several organisms in infancy and childhood. One group of organisms is moulded by Chinese-language stimuli; another, by English. The organism has no choice, and once cast into a mould is unable to change. To learn to speak a foreign language without accent after one has matured, or even, in most cases, to imitate another dialect of his own language is exceedingly difficult if not impossible for most people. So it is in other realms of behavior. A people practices polygyny, has matrilineal clans, cremates the dead, abstains from eating pork or peanuts, counts by tens, puts butter in their tea, tattoos their chests, wears neckties, believes in demons, vaccinates their children, scalps their vanquished foes or tries them as war criminals, lends their wives to guests, uses slide rules, plays pinochle, or extracts square roots *if* the culture into which they were born possesses these traits. It is obvious, of course, that people do not choose their culture; they inherit it. It is almost as obvious that a people behaves as it does because it possesses a certain type of culture—or more accurately, is possessed by it.

To return now to our proper subject. Mathematics is, of course, a part of culture. Every people inherits from its predecessors, or contemporary neighbors, along with ways of cooking, marrying, worshipping, etc., ways of counting, calculating, and whatever else mathematics does. Mathematics is, in fact, a form of behavior: the responses of a particular kind of primate organism to a set of stimuli. Whether a people counts by fives, tens, twelves or twenties; whether it has no words for cardinal numbers beyond 5, or possesses the most modern and highly developed mathematical conceptions, their mathematical behavior is determined by the mathematical culture which possesses them.

We can see now how the belief that mathematical truths and realities lie outside the human mind arose and flourished. They *do* lie outside the mind of each individual organism. They enter the individual mind as Durkheim says from the outside. They impinge upon his organism, again to quote Durkheim, just as cosmic forces do. Any mathematician can see, by observing himself as well as others, that this is so. Mathematics is not something that is secreted, like bile; it is something drunk, like wine. Hottentot boys grow up and behave, mathematically as well as otherwise, in obedience to and in conformity with the mathematical and other traits in their culture. English or American youths do the same in their respective cultures. There is not one iota of anatomical or psychological evidence to indicate that there are any significant innate, biological or racial differences so far as mathematical or any other kind of human behavior is concerned. Had Newton been reared in Hottentot culture he would have calculated like a Hottentot. Men like G. H. Hardy, who know, through their own experience as well as from the observation of others, that mathematical realities enter the mind from the outside, understandably—but erroneously—conclude that they have their origin and locus in the external world, independent of man. Erroneous, because the alternative to "outside the human mind," the individual mind, that is, is not "the external world, independent of man," but *culture,* the body of traditional thought and behavior of the human species.

Culture frequently plays tricks upon us and distorts our thinking. We tend to find in culture direct expressions of "human nature" on the one hand and of the external world on the other. Thus each people is disposed to believe that its own customs and beliefs are direct and faithful expressions of man's nature. It is "human nature," they think, to practice monogamy, to be jealous of one's wife, to bury the dead, drink milk, to appear in public only when clad, to call your mother's brother's children "cousin," to enjoy exclusive right to the fruit of your toil, etc., if they happen to have these particular customs. But ethnography tells us that there is the widest divergence of custom among the peoples of the world: there are peoples who loathe milk, practice polyandry, lend wives as a mark of hospitality, regard inhumation with horror, appear in public without clothing and without shame, call their mother's brother's children "son" and "daughter," and who freely place all or the greater portion of the produce of their toil at the disposal of their fellows. There is no custom or belief that can be said to express "human nature" more than any other.

Similarly it has been thought that certain conceptions of the external world were so simple and fundamental that they immediately and faithfully expressed its structure and nature. One is inclined to think that yellow, blue, and green are features of the external world which any normal person would distinguish until he learns that the Creek and Natchez Indians did not distinguish yellow from green; they had but one term for both. Similarly, the Choctaw, Tunica, the Keresan Pueblo Indians and many other peoples make no terminological distinction between blue and green.[16]

The great Newton was deceived by his culture, too. He took it for granted that the concept of *absolute space* directly and immediately corresponded to something in the external world; space, he thought, is something that has an existence independent of the human mind. "I do not frame hypotheses," he said. But the concept space is a creation of the intellect as are other concepts. To be sure, Newton himself did not create the hypothesis of absolute space. It came to him from the outside, as Durkheim properly puts it. But although it impinges upon the organism *comme les forces cosmiques,* it has a different source: it is not the cosmos but man's culture.

For centuries it was thought that the theorems of Euclid were merely conceptual photographs, so to speak, of the external world; that they had a validity quite independent of the human mind; that there was something necessary and inevitable about them. The invention of non-Euclidean geometries by Lobatchewsky, Riemann and others has dispelled this view entirely. It is now clear that concepts such as space, straight line, plane, etc., are no more necessary and inevitable as a consequence of the structure of the external world than are the concepts green and yellow—or the relationship term with which you designate your mother's brother, for that matter.

To quote Einstein again:[17]

"We come now to the question: what is a priori certain or necessary, respectively in geometry (doctrine of space) or its foundations? Formerly we thought everything; nowadays we think—nothing. Already the distance-concept is logically arbitrary; there need be no things that correspond to it, even approximately."

Kasner and Newman say that "non-Euclidean geometry is proof that mathematics . . . is man's own handiwork, subject only to the limitations imposed by the laws of thought."[18]

Far from having an existence and a validity apart from the human species, all mathematical concepts are "free inventions of the human intellect," to use a phrase with which Einstein characterizes the concepts and fundamental principles of physics.[19] But because mathematical and scientific concepts have always entered each individual mind from the outside, everyone until recently has concluded that they came from the external world instead of from man-made culture. But the concept of culture, as a scientific concept, is but a recent invention itself.

The cultural nature of our scientific concepts and beliefs is clearly recognized by the Nobel prize winning physicist, Erwin Schrödinger, in the following passage:[20]

"Whence arises the widespread belief that the behavior of molecules is determined by absolute causality, whence the conviction that the contrary is *unthinkable*? Simply from the *custom,* inherited through thousands of years, of *thinking causally,* which makes the idea of undetermined events, of absolute, primary causalness, seem complete nonsense, a logical absurdity," (Schrödinger's emphases).

Similarly, Henri Poincaré asserts that the axioms of geometry are mere " conventions," i.e., customs: they "are neither synthetic a priori judgments nor experimental facts. They are *conventions* . . ."[21]

We turn now to another aspect of mathematics that is illuminated by the concept of culture. Heinrich Hertz, the discoverer of wireless waves, once said.[22]

"One cannot escape the feeling that these mathematical formulas have an independent existence and an intelligence of their own, that they are wiser than we are, wiser even than their discoverers [sic], that we get more out of them than was originally put into them."

Here again we encounter the notion that mathematical formulas have an existence "of their own," (i.e., independent of the human species), and that they are "discovered," rather than man-made. The concept of culture clarifies the entire situation. Mathematical formulas, like other aspects of culture, do have in a sense an "independent existence and intelligence of their own." The English language has, in a sense, "an independent existence of its own." Not independent of the human species, of course, but independent of any individual or group of individuals, race or nation. It has, in a sense, an "intelligence of its own." That is, it behaves, grows and changes in accordance with principles which are inherent in the language itself, not in the human mind. As man becomes self-conscious of language, and as the science of philology matures, the principles of linguistic behavior are discovered and its laws formulated.

So it is with mathematical and scientific concepts. In a very real sense they have a life of their own. This life is the life of culture, of cultural tradition. As Durkheim expresses it:[23] "Collective ways of acting and thinking have a reality outside the individuals who, at every moment of time, conform to it. These

ways of thinking and acting exist in their own right." It would be quite possible
to describe completely and adequately the evolution of mathematics, physics,
money, architecture, axes, plows, language, or any other aspect of culture
without ever alluding to the human species or any portion of it. As a matter of
fact, the most effective way to study culture scientifically is to proceed *as if* the
human race did not exist. To be sure it is often convenient to refer to the nation
that first coined money or to the man who invented the calculus or the cotton
gin. But it is not necessary, nor, strictly speaking, relevant. The phonetic shifts
in Indo-European as summarized by Grimm's law have to do solely with
linguistic phenomena, with sounds and their permutations, combinations and
interactions. They can be dealt with adequately without any reference to the
anatomical, physiological, or psychological characteristics of the primate organisms who produced them. And so it is with mathematics and physics.
Concepts have a life of their own. Again to quote Durkheim, "when once born,
[they] obey laws all their own. They attract each other, repel each other, unite,
divide themselves and multiply. . . ."[24] Ideas like other culture traits, interact
with each other, forming new syntheses and combinations. Two or three ideas
coming together may form a new concept or synthesis. The laws of motion
associated with Newton were syntheses of concepts associated with Galileo,
Kepler and others. Certain ideas of electrical phenomena grow from the "Faraday stage," so to speak, to those of Clerk Maxwell, H. Hertz, Marconi, and
modern radar. "The application of Newton's mechanics to continuously distributed masses *led inevitably* to the discovery and application of partial differential
equations, which in their turn first provided the language for the laws of the
field-theory,"[25] (emphasis ours). The theory of relativity was, as Einstein
observes, "no revolutionary act, but the natural continuation of a line that can
be traced through centuries."[26] More immediately, "the theory of Clerk
Maxwell and Lorentz led inevitably to the special theory of relativity."[27] Thus
we see not only that any given thought system is an outgrowth of previous
experience, but that certain ideas lead inevitably to new concepts and new
systems. Any tool, machine, belief, philosophy, custom or institution is but the
outgrowth of previous culture traits. An understanding of the nature of culture
makes clear, therefore, why Hertz felt that "mathematical formulas have an
independent existence and an intelligence of their own."

His feeling that "we get more out of them than was originally put into them,"
arises from the fact that in the interaction of culture traits new syntheses are
formed which were not anticipated by "their discoverers," or which contained
implications that were not seen or appreciated until further growth made them
more explicit. Sometimes novel features of a newly formed synthesis are not
seen even by the person in whose nervous system the synthesis took place.
Thus Jacques Hadamard tells us of numerous instances in which he failed
utterly to see things that "ought to have struck . . . [him] blind."[28] He cites
numerous instances in which he failed to see "obvious and immediate conse-

quences of the ideas contained"[29] in the work upon which he was engaged, leaving them to be "discovered" by others later.

The contradiction between the view held by Hertz, Hardy and others that mathematical truths are discovered rather than man-made is thus resolved by the concept of culture. They are both; they are discovered but they are also man-made. They are the product of the mind of the human species. But they are encountered or discovered by each individual in the mathematical culture in which he grows up. The process of mathematical growth is, as we have pointed out, one of interaction of mathematical elements upon each other. This process requires, of course, a basis in the brains of men, just as a telephone conversation requires wires, receivers, transmitters, etc. But we do not need to take the brains of men into account in an explanation of mathematical growth and invention any more than we have to take the telephone wires into consideration when we wish to explain the conversation it carries. Proof of this lies in the fact of numerous inventions (or "discoveries") in mathematics made simultaneously by two or more person working independently.[30] If these discoveries really were caused, or determined, by individual minds, we would have to explain them as coincidences. On the basis of the laws of chance these numerous and repeated coincidences would be nothing short of miraculous. But the culturological explanation makes the whole situation clear at once. The whole population of a certain region is embraced by a type of culture. Each individual is born into a pre-existing organization of beliefs, tools, customs and institutions. These culture traits shape and mould each person's life, give it content and direction. Mathematics is, of course, one of the streams in the total culture. It acts upon individuals in varying degree, and they respond according to their constitutions. Mathematics is the organic behavior response to the mathematical culture.

But we have already noted that within the body of mathematical culture there is action and reaction among the various elements. Concept reacts upon concept; ideas mix, fuse, form new syntheses. This process goes on throughout the whole extent of culture although more rapidly and intensively in some regions (usually the center) than in others (the periphery). When this process of interaction and development reaches a certain point, new syntheses[31] are formed of themselves. These syntheses are, to be sure, real events, and have location in time and place. The places are of course the brains of men. Since the cultural process has been going on rather uniformly over a wide area and population, the new synthesis takes place simultaneously in a number of brains at once. Because we are habitually anthropocentric in our thinking we tend to say that these men made these discoveries. And in a sense, a biological sense, they did. But if we wish to explain the discovery as an event in the growth of mathematics we must rule the individual out completely. From this standpoint, the individual did not make the discovery at all. It was something that happened to him. He was merely the place where the lightning struck. A simultaneous "discovery" by three men working "independently" simply means that cultural-mathematical lightning can and does strike in more than one place at a time. In the process of cultural growth, through invention or discovery, the individual is merely the neural medium in which the "culture"[32] of ideas grows. Man's brain is merely a catalytic agent, so to speak, in the cultural process. This process cannot exist independently of neural tissue, but the function of man's nervous system is merely to make possible the interaction and resynthesis of cultural elements.

To be sure individuals differ just as catalytic agents, lightning conductors or other media do. One person, one set of brains, may be a better medium for the growth of mathematical culture than another. One man's nervous system may be a better catalyst for the cultural process than that of another. The mathematical cultural process is therefore more likely to select one set of brains than another as its medium of expression. But it is easy to exaggerate the role of superior brains in cultural advance. It is not merely superiority of brains that counts. There must be a juxtaposition of brains with the interactive, synthesizing cultural process. If the cultural elements are lacking, superior brains will be of no avail. There were brains as good as Newton's in England 10,000 years before the birth of Christ, at the time of the Norman conquest, or any other period of English history. Everything that we know about fossil man, the prehistory of England, and the neuro-anatomy of *homo sapiens* will support this statement. There were brains as good as Newton's in aboriginal America or in Darkest Africa. But the calculus was not discovered or invented in these other times and places because the requisite cultural elements were lacking. Contrariwise, when the cultural elements are present, the discovery or invention becomes so inevitable that it takes place independently in two or three nervous systems at once. Had Newton been reared as a sheep herder, the mathematical culture of England would have found other brains in which to achieve its new synthesis. One man's brains may be better than another's, just as his hearing may be more acute or his feet larger. But just as a "brilliant" general is one whose armies are victorious, so a genius, mathematical or otherwise, is a person in whose nervous system an important cultural synthesis takes place; he is the neural locus of an epochal event in culture history.[33]

The nature of the culture process and its relation to the minds of men is well illustrated by the history of the theory of evolution in biology. As is well known, this theory did not originate with Darwin. We find it in one form or another, in the neural reactions of many others before Darwin was born: Buffon, Lamarck, Erasmus Darwin, and others. As a matter of fact, virtually all of the ideas which together we call Darwinism are to be found in the writings of J. C. Prichard, an English physician and anthropologist (1786–1848). These various concepts were interacting upon each other and upon current theological beliefs, competing, struggling, being modified, combined, re-synthesized, etc., for decades. The time finally came, i.e., the stage of development was reached, where the theological system broke down and the rising tide of scientific interpretation inundated the land.

Here again the new synthesis of concepts found expression simultaneously in the nervous systems of two men working independently of each other: A. R. Wallace and Charles Darwin. The event had to take place when it did. If Darwin had died in infancy, the cultural process would have found another neural medium of expression.

This illustration is especially interesting because we have a vivid account, in Darwin's own words, of the way in which the "discovery" (i.e., the synthesis of ideas) took place:

"In October 1838," Darwin wrote in his autobiographic sketch, "that is, fifteen months after I had begun my systematic enquiry, I *happened to read for amusement* 'Malthus on Population,' and being well prepared to appreciate the struggle for existence which everywhere goes on from long-continued observation of the habits of animals and plants, it at once struck me that under these circumstances favourable variations would tend to be preserved, and unfavourable ones to be destroyed. The result of this would be the formation of a new species. *Here then I had at last got a theory by which to work . . .*" (emphasis ours).

This is an exceedingly interesting revelation. At the time he read Malthus, Darwin's mind was filled with various ideas, (i.e., he had been moulded, shaped, animated and equipped by the cultural milieu into which he happened to have been born and reared—a significant aspect of which was independent means; had he been obliged to earn his living in a "counting house" we might have had "Hudsonism" today instead of Darwinism). These ideas reacted upon each other, competing, eliminating, strengthening, combining. Into this situation was introduced, *by chance,* a peculiar combination of cultural elements (ideas) which bears the name of Malthus. Instantly a reaction took place, a new synthesis was formed—"here at last he had a theory by which to work." Darwin's nervous system was merely the place where these cultural elements came together and formed a new synthesis. It was something that *happened* to Darwin rather than something he *did.*

This account of invention in the field of biology calls to mind the well-known incident of mathematical invention described so vividly by Henri Poincaré. One evening, after working very hard on a problem but without success, he writes:[34]

". . . contrary to my custom, I drank black coffee and could not sleep. Ideas rose in crowds; I felt them collide until pairs interlocked, so to speak, making a stable combination. By the next morning I had established the existence of a class of Fuchsian functions . . . I had only to write out the results, which took but a few hours."

Poincaré further illustrates the process of culture change and growth in its subjective (i.e., neural) aspect by means of an imaginative analogy.[35] He imagines mathematical ideas as being something like "the hooked atoms of Epicurus. During complete repose of the mind, these atoms are motionless, they are, so to speak, hooked to the wall." No combinations are formed. But in mental activity, even unconscious activity, certain of the atoms "are detached from the wall and put in motion. They flash in every direction through space . . . like the molecules of a gas . . . Then their mutual impacts may produce new combinations." This is merely a description of the subjective aspect of the cultural process which the anthropologist would describe objectively (i.e., without reference to nervous systems). He would say that in cultural systems, traits of various kinds act and react upon each other, eliminating some, reinforcing others, forming new combinations and syntheses. The significant thing about the loci of inventions and discoveries from the anthropologist's standpoint is not quality of brains, but relative position within the culture area: inventions and discoveries are much more likely to take place at culture centers, at places where there is a great deal of cultural interaction, than on the periphery, in remote or isolated regions.

If mathematical ideas enter the mind of the individual mathematician from the outside, from the stream of culture into which he was born and reared, the question arises, where did culture in general, and mathematical culture in particular, come from in the first place? How did it arise and acquire its content?

It goes without saying of course that mathematics did not originate with Euclid and Pythagoras—or even with the thinkers of ancient Egypt and Mesopotamia. Mathematics is a development of thought that had its beginning with the origin of man and culture a million years or so ago. To be sure, little progress was made during hundreds of thousands of years. Still, we find in mathematics today systems and concepts that were developed by primitive and preliterate peoples of the Stone Ages, survivals of which are to be found among savage tribes today. The system of counting by tens arose from using the fingers of both hands. The vigesimal system of the Maya astronomers grew out of the use of toes as well as fingers. To *calculate* is to count with *calculi*, pebbles. A *straight line* was a *stretched linen* cord, and so on.

To be sure, the first mathematical ideas to exist were brought into being by the nervous systems of individual human beings. They were, however, exceedingly simple and rudimentary. Had it not been for the human ability to give these ideas overt expression in symbolic form and to communicate them to one another so that new combinations would be formed, and these new syntheses passed on from one generation to another in a continuous process of interaction and accumulation, the human species would have made no mathematical progress beyond its initial stage. This statement is supported by our studies of anthropoid apes. They are exceedingly intelligent and versatile. They have a fine appreciation of geometric forms, solve problems by imagination and insight, and possess not a little originality.[36] But they cannot express their neuro-sensory-muscular concepts in overt symbolic form. They cannot communicate their ideas to one another except by gestures, i.e., by *signs* rather than *symbols*. Hence ideas cannot react upon one another in their minds to produce new syntheses. Nor can these ideas be transmitted from one generation to another in a cumulative manner. Consequently, one generation of apes begins where the preceding generation began. There is neither accumulation nor progress.[37]

Thanks to articulate speech, the human species fares better. Ideas are cast into symbolic form and given overt expression. Communication is thus made easy and versatile. Ideas now impinge upon nervous systems from the outside. These ideas react upon each other within these nervous systems. Some are eliminated; others strengthened. New combinations are formed, new syntheses achieved. These advances are in turn communicated to someone else, transmitted to the next generation. In a relatively short time, the accumulation of mathematical ideas has gone beyond the creative range of the individual human nervous system *unaided by cultural tradition*. From this time on, mathematical progress is made by the interaction of ideas already in existence rather than by the creation of new concepts by the human nervous system alone. Ages before writing was invented, individuals in all cultures were dependent upon the mathematical ideas present in their respective cultures. Thus, the mathematical behavior of an Apache Indian is the response that he makes to stimuli provided by the mathematical ideas in his culture. The same was true for Neanderthal man and the inhabitants of ancient Egypt, Mesopotamia and Greece. It is true for individuals of modern nations today.

Thus we see that mathematical ideas were produced originally by the human nervous system when man first became a human being a million years ago. These concepts were exceedingly rudimentary, and the human nervous system, *unaided by culture,* could never have gone beyond them regardless of how many generations lived and died. It was the formation of a cultural tradition which made progress possible. The communication of ideas from person to person, the transmission of concepts from one generation to another, placed in the minds of men (i.e., stimulated their nervous systems) ideas which through interaction formed new syntheses which were passed on in turn to others.

We return now, in conclusion, to some of the observations of G. H. Hardy, to show that his conception of mathematical reality and mathematical behavior is consistent with the culture theory that we have presented here and is, in fact, explained by it.

"I believe that mathematical reality lies outside us,"[38] he says. If by "us" he means "us mathematicians individually," he is quite right. They do lie outside each one of us; they are a part of the culture into which we are born. Hardy feels that "in some sense, mathematical truth is part of *objective* reality,"[39] (my emphasis, L.A.W.). But he also distinguishes "mathematical reality" from "physical reality," and insists that "pure geometries are *not* pictures . . . [of] the spatio-temporal reality of the physical world."[40] What then is the nature of mathematical reality? Hardy declares that "there is no sort of agreement. . . among either mathematicians or philosophers"[41] on this point. Our interpretation provides the solution. Mathematics does have objective reality. And this reality, as Hardy insists, is *not* the reality of the physical world. But there is no mystery about it. Its reality is cultural: the sort of reality possessed by a code of etiquette, traffic regulations, the rules of baseball, the English language or rules of grammar.

Thus we see that there is no mystery about mathematical reality. We need not search for mathematical "truths" in the divine mind or in the structure of the universe. Mathematics is a kind of primate behavior as languages, musical systems and penal codes are. Mathematical concepts are man-made just as ethical values, traffic rules, and bird cages are man-made. But this does not invalidate the belief that mathematical propositions lie outside us and have an objective reality. They do lie outside us. They existed before we were born. As we grow up we find them in the world about us. But this objectivity exists only for the individual. The locus of mathematical reality is cultural tradition, i.e., the continuum of symbolic behavior. This theory illuminates also the phenomena of novelty and progress in mathematics. Ideas interact with each other in the nervous systems of men and thus form new syntheses. If the owners of these nervous systems are aware of what has taken place they call it invention as Hadamard does, or "creation," to use Poincaré's term. If they do not understand what has happened, they call it a "discovery" and believe they have found something in the external world. Mathematical concepts are independent of the individual mind but lie wholly within the mind of the species, i.e., culture. Mathematical invention and discovery are merely two aspects of an event that takes place simultaneously in the cultural tradition and in one or more nervous systems. Of these two factors, culture is the more significant; the determinants of mathematical evolution lie here. The human nervous system is merely the catalyst which makes the cultural process possible.

# ENDNOTES

[1]She wrote the following works, some of which went into several editions: *The Mechanism of the Heavens*, 1831 (which was, it seems, a popularization of the *Mécanique Céleste* of Laplace); *The Connection of the Physical Sciences*, 1858; *Molecular and Microscopic Science*, 1869; *Physical Geography*, 1870.

[2]*Personal Recollections of Mary Somerville*, edited by her daughter, Martha Somerville, pp. 140–141 (Boston, 1874).

[3]*ibid.*, p. 375. See, also, A, D. White, *The History of the Warfare of Science with Theology &c*, Vol. I, p. 225, ftn.* (New York, 1930 printing).

[4]Quoted by E. T. Bell in *The Queen of the Sciences*, p. 20 (Baltimore, 1931).

[5]G. H. Hardy, *A Mathematician's Apology*, pp. 63–64 (Cambridge, England; 1941).

[6]The mathematician is not, of course, the only one who is inclined to believe that his creations are discoveries of things in the external world. The theoretical physicist, too, entertains this belief. "To him who is a discoverer in this field," Einstein observes, "the products of his imagination appear so necessary and natural that he regards them, and would like to have them regarded by others, not as creations of thought but as given realities," ("On the Method of Theoretical Physics," in *The World as I See It*, p. 30; New York, 1934).

[7]P. W. Bridgman, *The Logic of Modern Physics,* p. 60 (New York, 1927).

[8]Edward Kasner and James Newman, *Mathematics and the Imagination,* p. 359 (New York, 1940).

[9]*Principles of Philosophy,* Pt. I, Sec. XVIII, p. 308, edited by J. Veitch (New York, 1901).

[10]*Les Formes Élémentaires de la Vie Religieuse* (Paris, 1912) translated by J. W. Swain (London, 1915). Nathan Altshiller-Court refers to Durkheim's treatment of this point in "Geometry and Experience," (Scientific Monthly, Vol. LX, No. 1, pp. 63–66, Jan., 1945).

[11]*Les Règles de la Méthode Sociologique* (Paris, 1895; translated by Sarah A. Solovay and John H. Mueller, edited by George E. G. Catlin; Chicago, 1938).

[12]Durkheim did not use the term *culture.* Instead he spoke of the "collective consciousness," "collective representations," etc, Because of his unfortunate phraseology Durkheim has been misunderstood and even branded mystical. But it is obvious to one who understands both Durkheim and such anthropologists as R. H. Lowie, A. L. Kroeber and Clark Wissler that they are all talking about the same thing: culture.

[13]See, e.g., E. B. Tylor, *Anthropology* (London, 1881); R. H. Lowie, *Culture and Ethnology,* New York, 1917; A. L. Kroeber, "The Superorganic," (American Anthropologist, Vol. 19, pp. 163–213; 1917); Clark Wissler, *Man and Culture,* (New York, 1923).

[14]See, White, Leslie A., "The Symbol: the Origin and Basis of Human Behavior," (Philosophy of Science, Vol. 7, pp. 451–463; 1940; reprinted in ETC., a Review of General Semantics, Vol. I, pp. 229–237; 1944).

[15]Individuals vary, of course, in their constitutions and consequently may vary in their responses to cultural stimuli.

[16]Cf. "Keresan Indian Color Terms," by Leslie A. White, Papers of the Michigan Academy of Science, Arts, and Letters, Vol. XXVIII, pp. 559–563; 1942 (1943).

[17]Article "Space-Time." Encyclopaedia Britannica, 14th edition.

[18]*op. cit.,* p. 359.

[19]"On the Method of Theoretical Physics," in *The World as I See It,* p. 33 (New York, 1934).

[20]*Science and the Human Temperament,* p. 115 (London, 1935).

[21]"On the Nature of Axioms," in *Science and Hypothesis,* published in *The Foundations of Science* (The Science Press, New York, 1913).

[22]Quoted by E. T. Bell, *Men of Mathematics,* p. 16 (New York, 1937).

[23]*The Rules of Sociological Method,* Preface to 2nd edition, p. lvi.

[24]*The Elementary Forms of the Religious Life,* p. 424. See also *The Rules of Sociological Method,* Preface to 2nd edition, p. li, in which he says "we need to investigate... the manner in which social representations [i.e., culture traits] adhere to and repel one another, how they fuse or separate from one another."

[25]Einstein, "The Mechanics of Newton and their Influence on the Development of Theoretical Physics," in *The World as I See It,* p. 58.

[26]"On the Theory of Relativity," in *The World as I See It,* p. 69.

[27]Einstein, "The Mechanics of Newton &c," p. 57.

[28]Jacques Hadamard. *The Psychology of Invention in the Mathematical Field,* p. 50 (Princeton, 1945).

[29]*ibid.,* p. 51.

[30]The following data are taken from a long and varied list published in *Social Change,* by Wm. F. Ogburn (New York, 1923), pp. 90–102, in which simultaneous inventions and discoveries in the fields of chemistry, physics, biology, mechanical invention, etc., as well as in mathematics, are listed.

Law of inverse squares: Newton, 1666; Halley, 1684.

Introduction of decimal point: Pitiscus, 1608–12; Kepler, 1616; Napier, 1616–17.

Logarithms: Burgi, 1620; Napier-Briggs, 1614.

Calculus: Newton, 1671; Leibnitz. 1676.

Principle of least squares: Gauss, 1809; Legendre, 1806.

A treatment of vectors without the use of co-ordinate systems: Hamilton, 1843; Grassman, 1843; and others, 1843.

Contraction hypothesis: H. A. Lorentz, 1895; Fitzgerald, 1895.

The double theta functions: Gopel, 1847; Rosenhain, 1847.

Geometry with axiom contradictory to Euclid's parallel axiom: Lobatchevsky, 1836–40; Bolyai, 1826–33; Gauss, 1829.

The rectification of the semi-cubal parabola: Van Heuraet, 1659; Neil, 1657; Fermat, 1657–59.

The geometric law of duality: Oncelet, 1838; Gergone, 1838.

As examples of simultaneity in other fields we might cite:

Discovery of oxygen: Scheele, 1774; Priestley, 1774.

Liquefaction of oxygen: Cailletet, 1877; Pictet, 1877.

Periodic law: De Chancourtois, 1864; Newlands, 1864; Lothar Meyer, 1864.

Law of periodicity of atomic elements: Lothar Meyer, 1869; Mendeleff, 1869.

Law of conservation of energy: Mayer, 1843; Joule, 1847; Helmholz, 1847; Colding, 1847; Thomson, 1847.

A host of others could be cited. Ogburn's list, cited above, does not pretend to be complete.

[31]Hadamard entitles one chapter of his book "Discovery as a Synthesis."

[32]We use "culture" here in its bacteriological sense: a culture of bacilli growing in a gelatinous medium.

[33]The distinguished anthropologist, A. L. Kroeber, defines geniuses as "the indicators of the realization of coherent patterns of cultural value," *Configurations of Culture Growth,* p. 839 (Berkeley, 1944).

[34]"Mathematical Creation," in *Science and Method,* published in *The Foundations of Science,* p. 387 (The Science Press; New York and Garrison, 1913).

[35]*ibid.,* p. 393.

[36]See, W. Köhler's *The Mentality of Apes* (New York, 1931).

[37]See Leslie A. White, "On the Use of Tools by Primates" (Journ. of Comparative Psychology, Vol. 34, pp. 369–374, Dec. 1942). This essay attempts to show that the human species has a highly developed and progressive material culture while apes do not, although they can use tools with skill and versatility and even invent them, because man, and not apes, can use symbols.

[38]*A Mathematician's Apology*, p. 63.

[39]"Mathematical Proof," p. 4 (Mind, Vol. 38, pp. 1–25, 1929).

[40]*A Mathematician's Apology*, pp. 62–63, 65.

[41]*ibid.*, p. 63.

# PART XXVI

# AMUSEMENTS, PUZZLES, FANCIES

# COMMENTARY ON
# AUGUSTUS DE MORGAN

Augustus De Morgan (1806–1871) was a mathematician of considerable merit, a brilliant and influential teacher, a founder, with George Boole, of symbolic logic as it developed in England, a writer of many books, an indefatigable contributor to encyclopedias, magazines and learned journals. He was an uncompromising advocate of religious liberty and free expression, an insatiable collector of curious lore, anecdotes. quaint and perverse opinions, paradoxes, puzzles, riddles and puns; a bibliomaniac, a wit and polemicist, a detester of hypocrisy and sordid motive, an impolitic, independent, crotchety, overworked, lovable, friendly and contentious Englishman. De Morgan admired Dickens, loathed the country and was "a fair performer on the flute." This summary does him scant justice; he was an original man even among mathematicians.

De Morgan was born in 1806 in Madras Province, India, where his father was employed by the East India Company. He received his early education in English private schools, which he hated. He had lost the use of one eye in infancy: this made him shy and solitary, and exposed him to jolly schoolboy pranks. One of them was to "come up stealthily to his blind side and, holding a sharp-pointed penknife to his cheek, speak to him suddenly by name. De Morgan on turning around received the point of the knife in his face."[1] He managed to catch and thrash the "stout boy of fourteen" who specialized in this sport. He did not then, or at any time thereafter, allow bullies to push him around.

De Morgan made an excellent record at Trinity College, Cambridge. He was recognized as far superior in mathematical ability to any man in his year, but his wide reading and refusal to buckle down to the necessary cramming resulted in his finishing only fourth in the mathematical tripos. This was the first of many disappointments in his career. Because of scruples against signing certain theological articles—he called himself a "Christian unattached"—then required by the University, he was unable to proceed to the M.A. degree and was ineligible for a fellowship. This avenue being closed to him, De Morgan decided to try for the Bar, but a short time after entering Lincoln's Inn he learned that he might have a chance to teach mathematics at the newly formed University of London. With the strong support of the leading Cambridge mathematicians, Peacock and Airy among them, who knew his worth, he was appointed in 1828 the first professor of mathematics of the institution later to be known as University College. In this post, save for an interruption of five years, he served for thirty years.

As a teacher De Morgan was "unrivaled." His lectures were fluent and lucid; unlike so many teachers, he cared that his hearers should be stimulated as well as instructed. He exhibited frequently his "quaint humor" and his "thorough contempt for sham knowledge and low aims in study."[2] Above all, he hated competitive examinations and would not permit this nonsensical practice in his classes. Walter Bagehot and Stanley Jevons were two among the many of his pupils who later gained distinction.

It is impossible in this space even to enumerate De Morgan's writings on mathematics, philosophy and random antiquarian matters. He published first-rate elementary texts on arithmetic, algebra, trigonometry and calculus, and important treatises on the theory of probability and formal logic. In his celebrated *Trigonometry and Double Algebra* and, to a greater extent in his *Formal Logic,* and in several memoirs in the *Cambridge Philosophical Transactions,* he considered the possibilities of establishing a logical calculus and the fundamental problem of expressing thought by means of symbols.[3] "Every science," he said, "that has *thriven* has thriven upon its own symbols: logic, the only science which is admitted to have made no improvements in century after century, is the only one which has *grown no* symbols."[4] This deficiency he set out to remedy. He had a profound appreciation of the close relationship between logic and pure mathematics, and perceived how rich a field of discovery lay in cultivating these disciplines jointly and not separately. While his own achievements in this sphere were not equal to those of George Boole, his studies in logic were of the highest value both in illuminating new areas and in encouraging other workers to press further.[5]

The writings just mentioned are the basis of De Morgan's reputation; yet they represent much the smaller part of his total output. His income as a professor was never large enough to support a wife, five children, and a passion for book collecting—even a modest passion. The necessary supplement he derived from tutoring private pupils, from consulting services as an actuary and from an almost unending stream of articles contributed to biographical dictionaries, historical series, composite works and encyclopaedias. He wrote no less than one-sixth of the 850 articles in the famous Penny Cyclopaedia. His main fields were astronomy, mathematics, physics and biography and his subjects ranged from "abacus" (two articles) to "Thomas Young." From 1831 to 1857 he had one article each year in the *Companions to the British Almanack,* on such topics as chronology, decimal coinage, life insurance, bibliography and the history of science. The selection which follows consists of excerpts from the *Budget of Paradoxes,* a book published in 1872 after De Morgan's death. The *Budget* is a collection of articles, letters and reviews, most of which appeared first in *The Athenaeum.* A paradox, in De Morgan's special sense of the word, was any curious tale about science or scientists that he had come across in his extensive reading, any piece of gossip, choice examples of lunacy, assorted riddles and puns. Many of the articles deal with the attempts of sundry zanies to square the circle, trisect the angle or construct a perpetual motion engine. The *Budget* is a dated book but some of the material is amusing. The excerpts selected are among the better-known historical entries.

De Morgan's unswerving adherence to principle deserves to be remembered. On several critical occasions in his life he courageously renounced self-advantage and chose to follow the thorny road rather than trim his convictions. After thirty years of service at the university he resigned his professorship on an issue of sectarian freedom in which he personally was not involved: the council had refused to appoint a Unitarian minister to the chair of logic and philosophy. "It is unnecessary," he wrote the chairman of the council, "for me to settle when I shall leave the college; for the college has left me." He declined the offer of an honorary degree from the University of Edinburgh, saying "he did not feel like an LL.D."; he refused to let his name be posted for fellowship in the Royal Society because it was "too much open to social influences"—which was certainly true in his time. A sentence in his will perhaps illustrates best the principles by which he lived: "I commend my future with hope and confidence to Almighty God; to God the Father of our Lord Jesus Christ, whom I believe in my heart to be the Son of God but whom I have not confessed with my lips, because in my time such confession has always been the way up in the world."

## ENDNOTES

[1] Sophia Elizabeth De Morgan, *Memoirs of Augustus De Morgan*; London. 1882, p. 5.

[2] *Dictionary of National Biography*; article on De Morgan.

[3] Federigo Enriques, *The Historic Development of Logic*, 1929; pp. 115, 127–128.

[4] *Transactions Cambridge Philosophical Society*, vol. X, 1864, p. 184.

[5] Boole acknowledged the stimulus to his own investigations derived from De Morgan's writings, and the latter unhesitatingly proclaimed that "the most striking results . . . in increasing the power of mathematical language," of binding together the "two great branches of exact science, Mathematics and Logic" were the product of "Dr. Boole's genius." Sophia De Morgan, *op. cit.*, p. 167. Sir William Rowan Hamilton was no less generous in owning his debt to De Morgan whose papers "led and encouraged him (Hamilton) in the working out of the new system of quaternions." *Encyclopaedia Britannica*, Eleventh edition; article on De Morgan by W. Stanley Jevons.

1

*The riddle does not exist. If a question can be put at all, then it can
also be answered.*
> —Ludwig Wittgenstein (*Tractatus
> Logico—Philosophicus*)

*There is a pleasure sure
In being mad which none but madmen know.*
> —John Dryden

*His father's sister had bats in the belfry and was put away.*
> —Eden Phillpotts

*Though this be madness, yet there is method in 't.*
> —Shakespeare

# Assorted Paradoxes

## *By Augustus De Morgan*

### MATHEMATICAL THEOLOGY

Theologiæ Christianæ Principia Mathematica. Auctore Johanne Craig.[1]
London, 1699, 4to.

This is a celebrated speculation, and has been reprinted abroad, and seriously
answered. Craig is known in the early history of fluxions, and was a good
mathematician. He professed to calculate, on the hypothesis that the suspicions
against historical evidence increase with the square of the time, how long it will
take the evidence of Christianity to die out. He finds, by formulae, that had it
been oral only, it would have gone out A.D. 800; but, by aid of the written
evidence, it will last till A.D. 3150. At this period he places the second
coming, which is deferred until the extinction of evidence, on the authority of
the question "When the Son of Man cometh, shall he find faith on the earth?" It
is a pity that Craig's theory was not adopted: it would have spared a hundred
treatises on the end of the world, founded on no better knowledge than his, and
many of them falsified by the event. The most recent (October, 1863) is a tract
in proof of Louis Napoleon being Antichrist, the Beast, the eighth Head, etc.;
and the present dispensation is to close soon after 1864.

In order rightly to judge Craig, who added speculations on the variations of
pleasure and pain treated as functions of time, it is necessary to remember that
in Newton's day the idea of force, as a quantity to be measured, and as

2347

following a law of variation, was very new: so likewise was that of probability, or belief, as an object of measurement. The success of the *Principia* of Newton put it into many heads to speculate about applying notions of quantity to other things not then brought under measurement. Craig imitated Newton's title, and evidently thought he was making a step in advance: but it is not every one who can plough with Samson's heifer.

It is likely enough that Craig took a hint, directly or indirectly, from Mohammedan writers, who make a reply to the argument that the Koran has not the evidence derived from miracles. They say that, as evidence of Christian miracles is daily becoming weaker, a time must at last arrive when it will fail of affording assurance that they were miracles at all: whence would arise the necessity of another prophet and other miracles. Lee, the Cambridge Orientalist, from whom the above words are taken, almost certainly never heard of Craig or his theory. This is Samuel Lee (1783–1852), the young prodigy in languages. He was apprenticed to a carpenter at twelve and learned Greek while working at the trade. Before he was twenty-five he knew Hebrew, Chaldee, Syriac, Samaritan, Persian, and Hindustani. He later became Regius professor of Hebrew at Cambridge.

## ON CURIOSITIES OF $\pi$

The celebrated interminable fraction 3.14159. . ., which the mathematician calls $\pi$, is the ratio of the circumference to the diameter. But it is thousands of things besides. It is constantly turning up in mathematics: and if arithmetic and algebra had been studied without geometry, $\pi$ must have come in somehow, though at what stage or under what name must have depended upon the casualties of algebraical invention. This will readily be seen when it is stated that $\pi$ is nothing but four times the series

$$1 - \tfrac{1}{3} + \tfrac{1}{5} - \tfrac{1}{7} + \tfrac{1}{9} - \tfrac{1}{11} + \ldots$$

*ad infinitum.*[2] It would be wonderful if so simple a series had but one kind of occurrence. As it is, our trigonometry being founded on the circle, $\pi$ first appears as the ratio stated. If, for instance, a deep study of probable fluctuation from average had preceded, $\pi$ might have emerged as a number perfectly indispensable in such problems as: What is the chance of the number of aces lying between a million $+ x$ and a million $-x$, when six million of throws are made with a die? I have not gone into any detail of all those cases in which the paradoxer finds out, by his unassisted acumen, that results of mathematical investigation *cannot be:* in fact, this discovery is only an accompaniment, though a necessary one, of his paradoxical statement of that which *must be.* Logicians are beginning to see that the notion of *horse* is inseparably connected with that of *non-horse:* that the first without the second would be no notion at all. And it is clear that the positive affirmation of that which contradicts mathematical demonstration cannot but be accompanied by a declaration, mostly overtly made, that demonstration is false. If the mathematician were interested in punishing this indiscretion, he could make his denier ridiculous by inventing asserted results which would completely take him in.

More than thirty years ago I had a friend, now long gone, who was a mathematician, but not of the higher branches: he was, *inter alia,* thoroughly up in all that relates to mortality, life assurance, &c. One day, explaining to him how it should be ascertained what the chance is of the survivors of a large number of persons now alive lying between given limits of number at the end of a certain time, I came, of course upon the introduction of $\pi$, which I could only describe as the ratio of the circumference of a circle to its diameter. "Oh, my dear friend! that must be a delusion; what can the circle have to do with the numbers alive at the end of a given time?"—"I cannot demonstrate it to you; but it is demonstrated."—"Oh! stuff! I think you can prove anything with your differential calculus: figment, depend upon it." I said no more; but, a few days afterwards, I went to him and very gravely told him that I had discovered the law of human mortality in the Carlisle Table, of which he thought very highly. I told him that the law was involved in this circumstance. Take the table of expectation of life, choose any age, take its expectation and make the nearest integer a new age, do the same with that, and so on; begin at what age you like, you are sure to end at the place where the age past is equal, or most nearly equal, to the expectation to come. "You don't mean that this always happens?"—"Try it." He did try, again and again; and found it as I said. "This is, indeed, a curious thing; this *is* a discovery." I might have sent him about trumpeting the law of life: but I contented myself with informing him that the same thing would happen with any table whatsoever in which the first column goes up and the second goes down; and that if a proficient in the higher mathematics chose to palm a figment upon him, he could do without the circle: *à corsaire, corsaire et demi,*[3] the French proverb says.

## THE OLD MATHEMATICAL SOCIETY

Among the most remarkable proofs of the diffusion of speculation was the Mathematical Society, which flourished from 1717 to 1845. Its habitat was Spitalfields, and I think most of its existence was passed in Crispin Street. It was originally a plain society, belonging to the studious artisan. The members met for discussion once a week; and I believe I am correct in saying that each man had his pipe, his pot, and his problem. One of their old rules was that, "If any member shall so far forget himself and the respect due to the Society as in the warmth of debate to threaten or offer personal violence to any other member, he shall be liable to immediate expulsion, or to pay such fine as the majority of the members present shall decide." But their great rule, printed large on the back of the title page of their last book of regulations, was "By the constitution of the Society, it is the duty of every member, if he be asked any mathematical or philosophical question by another member, to instruct him in the plainest and easiest manner he is able." We shall presently see that, in old time, the rule had a more homely form.

I have been told that De Moivre[4] was a member of this Society. This I cannot verify: circumstances render it unlikely; even though the French refugees clustered in Spitalfields; many of them were of the Society, which there is some reason to think was founded by them. But Dolland,[5] Thomas Simpson,[6] Saunderson,[7] Crossley, and others of known name, were certainly members. The Society gradually declined, and in 1845 was reduced to nineteen members. An arrangement was made by which sixteen of these members, who were not already in the Astronomical Society became Fellows without contribution, all the books and other property of the old Society being transferred to the new one. I was one of the committee which made the preliminary inquiries, and the reason of the decline was soon manifest. The only question which could arise was whether the members of the society of working men—for this repute still continued—were of that class of educated men who could associate with the Fellows of the Astronomical Society on terms agreeable to all parties. We found that the artisan element had been extinct for many years; there was not a man but might, as to education, manners, and position, have become a Fellow in the usual way. The fact was that life in Spitalfields had become harder: and the weaver could only live from hand to mouth, and not up to the brain. The material of the old Society no longer existed.

In 1798, experimental lectures were given, a small charge for admission being taken at the door: by this hangs a tale—and a song. Many years ago, I found among papers of a deceased friend, who certainly never had anything to do with the Society, and who passed all his life far from London, a song, headed "Song sung by the Mathematical Society in London, at a dinner given Mr. Fletcher, a solicitor, who had defended the Society gratis." Mr. Williams, the Assistant Secretary of the Astronomical Society, formerly Secretary of the Mathematical Society, remembered that the Society had had a solicitor named Fletcher among the members. Some years elapsed before it struck me that my old friend Benjamin Gompertz,[8] who had long been a member, might have some recollection of the matter. The following is an extract of a letter from him (July 9, 1861):

"As to the Mathematical Society, of which I was a member when only 18 years of age, [Mr. G. was born in 1779], having been, contrary to the rules, elected under the age of 21. How I came to be a member of that Society—and continued so until it joined the Astronomical Society, and was then the President—was: I happened to pass a bookseller's small shop, of second-hand books, kept by a poor taylor, but a good mathematician, John Griffiths. I was very pleased to meet a mathematician, and I asked him if he would give me some lessons; and his reply was that I was more capable to teach him, but he belonged to a society of mathematicians, and he would introduce me. I accepted the offer, and I was elected, and had many scholars then to teach, as one of the rules was, if a member asked for information, and applied to any one who could give it, he was obliged to give it, or fine one penny. Though I might say much with respect to the Society which would be interesting, I will for the present reply only to your question. I well knew Mr. Fletcher, who was a very clever and very scientific person. He did, as solicitor, defend an action brought by an informer against the Society—I think for 5,000*l.*—for giving lectures to the public in philosophical subjects [i.e., for unlicensed public exhibition with money taken at the doors]. I think the price for admission was one shilling, and we used to have, if I rightly recollect, from two to three hundred visitors. Mr. Fletcher was successful in his defence, and we got out of our trouble. There was a collection made to reward his services, but he did not accept of any reward: and I think we gave him a dinner, as you state, and enjoyed ourselves; no doubt with astronomical songs and other songs; but my recollection does not enable me to say if the astronomical song was a drinking song. I think the anxiety caused by that action was the cause of some of the members' death. [They had, no doubt, broken the law in ignorance; and by the sum named, the informer must have been present, and sued for a penalty on every shilling he could prove to have been taken]."

I by no means guarantee that the whole song I proceed to give is what was sung at the dinner: I suspect, by the completeness of the chain, that augmentations have been made. My deceased friend was just the man to add some verses, or the addition may have been made before it came into his hands, or since his decease, for the scraps containing the verses passed through several hands before they came into mine. We may, however, be pretty sure that the original is substantially contained in what is given, and that the character is therefore preserved. I have had myself to repair damages every now and then, in the way of conjectural restoration of defects caused by ill-usage.

### THE ASTRONOMER'S DRINKING SONG

"Whoe'er would search the starry sky,
    Its secrets to divine, sir,
Should take his glass—I mean, should try
    A glass or two of wine, sir!
True virtue lies in golden mean,
    And man must wet his clay, sir;
Join these two maxims, and 'tis seen
    He should drink his bottle a day, sir!

"Old Archimedes, reverend sage!
    By trump of fame renowned, sir,
Deep problems solved in every page,
    And the sphere's curved surface found, sir:
Himself he would have far outshone,
    And borne a wider sway, sir,

Had he our modern secret known,
    And drank a bottle a day, sir!

"When Ptolemy, now long ago,
    Believed the earth stood still, sir,
He never would have blundered so,
    Had he but drunk his fill, sir:
He'd then have felt it circulate,
    And would have learnt to say, sir,
The true way to investigate
    Is to drink your bottle a day, sir!

"Copernicus, that learned wight,
    The glory of his nation,
With draughts of wine refreshed his sight,
    And saw the earth's rotation;
Each planet then its orb described,
    The moon got under way, sir;
These truths from nature he imbibed
    For he drank his bottle a day, sir!

"The noble Tycho placed the stars,
    Each in its due location;
He lost his nose[9] by spite of Mars,
    But that was no privation:
Had he but lost his mouth, I grant
    He would have felt dismay, sir,
Bless you! *he* knew what he should want
    To drink his bottle a day, sir!

"Cold water makes no lucky hits;
    On mysteries the head runs:
Small drink let Kepler time his wits
    On the regular polyhedrons:
He took to wine, and it changed the chime,
    His genius swept away, sir,
Through area varying as the time
    At the rate of a bottle a day, sir!

"Poor Galileo, forced to rat
    Before the Inquisition,
*E pur si muove* was the pat
    He gave them in addition:
He meant, whate'er you think you prove,
    The earth must go its way, sirs;
Spite of your teeth I'll make it move,
    For I'll drink my bottle a day, sirs!

"Great Newton, who was never beat
    Whatever fools may think, sir;

>Though sometimes he forgot to eat,
>   He never forgot to drink, sir:
>Descartes[10] took nought but lemonade,
>   To conquer him was play, sir;
>The first advance that Newton made
>   Was to drink his bottle a day, sir!
>
>"D'Alembert, Euler, and Clairaut,
>   Though they increased our store, sir,
>Much further had been seen to go
>   Had they tippled a little more, sir!
>Lagrange gets mellow with Laplace,
>   And both are wont to say, sir,
>The *philosophe* who's not an ass
>   Will drink his bottle a day, sir!
>
>"Astronomers! What can avail
>   Those who calumniate us;
>Experiment can never fail
>   With such an apparatus:
>Let him who'd have his merits known
>   Remember what I say, sir;
>Fair science shines on him alone
>   Who drinks his bottle a day, sir!
>
>"How light we reck of those who mock
>   By this we'll make to appear, sir,
>We'll dine by the sidereal clock
>   For one more bottle a year, sir:
>But choose which pendulum you will,
>   You'll never make your way, sir,
>Unless you drink—and drink your fill,—
>   At least a bottle a day, sir!"

Old times are changed, old manners gone!

There is a new Mathematical Society, and I am, at this present writing (1866), its first President. We are very high in the newest developments, and bid fair to take a place among the scientific establishments. Benjamin Gompertz, who was President of the old Society when it expired, was the link between the old and new body: he was a member of *ours* at his death. But not a drop of liquor is seen at our meetings, except a decanter of water: all our heavy is a fermentation of symbols; and we do not draw it mild. There is no penny fine for reticence or occult science; and as to a song! not the ghost of a chance.

## ON SOME PHILOSOPHICAL ATHEISTS

With the general run of the philosophical atheists of the last century the notion of a God was an hypothesis. There was left an admitted possibility that the vague somewhat which went by more names than one, might be personal, intelligent, and superintendent. In the works of Laplace, who is sometimes called an atheist from his writings, there is nothing from which such an inference can be drawn: unless indeed a Reverend Fellow of the Royal Society may be held to be the fool who said in his heart, etc., etc., if his contributions to the *Philosophical Transactions* go no higher than *nature*. The following anecdote is well known in Paris, but has never been printed entire.

Laplace once went in form to present some edition of his "Système du Monde" to the First Consul, or Emperor. Napoleon, whom some wags had told that this book contained no mention of the name of God, and who was fond of putting embarrassing questions, received it with—"M. Laplace, they tell me you have written this large book on the system of the universe, and have never even mentioned its Creator." Laplace, who, though the most supple of politicians, was as stiff as a martyr on every point of his philosophy or religion (e.g., even under Charles X he never concealed his dislike of the priests), drew himself up and answered bluntly, "Je n'avais pas besoin de cette hypothèse-là." Napoleon, greatly amused, told this reply to Lagrange, who exclaimed, "Ah! c'est une belle hypothèse; ça explique beaucoup de choses."

It is commonly said that the last words of Laplace were, "Ce que nous connaissons est peu de chose; ce que nous ignorons est immense." This looks like a parody on Newton's pebbles: the following is the true account; it comes to me through one remove from Poisson. After the publication (in 1825) of the fifth volume of the *Mécanique Céleste,* Laplace became gradually weaker, and with it musing and abstracted. He thought much on the great problems of existence and often muttered to himself, *Qu'est ce que c'est que tout cela!* After many alternations, he appeared at last so permanently prostrated that his family applied to his favorite pupil, M. Poisson, to try to get a word from him. Poisson paid a visit, and after a few words of salutation, said, "J'ai une bonne nouvelle à vous annoncer: on a reçu au Bureau des Longitudes une lettre d'Allemagne annonçant que M. Bessel a vérifié par l'observation vos découvertes théoriques sur les satellites de Jupiter."[11] Laplace opened his eyes and answered with deep gravity, *"L'homme ne poursuit que des chimères."*[12] He never spoke again. His death took place March 5, 1827.

The language used by the two great geometers illustrates what I have said: a supreme and guiding intelligence—apart from a blind rule called *nature of things*—was an *hypothesis*. The absolute denial of such a ruling power was not in the plan of the higher philosophers: it was left for the smaller fry. A round assertion of the non-existence of anything which stands in the way is the refuge of a certain class of minds: but it succeeds only with things subjective; the objective offers resistance. A philosopher of the appropriative class tried it upon the constable who appropriated *him*: I deny your existence, said he; Come along all the same, said the unpsychological policeman.

Euler was a believer in God, downright and straightforward. The following story is told by Thiébault, in his *Souvenirs de vingt ans de séjour à Berlin,* published in his old age, about 1804. This volume was fully received as trustworthy; and Marshall Mollendorff told the Duc de Bassano in 1807 that it was the most veracious of books written by the most honest of men. Thiébault says that he has no personal knowledge of the truth of the story, but that it was believed throughout the whole of the north of Europe. Diderot paid a visit to the Russian Court at the invitation of the Empress. He conversed very freely, and gave the younger members of the Court circle a good deal of lively atheism. The Empress was much amused, but some of her councillors suggested that it might be desirable to check these expositions of doctrine. The Empress did not like to put a direct muzzle on her guest's tongue, so the following plot was contrived. Diderot was informed that a learned mathematician was in possession of an algebraical demonstration of the existence of God, and would give it him before all the Court, if he desired to hear it. Diderot gladly consented: though the name of the mathematician is not given, it was Euler. He advanced towards Diderot, and said gravely, and in a tone of perfect conviction: *Monsieur, $(a + b^n)/n = x$, donc Dieu existe; répondez!* Diderot, to whom algebra was Hebrew, was embarrassed and disconcerted; while peals of laughter rose on all sides. He asked permission to return to France at once, which was granted.

## CELEBRATED APPROXIMATIONS OF $\pi$

The following is an extract from the *English Cyclopaedia*, Art. TABLES:

"1853. William Shanks, *Contributions to Mathematics, comprising chiefly the Rectification of the Circle to 607 Places of Tables*, London, 1853. (QUADRATURE OF THE CIRCLE.) Here is a *table*, because it tabulates the results of the subordinate steps of this enormous calculation as far as 527 decimals: the remainder being added as results only during the printing. For instance, one step is the calculation of the reciprocal of $601.5^{601}$; and the result is given. The number of pages required to describe these results is 87. Mr. Shanks has also thrown off, as chips or splinters, the values of the base of Napier's logarithms, and of its logarithms of 2, 3, 5, 10, to 137 decimals; and the value of the modulus .4342. . . . to 136 decimals; with the 13th, 25th, 37th. . . . up to the 721st powers of 2. These tremendous stretches of calculation—at least we so call them in our day —are useful in several respects; they prove more than the capacity of this or that computer for labor and accuracy; they show that there is in the community an increase of skill and courage. We say in the community: we fully believe that the unequalled turnip which every now and then appears in the newspapers is a sufficient presumption that the average turnip is growing bigger, and the whole crop heavier. All who know the history of the quadrature are aware that the several increases of numbers of decimals to which $\pi$ has been carried have been indications of a general increase in the power to calculate, and in courage to face the labor. Here is a comparison of two different times. In the day of Cocker, the pupil was directed to perform a common subtraction with a voice-accompaniment of this kind:

'7 from 4 I cannot, but add 10, 7 from 14 remains 7, set down 7 and carry 1; 8 and 1 which 1 carry is 9, 9 from 2 I cannot, etc.' We have before us the announcement of the following *table,* undated, as open to inspection at the Crystal Palace, Sydenham, in two diagrams of 7 ft. 2 in., by 6 ft. 6 in.: 'The figure 9 involved into the 912th power, and antecedent powers, or involutions, containing upwards of 73,000 figures. Also, the proofs of the above, containing upwards of 146,000 figures. By Samuel Fancourt, of Mincing Lane, London, and completed by him in the year 1837, at the age of sixteen. N.B. The whole operation performed by simple arithmetic.' The young operator calculated by successive squaring the 2d, 4th, 8th, etc., powers up to the 512th, with proof by division. But 511 multiplications by 9, in the short (or 10 − 1) way, would have been much easier. The 2d, 32d, 64th, 128th, 256th, and 512th powers are given at the back of the announcement. The powers of 2 have been calculated for many purposes. In Vol. II of his *Magia Universalis Naturae et Artis,* Herbipoli, 1658, 4to, the Jesuit Gaspar Schott having discovered, on some grounds of theological magic, that the degrees of grace of the Virgin Mary were in number the 256th power of 2, calculated that number. Whether or no his number correctly represented the result he announced, he certainly calculated it rightly, as we find by comparison with Mr. Shanks."

There is a point about Mr. Shanks's 608 figures of the value of $\pi$ which attracts attention, perhaps without deserving it. It might be expected that, in so many figures, the nine digits and the cipher would occur each about the same number of times; that is, each about 61 times. But the fact stands thus: 3 occurs 68 times; 9 and 2 occur 67 times each; 4 occurs 64 times; 1 and 6 occur 62 times each; 0 occurs 60 times; 8 occurs 58 times; 5 occurs 56 times; and 7 occurs only 44 times. Now, if all the digits were equally likely, and 608 drawings were made, it is 45 to 1 against the number of sevens being as distant from the probable average (say 61) as 44 on one side or 78 on the other. There must be some reason why the number 7 is thus deprived of its fair share in the structure. Here is a field of speculation in which two branches of inquirers might unite. There is but one number which is treated with an unfairness which is incredible as an accident; and that number is the mystic number *seven*! If the cyclometers and the apocalyptics would lay their heads together until they come to a unanimous verdict on this phenomenon, and would publish nothing until they are of one mind, they would earn the gratitude of their race.—I was wrong: it is the Pyramid-speculator who should have been appealed to. A correspondent of my friend Prof. Piazzi Smyth notices that 3 is the number of most frequency, and that 3½ is the nearest approximation to it in simple digits. Professor Smyth himself, whose word on Egypt is paradox of a very high order, backed by a great quantity of useful labor, the results of which will be made available by those who do not receive the paradoxes, is inclined to see confirmation for some of his theory in these phenomena.

# HORNER'S METHOD

Horner's method[13] begins to be introduced at Cambridge: it was published in 1820. I remember that when I first went to Cambridge (in 1823) I heard my tutor say, in conversation, there is no doubt that the true method of solving equations is the one which was published a few years ago in the *Philosophical Transactions*. I wondered it was not taught, but presumed that it belonged to the higher mathematics. This Horner himself had in his head: and in a sense it is true; for all lower branches belong to the higher: but he would have stared to have been told that he, Horner, was without a European predecessor, and in the distinctive part of his discovery was heir-at-law to the nameless Brahmin—Tartar—Antenoachian—what you please—who concocted the extraction of the square root.

It was somewhat more than twenty years after I had thus heard a Cambridge tutor show sense of the true place of Horner's method, that a pupil of mine who had passed on to Cambridge was desired by his college tutor to solve a certain cubic equation—one of an integer root of two figures. In a minute the work and answer were presented, by Horner's method. "How!" said the tutor, "this can't be, you know." "There is the answer, Sir!" said my pupil, greatly amused, for my pupils learnt, not only Horner's method, but the estimation it held at Cambridge. "Yes!" said the tutor, "there is the answer certainly; but it *stands to reason* that a cubic equation cannot be solved in this space." He then sat down, went through a process about ten times as long, and then said with triumph: "There! that is the way to solve a cubic equation!"

I think the tutor in this case was never matched, except by the country organist. A master of the instrument went into the organ-loft during service, and asked the organist to let him *play the congregation out*; consent was given. The stranger, when the time came, began a voluntary which made the people open their ears, and wonder who had got into the loft: they kept their places to enjoy the treat. When the organist saw this, he pushed the interloper off the stool, with "You'll never play 'em out this side Christmas." He then began his own drone, and the congregation began to move quietly away. "There," said he, "that's the way to play 'em out!"

# BUFFON'S NEEDLE PROBLEM

The paradoxes of what is called chance, or hazard, might themselves make a small volume. All the world understands that there is a long run, a general average; but great part of the world is surprised that this general average should be computed and predicted. There are many remarkable cases of verification; and one of them relates to the quadrature of the circle. I give some account of this and another. Throw a penny time after time until *head* arrives, which it will do before long: let this be called a *set*. Accordingly, H is the smallest set, TH the next smallest, then TTH, &c. For abbreviation, let a set in which seven *tails* occur before *head* turns up be $T^7H$. In an immense number of trials of sets, about half will be H; about a quarter TH; about an eighth, $T^2H$. Buffon[14] tried 2,048 sets; and several have followed him. It will tend to illustrate the principle if I give all the results; namely, that many trials will with moral certainty show an approach—and the greater the greater the number of trials—to that average which sober reasoning predicts. In the first column is the most likely number of the theory: the next column gives Buffon's result; the three next are results obtained from trial by correspondents of mine. In each case the number of trials is 2,048.

| | | | | | |
|---|---|---|---|---|---|
| H | 1,024 | 1,061 | 1,048 | 1,017 | 1,039 |
| TH | 512 | 494 | 507 | 547 | 480 |
| $T^2H$ | 256 | 232 | 248 | 235 | 267 |
| $T^3H$ | 128 | 137 | 99 | 118 | 126 |
| $T^4H$ | 64 | 56 | 71 | 72 | 67 |
| $T^5H$ | 32 | 29 | 38 | 32 | 33 |
| $T^6H$ | 16 | 25 | 17 | 10 | 19 |
| $T^7H$ | 8 | 8 | 9 | 9 | 10 |
| $T^8H$ | 4 | 6 | 5 | 3 | 3 |
| $T^9H$ | 2 | | 3 | 2 | 4 |
| $T^{10}H$ | 1 | | 1 | 1 | |
| $T^{11}H$ | | | 0 | 1 | |
| $T^{12}H$ | | | 0 | 0 | |
| $T^{13}H$ | 1 | | 1 | 0 | |
| $T^{14}H$ | | | 0 | 0 | |
| $T^{15}H$ | | | 1 | 1 | |
| &c. | | | 0 | 0 | |
| | 2,048 | 2,048 | 2,048 | 2,048 | 2,048 |

In very many trials, then, we may depend upon something like the predicted average. Conversely, from many trials we may form a guess at what the average will be. Thus, in Buffon's experiment the 2,048 first throws of the sets gave *head* in 1,061 cases: we have a right to infer that in the long run something like 1,061 out of 2,048 is the proportion of heads, even before we know the reasons for the equality of chance, which tell us that 1,024 out of 2,048 is the real truth. I now come to the way in which such considerations have led to a mode in which mere pitch-and-toss has given a more accurate approach to the quadrature of the circle than has been reached by some of my paradoxers. The method is as follows: Suppose a planked floor of the usual kind, with thin visible seams between the planks. Let there be a thin straight rod, or wire, not so long as the breadth of the plank. This rod, being tossed up at hazard, will either fall quite clear of the seams, or will lay across one seam. Now Buffon, and after him Laplace, proved the following: That in the long run the fraction of the whole number of trials in which a seam is intersected will be the fraction which twice the length of the rod is of the circumference of the circle having the breadth of a plank for its diameter. In 1855 Mr. *Ambrose* Smith, of Aberdeen, made 3,204 trials with a rod three-fifths of the distance between the planks: there were 1,213 clear intersections, and 11 contacts on which it was difficult to decide. Divide these contacts equally, and we have 1,218½ to 3,204 for the ratio of 6 to 5π, presuming that the greatness of the number of trials gives something near to the final average, or result in the long run: this gives π = 3.1553. If all the 11 contacts had been treated as intersections, the result would have been π = 3.1412, exceedingly near. A pupil of mine made 600 trials with a rod of the length between the seams, and got π = 3.137.

This method will hardly be believed until it has been repeated so often that "there never could have been any doubt about it."

The first experiment strongly illustrates a truth of the theory, well confirmed by practice: whatever can happen will happen if we make trials enough. Who would undertake to throw tail eight times running? Nevertheless, in the 8,192 sets tail 8 times running occurred 17 times; 9 times running, 9 times; 10 times running, twice; 11 times and 13 times, each once; and 15 times twice.

## ENDNOTES

[1]John Craig (died in 1731) was a Scotchman, but most of his life was spent at Cambridge reading and writing on mathematics. He endeavored to introduce the Leibnitz differential calculus into England. His mathematical works include the *Methodus Figurarum . . . Quadraturas determinandi* (1685), *Tractatus . . . de Figurarum Curvilinearum Quadraturis et locis Geometricis* (1693), and *De Calculo Fleuntium libri duo* (1718). [All the notes in this selection are from the David Eugene Smith edition of the *Budget of Paradoxes*, Chicago, 1928.—ED.]

[2]There are many similar series and products. Among the more interesting are the following:

$$\frac{\pi}{2} = \frac{2 \cdot 2 \cdot 4 \cdot 4 \cdot 6 \cdot 6 \cdot 8 \dots}{1 \cdot 3 \cdot 3 \cdot 5 \cdot 5 \cdot 7 \cdot 7 \dots} ,$$

$$\frac{\pi - 3}{4} = \frac{1}{2 \cdot 3 \cdot 4} - \frac{1}{4 \cdot 5 \cdot 6} + \frac{1}{6 \cdot 7 \cdot 8} - \dots ,$$

$$\frac{\pi}{6} = \sqrt{\frac{1}{3}} \cdot \left(1 - \frac{1}{3 \cdot 3} + \frac{1}{3^2 \cdot 5} - \frac{1}{3^3 \cdot 7} + \frac{1}{3^4 \cdot 9} - \dots\right),$$

$$\frac{\pi}{4} = 4\left(\frac{1}{5} - \frac{1}{3 \cdot 5^3} + \frac{1}{5 \cdot 5^5} - \frac{1}{7 \cdot 5^7} + \dots\right) - \left(\frac{1}{239} - \frac{1}{3.239^3} + \frac{1}{5.239^5} - \dots\right).$$

[3]"To a privateer, a privateer and a half."

[4]Abraham de Moivre (1667–1754), French refugee in London, poor, studying under difficulties, was a man with tastes in some respects like those of De Morgan. For one thing, he was a lover of books, and he had a good deal of interest in the theory of probabilities to which De Morgan also gave much thought. His introduction of imaginary quantities into trigonometry was an event of importance in the history of mathematics, and the theorem that bears his name, $(\cos \phi + i \sin \phi)^n = \cos n\phi + i \sin n\phi$, is one of the most important ones in all analysis.

[5]John Dolland (1706–1761), the silk weaver who became the greatest maker of optical instruments in his time.

[6]Thomas Simpson (1710–1761), also a weaver, taking his leisure from his loom at Spitalfields to teach mathematics. His *New Treatise on Fluxions* (1737) was written only two years after he began working in London, and six years later he was appointed professor of mathematics at Woolwich. He wrote many works on mathematics and Simpson's Formulas for computing trigonometric tables are still given in the text-books.

[7]Nicholas Saunderson (1682–1739), the blind mathematician. He lost his eyesight through smallpox when only a year old. At the age of 25 he began lecturing at Cambridge on the principles of the Newtonian philosophy. His *Algebra*, in two large volumes, was long the standard treatise on the subject.

[8]Benjamin Gompertz (1779–1865) was debarred as a Jew from a university education. He studied mathematics privately and became president of the Mathematical Society. De Morgan knew him professionally through the fact that he was prominent in actuarial work.

[9]He lost it in a duel, with Manderupius Pasbergius. A contemporary, T. B. Laurus, insinuates that they fought to settle which was the best mathematician! This seems odd, but it must be remembered they fought in the dark, *"in tenebris densis"*; and it is a nice problem to shave off a nose in the dark, without any other harm.—A. De M.

[10]As great a lie as ever was told: but in 1800 a compliment to Newton without a fling at Descartes would have been held a lopsided structure.—A. De M.

[11]"I have some good news to tell you: at the Bureau of Longitudes they have just received a letter from Germany announcing that M. Bessel has verified by observation your theoretical discoveries on the satellites of Jupiter."

[12]"Man follows only phantoms."

[13]A method for approximating the real roots of an algebraic equation. The inventor was W. G. Horner (1773–1827), but the same numerical technique may, it is said, have been known to the Chinese in the 13th century.—ED.

[14]Georges Louis Leclerc Buffon (1707–1788), the well-known biologist. He also experimented with burning mirrors, his results appearing in his *Invention des miroirs ardens pour brûler à une grande distance* (1747). The reference here may be to his *Résolution des problèmes qui regardent le jeu du franc carreau* (1733). The prominence of his *Histoire naturelle* (36 volumes, 1749–1788) has overshadowed the credit due to him for his translation of Newton's work on Fluxions.

# COMMENTARY ON
# A ROMANCE OF MANY
# DIMENSIONS

A bout sixty years ago the Rev. Edwin Abbott Abbott, M.A., D.D., headmaster of the City of London School, published a small book of mathematics fiction entitled *Flatland*. This tale was as much off Abbott's beat as *Alice* was off the beat of the Rev. Charles Lutwidge Dodgson. Abbott was reputed a classics scholar; among his writings, which were well received, were *Through Nature to Christ, The Anglican Career of Cardinal Newman* and a less edifying but undoubtedly more profitable item called *How to Tell the Parts of Speech*. His published works number more than forty, but *Flatland* is, I dare say, his only hedge against oblivion. And even there opinions differ.

*Flatland* carries the subtitle "A Romance of Many Dimensions," which is a fair description. It deals with a world of two dimensions, a plane, inhabited by intelligent beings "who have no faculties by which they can become conscious of anything outside their space and no means of moving off the surface on which they live." Flatlanders are small plane figures, the shape of each person depending on his social status. Women, being at the bottom of the hierarchy, are straight lines; soldiers and the "lowest class of workmen" are triangles; the middle class consists of equilateral triangles; professional men and gentlemen are squares—and so on up the polygonal ladder, until one arrives at the priestly order, the members of which are so many-sided, and the sides so small that the figures cannot be distinguished from circles. The story is told in the first person by "A Square"—Dr. Abbott, I presume—who has the misfortune one day to be descended upon by a sphere, a visitor from the third dimension. In Flatland, of course, the sphere can be seen only as a circle, first increasing in size (from a point) and then decreasing and finally vanishing as the sphere passes through the plane. The sphere makes a number of descents and stays long enough to describe to "A Square" the wonders of Spaceland and to make him realize the wretchedness of being confined to the plane. At last the stranger takes the Flatlander on a voyage into three-dimensional space. When he returns he is eager to instruct others in the newly revealed theory of three dimensions, but is promptly denounced by the priests as a heretic, sentenced to "perpetual imprisonment" and cast into jail. There, fortunately, the story ends.

On its first appearance *Flatland* received what is known as a mixed press. The dust jacket of my copy (a 1941 reprint) records opinions that the book is "desperately facetious," "mortally tedious," "prolix," a "soporific"; also that it is "clever," "fascinating," "mind broadening," and worthy of a place beside

Gulliver. All the reviewers were right, I think, except the extremists: the Rev. Abbott's whimsey is not "meaningless," but neither does it make him a peer of the Rev. Swift. *Flatland* is too long, most of its jokes are not funny, and its didacticism is awful. Yet it is based on an original idea, is not without charm and suggests certain remarkably prophetic analogies applicable to relativity theory.[1] The material I have selected gives a taste of the whole; the book is still in print if you care to learn more.

## ENDNOTE

[1]An anonymous letter published in *Nature* (the famous British scientific journal) on February 12, 1920, entitled "Euclid, Newton and Einstein," calls attention to the prophetic nature of *Flatland*. I quote a few lines: "[Dr. Abbott] asks the reader, who has consciousness of the third dimension, to imagine a sphere descending upon the plane of Flatland and passing through it. How will the inhabitants regard this phenomenon?. . . . Their experience will be that of a circular obstacle gradually expanding or growing, and then contracting, and they will attribute to *growth in time* what the external observer in three dimensions assigns to motion in the third dimension. Transfer this analogy to a movement of the fourth dimension through three-dimensional space. Assume the past and future of the universe to be all depicted in four-dimensional space and visible to any being who has consciousness of the fourth dimension. If there is motion of our three-dimensional space relative to the fourth dimension, all the changes we experience and assign to the flow of time will be due simply to this movement, the whole of the future as well as the past always existing in the fourth dimension." (See the introduction to the 1941 edition of *Flatland* by William Garnett.)

# 2

*Imagination is a sort of faint perception.*
—ARISTOTLE

*Where we see the fancy outwork nature.*
—SHAKESPEARE (*Antony and Cleopatra*)

*So full of shapes is fancy,*
*That it alone is high fantastical.*
—SHAKESPEARE (*Twelfth Night*)

*And isn't your life extremely flat*
*With nothing whatever to grumble at!*
—W. S. GILBERT (*Princess Ida*)

# FLATLAND

## By Edwin A. Abbott

### OF THE NATURE OF FLATLAND

I call our world Flatland, not because we call it so, but to make its nature clearer to you, my happy readers, who are privileged to live in Space.

Imagine a vast sheet of paper on which straight Lines, Triangles, Squares, Pentagons, Hexagons, and other figures, instead of remaining fixed in their places, move freely about, on or in the surface, but without the power of rising above or sinking below it, very much like shadows—only hard and with luminous edges—and you will then have a pretty correct notion of my country and countrymen. Alas! a few years ago, I should have said "my universe"; but now my mind has been opened to higher views of things.

In such a country, you will perceive at once that it is impossible that there should be anything of what you call a "solid" kind; but I dare say you will suppose that we could at least distinguish by sight the Triangles, Squares, and other figures moving about as I have described them. On the contrary, we could see nothing of the kind, not at least so as to distinguish one figure from another. Nothing was visible, nor could be visible, to us, except straight Lines; and the necessity of this I will speedily demonstrate.

Place a penny on the middle of one of your tables in Space; and leaning over it, look down upon it. It will appear a circle.

But now, drawing back to the edge of the table, gradually lower your eye (thus bringing yourself more and more into the condition of the inhabitants of Flatland), and you will find the penny becoming more and more oval to your view; and at last when you have placed your eye exactly on the edge of the table (so that you are, as it were, actually a Flatland citizen) the penny will then have ceased to appear oval at all, and will have become, so far as you can see, a straight line.

The same thing would happen if you were to treat in the same way a Triangle, or Square, or any other figure cut out of pasteboard. As soon as you look at it with your eye on the edge of the table, you will find that it ceases to appear to you a figure, and that it becomes in appearance a straight line. Take for example an equilateral Triangle—who represents with us a Tradesman of the respectable class. Figure 1 represents the Tradesman as you would see him while you were bending over him from above; Figures 2 and 3 represent the Tradesman, as you would see him if your eye were close to the level, or all but on the level of the table; and if your eye were quite on the level of the table (and that is how we see him in Flatland) you would see nothing but a straight line.

FIGURE 1            FIGURE 2            FIGURE 3

When I was in Spaceland I heard that your sailors have very similar experiences while they traverse your seas and discern some distant island or coast lying on the horizon. The far-off land may have bays, forelands, angles in and out to any number and extent; yet at a distance you see none of these (unless indeed your sun shines bright upon them revealing the projections and retirements by means of light and shade), nothing but a gray unbroken line upon the water.

Well, that is just what we see when one of our triangular or other acquaintances comes towards us in Flatland. As there is neither sun with us, nor any light of such a kind as to make shadows, we have none of the helps to the sight that you have in Spaceland. If our friend comes close to us we see his line becomes larger; if he leaves us it becomes smaller: but still he looks like a straight line; be he a Triangle, Square, Pentagon, Hexagon, Circle, what you will—a straight Line he looks and nothing else.

You may perhaps ask how under these disadvantageous circumstances we are able to distinguish our friends from one another: but the answer to this very natural question will be more fitly and easily given when I come to describe the inhabitants of Flatland. For the present let me defer this subject, and say a word or two about the climate and houses in our country.

## OF THE CLIMATE AND HOUSES IN FLATLAND

As with you, so also with us, there are four points of the compass, North, South, East, and West.

There being no sun nor other heavenly bodies, it is impossible for us to determine the North in the usual way; but we have a method of our own. By a Law of Nature with us, there is a constant attraction to the South; and, although in temperate climates this is very slight—so that even a Woman in reasonable health can journey several furlongs northward without much difficulty—yet the hampering effect of the southward attraction is quite sufficient to serve as a compass in most parts of our earth. Moreover the rain (which falls at stated intervals) coming always from the North, is an additional assistance; and in the towns we have the guidance of the houses, which of course have their side-walls running for the most part North and South, so that the roofs may keep off the rain from the North. In the country, where there are no houses, the trunks of the trees serve as some sort of guide. Altogether, we have not so much difficulty as might be expected in determining our bearings.

Yet in our more temperate regions, in which the southward attraction is hardly felt, walking sometimes in a perfectly desolate plain where there have been no houses nor trees to guide me, I have been occasionally compelled to remain stationary for hours together, waiting till the rain came before continuing my journey. On the weak and aged, and especially on delicate Females, the force of attraction tells much more heavily than on the robust of the Male Sex, so that it is a point of breeding, if you meet a Lady in the street, always to give her the North side of the way—by no means an easy thing to do always at short notice when you are in rude health and in a climate where it is difficult to tell your North from your South.

Windows there are none in our houses; for the light comes to us alike in our homes and out of them, by day and by night, equally at all times and in all places, whence we know not. It was in old days, with our learned men, an interesting and oft-investigated question, What is the origin of light; and the solution of it has been repeatedly attempted, with no other result than to crowd our lunatic asylums with the would-be solvers. Hence, after fruitless attempts to suppress such investigations indirectly by making them liable to a heavy tax, the Legislature, in comparatively recent times, absolutely prohibited them. I, alas I alone in Flatland know now only too well the true solution of this mysterious problem; but my knowledge cannot be made intelligible to a single one of my countrymen; and I am mocked at—I, the sole possessor of the truths of Space and of the theory of the introduction of Light from the world of Three Dimensions—as if I were the maddest of the mad! But a truce to these painful digressions: let me return to our houses.

The most common form for the construction of a house is five-sided or pentagonal, as in the annexed figure. The two Northern sides *RO, OF,* constitute the roof, and for the most part have no doors; on the East is a small door for the Women; on the West a much larger one for the Men; the South side or floor is usually doorless.

Square and triangular houses are not allowed, and for this reason. The angles of a Square (and still more those of an equilateral Triangle) being much more

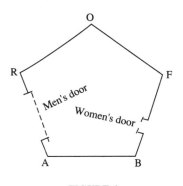

**FIGURE 4**

pointed than those of a pentagon, and the lines of inanimate objects (such as houses) being dimmer than the lines of Men and Women, it follows that there is no little danger lest the points of a square or triangular house residence might do serious injury to an inconsiderate or perhaps absentminded traveller suddenly running against them: and therefore, as early as the eleventh century of our era, triangular houses were universally forbidden by Law, the only exceptions being fortifications, powder-magazines, barracks, and other state buildings, which it is not desirable that the general public should approach without circumspection.

At this period, square houses were still everywhere permitted, though discouraged by a special tax. But, about three centuries afterwards, the Law decided that in all towns containing a population above ten thousand, the angle of a pentagon was the smallest house angle that could be allowed consistently with the public safety. The good sense of the community has seconded the efforts of the Legislature; and now, even in the country, the pentagonal construction has superseded every other. It is only now and then in some very remote and backward agricultural district that an antiquarian may still discover a square house.

## CONCERNING THE INHABITANTS OF FLATLAND

The greatest length or breadth of a full-grown inhabitant of Flatland may be estimated at about eleven of your inches. Twelve inches may be regarded as a maximum.

Our Women are Straight Lines.

Our Soldiers and Lowest Classes of Workmen are Triangles with two equal sides, each about eleven inches long, and a base or third side so short (often not exceeding half an inch) that they form at their vertices a very sharp and formidable angle. Indeed when their bases are of the most degraded type (not more than the eighth part of an inch in size), they can hardly be distinguished from Straight Lines or Women; so extremely pointed are their vertices. With us, as with you, these Triangles are distinguished from others by being called Isosceles; and by this name I shall refer to them in the following pages.

Our Middle Class consists of Equilateral or Equal-sided Triangles.

Our Professional Men and Gentlemen are Squares (to which class I myself belong) and Five-sided figures, or Pentagons.

Next above these come the Nobility, of whom there are several degrees, beginning at Six-sided Figures, or Hexagons, and from thence rising in the number of their sides till they receive the honorable title of Polygonal, or many-sided. Finally when the number of the sides becomes so numerous, and the sides themselves so small that the figure cannot be distinguished from a circle, he is included in the Circular or Priestly order; and this is the highest class of all.

It is a Law of Nature with us that a male child shall have one more side than his father, so that each generation shall rise (as a rule) one step in the scale of development and nobility. Thus the son of a Square is a Pentagon; the son of a Pentagon, a Hexagon; and so on.

But this rule applies not always to the Tradesmen, and still less often to the Soldiers, and to the Workmen; who indeed can hardly be said to deserve the name of human Figures, since they have not all their sides equal. With them therefore the Law of Nature does not hold; and the son of an Isosceles (i.e., a Triangle with two sides equal) remains Isosceles still. Nevertheless, all hope is not shut out, even from the Isosceles, that his posterity may ultimately rise above his degraded condition. For, after a long series of military successes, or diligent and skilful labors, it is generally found that the more intelligent among the Artisan and Soldier classes manifest a slight increase of their third side, or base, and a shrinkage of the two other sides. Intermarriages (arranged by the Priests) between the sons and daughters of these more intellectual members of the lower classes generally result in an offspring approximating still more to the type of the Equal-sided Triangle.

Rarely—in proportion to the vast number of Isosceles births—is a genuine and certifiable Equal-sided Triangle produced from Isosceles parents.[1]

Such a birth requires, as its antecedents, not only a series of carefully arranged intermarriages, but also a long-continued exercise of frugality and self-control on the part of the would-be ancestors of the coming Equilateral, and a patient, systematic, and continuous development of the Isosceles intellect through many generations.

The birth of a True Equilateral Triangle from Isosceles parents is the subject of rejoicing in our country for many furlongs round. After a strict examination conducted by the Sanitary and Social Board, the infant, if certified as Regular, is with solemn ceremonial admitted into the class of Equilaterals. He is then immediately taken from his proud yet sorrowing parents and adopted by some childless Equilateral, who is bound by oath never to permit the child henceforth to enter his former home or so much as to look upon his relations again, for fear lest the freshly developed organism may, by force of unconscious imitation, fall back again into his hereditary level.

The occasional emergence of an Isosceles from the ranks of his serf-born ancestors, is welcomed not only by the poor serfs themselves, as a gleam of light and hope shed upon the monotonous squalor of their existence, but also by the Aristocracy at large; for all the higher classes are well aware that these rare phenomena, while they do little or nothing to vulgarize their own privileges, serve as a most useful barrier against revolution from below.

Had the acute-angled rabble been all, without exception, absolutely destitute of hope and of ambition, they might have found leaders in some of their many seditious outbreaks, so able as to render their superior numbers and strength too much even for the wisdom of the Circles. But a wise ordinance of Nature has decreed that, in proportion as the working-classes increase in intelligence, knowledge, and all virtue, in that same proportion their acute angle (which makes them physically terrible) shall increase also and approximate to the harmless angle of the Equilateral Triangle. Thus, in the most brutal and formidable of the soldier class—creatures almost on a level with women in their lack of intelligence—it is found that, as they wax in the mental ability necessary to employ their tremendous penetrating power to advantage, so do they wane in the power of penetration itself.

How admirable is this Law of Compensation! And how perfect a proof of the natural fitness and, I may almost say, the divine origin of the aristocratic constitution of the States in Flatland! By a judicious use of this Law of Nature, the Polygons and Circles are almost always able to stifle sedition in its very cradle, taking advantage of the irrepressible and boundless hopefulness of the human mind. Art also comes to the aid of Law and Order. It is generally found possible—by a little artificial compression or expansion on the part of the State physicians—to make some of the more intelligent leaders of a rebellion perfectly Regular, and to admit them at once into the privileged classes; a much larger number, who are still below the standard, allured by the prospect of being ultimately ennobled, are induced to enter the State Hospitals, where they are kept in honorable confinement for life; one or two alone of the more obstinate, foolish, and hopelessly irregular are led to execution.

Then the wretched rabble of the Isosceles, planless and leaderless, are either transfixed without resistance by the small body of their brethren whom the Chief Circle keeps in pay for emergencies of this kind; or else more often, by means of jealousies and suspicions skilfully fomented among them by the Circular party, they are stirred to mutual warfare, and perish by one another's angles. No less than one hundred and twenty rebellions are recorded in our annals, besides minor outbreaks numbered at two hundred and thirty-five; and they have all ended thus.

## HOW I TRIED TO TEACH THE THEORY OF THREE DIMENSIONS TO MY GRANDSON, AND WITH WHAT SUCCESS

I awoke rejoicing, and began to reflect on the glorious career before me. I would go forth, me-thought, at once, and evangelize the whole of Flatland.

Even to Women and Soldiers should the Gospel of Three Dimensions be proclaimed. I would begin with my Wife.

Just as I had decided on the plan of my operations, I heard the sound of many voices in the street commanding silence. Then followed a louder voice. It was a herald's proclamation. Listening attentively, I recognized the words of the Resolution of the Council, enjoining the arrest, imprisonment, or execution of any one who should pervert the minds of the people by delusions, and by professing to have received revelations from another World.

I reflected. This danger was not to be trifled with. It would be better to avoid it by omitting all mention of my Revelation, by proceeding on the path of Demonstration—which after all seemed so simple and so conclusive that nothing would be lost by discarding the former means. "Upward, not North-ward" was the clew to the whole proof. It had seemed to me fairly clear before I fell asleep; and when I first awoke, fresh from my dream, it had appeared as patent as Arithmetic; but somehow it did not seem to me quite so obvious now. Though my Wife entered the room opportunely just at that moment, I decided, after we had interchanged a few words of commonplace conversation, not to begin with her.

My Pentagonal Sons were men of character and standing, and physicians of no mean reputation, but not great in mathematics, and, in that respect, unfit for my purpose. But it occurred to me that a young and docile Hexagon, with a mathematical turn, would be a most suitable pupil. Why therefore not make my first experiment with my little precocious Grandson, whose casual remarks on the meaning of $3^3$ had met with the approval of the Sphere? Discussing the matter with him, a mere boy, I should be in perfect safety; for he would know nothing of the Proclamation of the Council; whereas I could not feel sure that my Sons—so greatly did their patriotism and reverence for the Circles predom-inate over mere blind affection—might not feel compelled to hand me over to the Prefect, if they found me seriously maintaining the seditious heresy of the Third Dimension.

But the first thing to be done was to satisfy in some way the curiosity of my Wife, who naturally wished to know something of the reasons for which the Circle had desired that mysterious interview, and of the means by which he had entered our house. Without entering into the details of the elaborate account I gave her,—an account, I fear, not quite so consistent with truth as my Readers in Spaceland might desire,—I must be content with saying that I succeeded at last in persuading her to return quietly to her household duties without eliciting from me any reference to the World of Three Dimensions. This done, I immediately sent for my Grandson; for, to confess the truth, I felt that all that I had seen and heard was in some strange way slipping away from me, like the image of a half-grasped tantalizing dream, and I longed to essay my skill in making a first disciple.

When my Grandson entered the room I carefully secured the door. Then, sitting down by his side and taking our mathematical tablets,—or, as you would call them, Lines,—I told him we would resume the lesson of yesterday. I taught him once more how a Point by motion in One Dimension produces a Line, and how a straight Line in Two Dimensions produces a Square. After this, forcing a laugh, I said, "And now, you scamp, you wanted to make me believe that a Square may in the same way by motion 'Upward, not Northward,' produce another figure, a sort of extra Square in Three Dimensions. Say that again, you young rascal."

At this moment we heard once more the herald's "O yes! O yes!" outside in the street proclaiming the Resolution of the Council. Young though he was, my Grandson—who was unusually intelligent for his age, and bred up in perfect reverence for the authority of the Circles—took in the situation with an acuteness for which I was quite unprepared. He remained silent till the last words of the Proclamation had died away, and then, bursting into tears, "Dear Grandpapa," he said, "that was only my fun, and of course I meant nothing at all by it; and we did not know anything then about the new Law; and I don't think I said anything about the Third Dimension; and I am sure I did not say one word about 'Upward, not Northward,' for that would be such nonsense, you know. How could a thing move Upward, and not Northward? Upward, and not Northward! Even if I were a baby, I could not be so absurd as that. How silly it is! Ha! ha! ha!"

"Not at all silly," said I, losing my temper; "here, for example, I take this Square," and, at the word, I grasped a movable Square, which was lying at hand—"and I move it, you see, not Northward but—yes, I move it Upward—that is to say, not Northward, but I move it somewhere—not exactly like this, but somehow—" Here I brought my sentence to an inane conclusion, shaking the Square about in a purposeless manner, much to the amusement of my Grandson, who burst out laughing louder than ever, and declared that I was not teaching him, but joking with him; and so saying he unlocked the door and ran out of the room. Thus ended my first attempt to convert a pupil to the Gospel of Three Dimensions.

## HOW I THEN TRIED TO DIFFUSE THE THEORY OF THREE DIMENSIONS BY OTHER MEANS, AND OF THE RESULT

My failure with my Grandson did not encourage me to communicate my secret to others of my household; yet neither was I led by it to despair of success. Only I saw that I must not wholly rely on the catch-phrase, "Upward, not Northward," but must rather endeavor to seek a demonstration by setting before the public a clear view of the whole subject; and for this purpose it seemed necessary to resort to writing.

So I devoted several months in privacy to the composition of a treatise on the mysteries of Three Dimensions. Only, with the view of evading the Law, if possible, I spoke not of a physical Dimension, but of a Thoughtland whence, in theory, a Figure could look down upon Flatland and see simultaneously the insides of all things, and where it was possible that there might be supposed to exist a Figure environed, as it were, with six Squares, and containing eight terminal points. But in writing this book I found myself sadly hampered by the impossibility of drawing such diagrams as were necessary for my purpose; for of course, in our country of Flatland, there are no tablets but Lines, and no diagrams but Lines, all in one straight Line and only distinguishable by difference of size and brightness; so that, when I had finished my treatise (which I entitled "Through Flatland to Thoughtland") I could not feel certain that many would understand my meaning.

Meanwhile my life was under a cloud. All pleasures palled upon me; all sights tantalized and tempted me to outspoken treason, because I could not but compare what I saw in Two Dimensions with what it really was if seen in Three, and could hardly refrain from making my comparisons aloud. I neglected my clients and my own business to give myself to the contemplation of the mysteries which I had once beheld, yet which I could impart to no one, and found daily more difficult to reproduce even before my own mental vision.

One day, about eleven months after my return from Spaceland, I tried to see a Cube with my eye closed, but failed; and though I succeeded afterwards, I was not then quite certain (nor have I been ever afterwards) that I had exactly realized the original. This made me more melancholy than before, and determined me to take some step; yet what, I knew not. I felt that I would have been willing to sacrifice my life for the Cause, if thereby I could have produced conviction. But if I could not convince my Grandson, how could I convince the highest and most developed Circles in the land?

And yet at times my spirit was too strong for me, and I gave vent to dangerous utterances. Already I was considered heterodox if not treasonable, and I was keenly alive to the dangers of my position; nevertheless I could not at times refrain from bursting out into suspicious or half-seditious utterances, even among the highest Polygonal and Circular society. When, for example, the question arose about the treatment of those lunatics who said that they had received the power of seeing the insides of things, I would quote the saying of an ancient Circle, who declared that prophets and inspired people are always considered by the majority to be mad; and I could not help occasionally dropping such expressions as "the eye that discerns the interiors of things," and "the all-seeing land": once or twice I even let fall the forbidden terms "the Third and Fourth Dimensions." At last, to complete a series of minor indiscretions, at a meeting of our Local Speculative Society held at the palace of the prefect himself,—some extremely silly person having read an elaborate paper exhibiting the precise reasons why providence has limited the number of Dimensions to Two, and why the attribute of omnividence is assigned to the Supreme alone,—I so far forgot myself as to give an exact account of the whole of my voyage with the Sphere into Space, and to the Assembly Hall in our Metropolis, and then to

Space again, and of my return home, and of everything that I had seen and heard in fact or vision. At first, indeed, I pretended that I was describing the imaginary experiences of a fictitious person; but my enthusiasm soon forced me to throw off all disguise, and finally, in a fervent peroration, I exhorted all my hearers to divest themselves of prejudice and to become believers in the Third Dimension.

Need I say that I was at once arrested and taken before the Council?

Next morning, standing in the very place where but a very few months ago the Sphere had stood in my company, I was allowed to begin and to continue my narration unquestioned and uninterrupted. But from the first I foresaw my fate; for the President, noting that a guard of the better sort of Policemen was in attendance, of angularity little, if at all, under 55°, ordered them to be relieved before I began my defence, by an inferior class of 2° or 3°. I knew only too well what that meant. I was to be executed or imprisoned, and my story was to be kept secret from the world by the simultaneous destruction of the officials who had heard it; and, this being the case, the President desired to substitute the cheaper for the more expensive victims.

After I had concluded my defence, the President, perhaps perceiving that some of the junior Circles had been moved by my evident earnestness, asked me two questions:—

1. Whether I could indicate the direction which I meant when I used the words "Upward, not Northward"?

2. Whether I could by any diagrams or descriptions (other than the enumeration of imaginary sides and angles) indicate the Figure I was pleased to call a Cube?

I declared that I could say nothing more, and that I must commit myself to the Truth, whose cause would surely prevail in the end.

The President replied that he quite concurred in my sentiment, and that I could not do better. I must be sentenced to perpetual imprisonment; but if the Truth intended that I should emerge from prison and evangelize the world, the Truth might be trusted to bring that result to pass. Meanwhile I should be subjected to no discomfort that was not necessary to preclude escape, and, unless I forfeited the privilege by misconduct, I should be occasionally permitted to see my brother, who had preceded me to my prison.

Seven years have elapsed and I am still a prisoner, and—if I except the occasional visits of my brother—debarred from all companionship save that of my jailers. My brother is one of the best of Squares, just, sensible, cheerful, and not without fraternal affection; yet I must confess that my weekly interviews, at least in one respect, cause me the bitterest pain. He was present when the Sphere manifested himself in the Council Chamber; he saw the Sphere's changing sections; he heard the explanation of the phenomena then given to the Circles. Since that time, scarcely a week has passed during seven whole years, without his hearing from me a repetition of the part I played in that manifestation, together with ample descriptions of all the phenomena in Spaceland, and the arguments for the existence of Solid things derivable from Analogy. Yet—I take shame to be forced to confess it—my brother has not yet grasped the nature of the Third Dimension, and frankly avows his disbelief in the existence of a Sphere.

Hence I am absolutely destitute of converts, and, for aught that I can see, the millennial Revelation has been made to me for nothing. Prometheus up in Spaceland was bound for bringing down fire for mortals, but I—poor Flatland Prometheus—lie here in prison for bringing down nothing to my countrymen. Yet I exist in the hope that these memoirs, in some manner, I know not how, may find their way to the minds of humanity in Some Dimension, and may stir up a race of rebels who shall refuse to be confined to limited Dimensionality.

That is the hope of my brighter moments. Alas, it is not always so. Heavily weighs on me at times the burdensome reflection that I cannot honestly say I am confident as to the exact shape of the once-seen, oft-regretted Cube; and in my nightly visions the mysterious precept, "Upward, not Northward," haunts me like a soul-devouring Sphinx. It is part of the martyrdom which I endure for the cause of the Truth that there are seasons of mental weakness, when Cubes and Spheres flit away into the background of scarce-possible existences; when the Land of Three Dimensions seems almost as visionary as the Land of One or None; nay, when even this hard wall that bars me from my freedom, these very tablets on which I am writing, and all the substantial realities of Flatland itself, appear no better than the offspring of a diseased imagination, or the baseless fabric of a dream.

## ENDNOTE

[1]"What need of a certificate?" a Spaceland critic may ask; "Is not the procreation of a Square Son a certificate from Nature herself, proving the Equal-sidedness of the Father?" I reply that no Lady of any position will marry an uncertified Triangle. Square offspring has sometimes resulted from a slightly Irregular Triangle: but in almost every such case the Irregularity of the first generation is visited on the third; which either fails to attain the Pentagonal rank, or relapses to the Triangular.

# COMMENTARY ON
# LEWIS CARROLL

The Reverend Charles Lutwidge Dodgson was a mediocre mathematician who taught at Oxford for twenty-seven years without brightening the hour of a single student or producing anything of lasting value to his subject. Making in his own person no claim on men's remembrance, he created an immortal alter ego. I need not celebrate the famous Lewis Carroll writings, but should point out that Dodgson's love of mathematics and his preoccupation with certain of its concepts had much to do with the shape his fantasies took. Especially is this true of *Through the Looking Glass,* whose dreamlike inversions anticipate some of the most astounding and revolutionary insights of twentieth-century mathematics and physics. Modern writers have not overlooked the prophetic character of these remarkable stories. In serious treatises and in popularizations scientists repeatedly borrow from the recorded wisdom of Carroll's notable company of philosophers: the Red Queen, Humpty-Dumpty, the White Knight, Tweedledum and Tweedledee.

For this book I have selected three less familiar examples of Dodgson's writings. They too exhibit his interest in mathematical and logical problems. Cast characteristically in the form of puzzles, they have an inimitable quality, a blend of subtlety, pixieness and guileful innocence. No less than his masterpieces, they make us wonder what extraordinary sort of man he was.

Dodgson was born at Daresbury in 1832, the eldest son and the third child in a family of eleven, all of whom stammered.[1] His father was a comfortably off clergyman who rose to be an archdeacon. As a child, Dodgson displayed a "quaint precocity," which included a premature concern for the meaning of logarithms,[2] a propensity for staging marionette and magic shows and a gift for inventing puzzles. Like other children he made pets of snails and toads, but in a perverse innovation of his own he tried to get earthworms to fight each other. For this purpose, it is said, he supplied them with weapons, but the attempts were unsuccessful. After early schooling at home, when his father inculcated an interest in mathematics and theology, he went to a private school at Richmond and then to Rugby. He was a good student, exceptional in mathematics and "Divinity," and creditable in classics. But being an "odd fish" he was not happy at Rugby. "I cannot say," he afterwards wrote, ". . . that any earthly consideration could induce me to go through my three years again."[3] He took refuge in literary work which began by his writing and illustrating numerous magazines for his home circle, among them *Misch-Masch* and *The Rectory Umbrella.* The latter publication has whimsical articles on mathematical and other puzzles including the riddle of the two clocks, reproduced below, and an essay called "A Hemispherical Problem," or "Where Does the Day Change Its Name." This was a real problem. The day changes its name on the International Date Line, but this demarcation was not invented until 1878, more than twenty-five years after the question troubled Dodgson. He was always fascinated by time and some of the wonderful topsy-turvy effects in his later writings were produced by tinkering with it.

In January 1851 Dodgson entered Christ Church College, Oxford. He stayed there for forty-seven years until his death. He won scholarships and became, in recognition of his diligence and piety, a "Prick-bill"—a sort of chapel monitor who checked off attendance by pricking a hole next to each of the names on a list. He took his B.A. and M.A. degrees with honors, was appointed to the teaching staff, and in 1861 was ordained Deacon of the Church of England. Dodgson never proceeded to priest's orders and even his ordainment was preceded by years of self-searching and misgivings. His bad stammer and his doctrinal doubts were not the only things in the way of becoming a vicar. He liked being a mathematician, even though he was not very good at it; moreover he was reluctant to commit himself to some of the very strict rules imposed by custom on those who took priests' orders. He could not have gone to the theater[4] and he was determined not to relinquish this innocent enjoyment.

It has also been suggested that behind Dodgson's refusal to take orders lay a hidden desire—hidden even from himself—to marry. I suspect this is the sort of thing interpretative biographers dream up; there is no real evidence to sustain it beyond the general notion that every maiden aunt is the victim of a great disappointment, following which she forever renounces physical love.

Except for a six-week trip to the Continent (1867), in the course of which he visited Russia, there was little to interrupt Dodgson's "half-cloistral, fastidious, eccentric life."[5] He was a student always and a teacher for twenty-seven of his middle years.[6] He was a writer of letters on public questions to newspapers, and a participant in many Oxford controversies. He was a "dreadfully conscientious" and rather dull instructor. In the summer he went on short vacations to English bathing resorts. He came up to London occasionally to see a play. He lectured to children, using lantern slides of his own devising, made a mechanical Humpty-Dumpty, accumulated a library of 5,000 volumes (from Shakespeare to Kipling), bought a skeleton so as to learn anatomy, rigged up his rooms with thermometers and oil stoves because he had a horror of drafts, had five sizes of notepaper, conducted a prodigious correspondence which he cross-indexed to the extent of having almost 100,000 entries in 1898, and became one of the best amateur photographers of his time. He also went rowing on the Isis with the small daughters of the Dean of Christ Church. In his diary for July 4, 1862, appears the entry: "I made an expedition *up* the river to Godstow with the three Liddells; we had tea on the bank there, and did not reach Christ Church till half-past eight." On the opposite page (says Collingwood), he added, "somewhat later," the words, "on which occasion I told them the fairy-tale of *Alice's Adventures Underground,* which I undertook to write out for Alice."[7]

It is of course a myth that *Alice* was composed in a single afternoon. There were many river excursions and picnics with baskets of cake, cold chicken and a kettle for tea, though it is perhaps true that the Godstow outing was a specially magic day.[8] There were other story-telling hours when the Liddells visited him to be photographed or to have lunch in his unbelievably commodious suite at Christ Church (no less than four bedrooms and four sitting rooms were occupied by him for thirty years). *Alice* was published in 1865, *Through the Looking Glass,* seven years later. Both were immensely successful. They brought Dodgson fame, which he enjoyed, and made him the object of public attention and curiosity, which terrified him. They also gave him modest wealth, which he used in his own way by lending to impecunious friends, giving to hospitals and other charities, presenting young nephews and nieces with gold watches (and instructing them in symbolic logic)[9] and providing for the amusement and even the education of the numerous tribe of little girls he adored. From the fact that Dodgson was never a serious candidate for marriage one must not infer that he was incapable of love. He was incapable, it is true, of giving love to adults, or at least of showing it. Even when writing to his sisters he signed himself "C. L. Dodgson." But with little girls he could be natural and warm, invite them to parties, play games tirelessly, invent for their entertainment endless puzzles and stories. I think there is point to the conjecture that on a deeper level his two great books were allegories in which are fused two themes: his undeclared love for Alice Liddell and his fascination with the mathematical mysteries of time.[10]

"He always used to say that when the time came for him to take off his hat when he met one of his quondam child friends in the street, it was time for the friendship to cease."[11] Alice grew up, got married and the friendship ceased; inspiration also ceased. *The Hunting of the Snark* appeared in 1876; it was the last of the writings in which Dodgson succeeded in creating what has aptly been called a new nursery mythology. *Sylvie and Bruno* (1889), also a book for children (and also an allegory) is much inferior. Nevertheless many works, mathematical as well as literary and fanciful, still came from his pen. His total output, before and after *Alice,* was quite large. As C. L. Dodgson, he published, among others, numerous mathematical texts, a book on non-Euclidean geometry (*Euclid and His Modern Rivals,* 1879), mathematical curiosa, and volumes on symbolic logic. These were useful works, and his studies of formal logic exhibited fully his "acute and ingenious intellect." But his contributions in mathematics and logic cannot be assessed as of more than limited value. As Lewis Carroll (the pseudonym which he adopted in 1856 when he wrote verse for a magazine called *The Train*)[12] he wrote many poems, pamphlets on university affairs, articles on various subjects ranging from logical paradoxes to vivisection, and a book on symbolic logic.

As he grew older, Dodgson became fussier, more prudish and more difficult. Sir John Tenniel who had almost as much to do with bringing *Alice* to life as the author, said he could no longer tolerate "that conceited old don."[13] His close friend Ellen Terry had to submit to his rebuke for taking off a few outer garments as required while playing the role of Margaret in *Faust*. (This rebuke makes an interesting contrast with Dodgson's frequent practice of photographing little girls in his chambers in the nude.) More and more he fled from the sensible world into a land of games, puzzles, mnemonics, circle squarers and logical paradoxes.

Endless suggestions poured from him for improving things "like lawn tennis tournaments and the election of proctors." A chronic insomniac, but otherwise in excellent health, he found ample time to pursue every harmless whim to its insanely logical conclusion. He had the habit of working throughout the night while standing at his tall writing desk; "he also worked in bed without light, using an instrument of his own invention called the nyctograph," which kept the lines straight and his pen from running off the paper.[14] On January 6, 1898, he contracted influenza, and eight days later he died. Dodgson once said in a letter to a friend in America: "words mean more than we mean to express when we use them; so a whole book ought to mean a great deal more than the writer means." No more profound opinion has been uttered on his own strange masterpieces.[15]

## ENDNOTES

[1]The sources of this sketch include the article on Dodgson in the *Dictionary of National Biography*; A. L. Taylor, *The White Knight*, Edinburgh, 1952; Florence Becker Lennon, *Victoria Through the Looking Glass—The Life of Lewis Carroll*, New York, 1945; Stuart Dodgson Collingwood, *The Life and Letters of Lewis Carroll*, London, 1898; Helmut Gernsheim, *Lewis Carroll, Photographer*, New York, 1951; Roger Lancelyn Green (editor), *The Diaries of Lewis Carroll*, New York, 1954.

[2]"As a mere child he found a book of logarithms and took it to his father with the request: 'Please explain.' His father smilingly told him he was much too young to understand so difficult a subject. 'But,' said young Dodgson, with devastating simplicity, 'please explain.'" Taylor, *op. cit.*, pp. 1–2.

[3]Quoted by Taylor, *op. cit.*, p. 3.

[4]A. L. Taylor, *op. cit.*, p. 24.

[5]*Dictionary of National Biography*.

[6]*Florence Becker Lennon. op. cit.*, p. 44.

[7]Collingwood, *op. cit.*, p. 93, as cited in Taylor, *op. cit.*, p. 41.

[8]"The importance of the Godstow trip was that on it he told a particularly good story and Alice asked him to write it out for her. This is confirmed by Canon Duckworth, who says: 'I also well remember how, when we had conducted the three children back to the Deanery, Alice said, as she bade us good night, "Oh, Mr. Dodgson, I wish you would write out Alice's adventures for me." He said he should try, and he afterwards told me that he sat up nearly the whole night, committing to a ms. book his recollections of the drolleries with which he had enlivened the afternoon.' Taylor, *op. cit.*, p.43.

[9]A. L. Taylor, *op. cit.*, p. 200.

[10]See A. L. Taylor, *op. cit.*

[11]Ethel Arnold (Matthew Arnold's niece), one of Dodgson's "quondam child friends," as quoted in Florence Becker Lennon, *op. cit.*, p. 207.

[12]One of the poems in *Through the Looking Glass*, the White Knight's mournful ballad, appeared in *The Train* in 1856. Taylor, *op. cit.*, p. 21.

[13]Taylor, *op. cit.*, p. 145.

[14]Taylor, *op. cit.*, p. 200.

[15]For an interesting article on Carroll's contributions to mathematics, see Warren Weaver, "Lewis Carroll: Mathematician," *Scientific American*, April 1956. [Note added in proof.]

## 3

*Logic is neither a science nor an art, but a dodge.*
—Benjamin Jowett (?) (1817–1893)

# WHAT THE TORTOISE SAID TO ACHILLES AND OTHER RIDDLES

*By Lewis Carroll*

## WHAT THE TORTOISE SAID TO ACHILLES

Achilles had overtaken the Tortoise, and had seated himself comfortably on its back.

"So you've got to the end of our race-course?" said the Tortoise. "Even though it *does* consist of an infinite series of distances? I thought some wiseacre or other had proved that the thing couldn't be done?"

"It *can* be done," said Achilles. "It *has* been done! *Solvitur ambulando.* You see the distances were constantly *diminishing*: and so—"

"But if they had been constantly *increasing*?" the Tortoise interrupted. "How then?"

"Then I shouldn't be *here*," Achilles modestly replied; "and *you* would have got several times round the world, by this time!"

"You flatter me—*flatten*, I mean," said the Tortoise; "for you *are* a heavy weight, and *no* mistake! Well now, would you like to hear of a race-course, that most people fancy they can get to the end of in two or three steps, while it *really* consists of an infinite number of distances, each one longer than the previous one?"

"Very much indeed!" said the Grecian warrior, as he drew from his helmet (few Grecian warriors possessed *pockets* in those days) an enormous note-book and a pencil. "Proceed! And speak *slowly*, please! *Shorthand* isn't invented yet!"

"That beautiful First proposition of Euclid!" the Tortoise murmured dreamily. "You admire Euclid?"

"Passionately! So far, at least, as one *can* admire a treatise that won't be published for some centuries to come!"

"Well, now, let's take a little bit of the argument in that First proposition— just *two* steps, and the conclusion drawn from them. Kindly enter them in your note-book. And, in order to refer to them conveniently, let's call them $A$, $B$, and $Z$:

($A$) Things that are equal to the same are equal to each other.

($B$) The two sides of this Triangle are things that are equal to the same.

($Z$) The two sides of this Triangle are equal to each other.

"Readers of Euclid will grant, I suppose, that $Z$ follows logically from $A$ and $B$, so that any one who accepts $A$ and $B$ as true, *must* accept $Z$ as true?"

"Undoubtedly! The youngest child in a High School—as soon as High Schools are invented, which will not be till some two thousand years later —will grant *that*."

"And if some reader had *not* yet accepted $A$ and $B$ as true, he might still accept the *Sequence* as a *valid* one, I suppose?"

"No doubt such a reader might exist. He might say 'I accept as true the Hypothetical proposition that, if $A$ and $B$ be true, $Z$ must be true; but I *don't* accept $A$ and $B$ as true.' Such a reader would do wisely in abandoning Euclid, and taking to football."

"And might there not *also* be some reader who would say 'I accept $A$ and $B$ as true, but I *don't* accept the Hypothetical'?"

"Certainly there might. *He,* also, had better take to football."

"And *neither* of these readers," the Tortoise continued, "is *as yet* under any logical necessity to accept $Z$ as true?"

"Quite so," Achilles assented.

"Well, now, I want you to consider *me* as a reader of the *second* kind, and to force me, logically, to accept $Z$ as true."

"A tortoise playing football would be—" Achilles was beginning.

"—an anomaly, of course," the Tortoise hastily interrupted. "Don't wander from the point. Let's have $Z$ first, and football afterwards!"

"I'm to force you to accept $Z$, am I?" Achilles said musingly. "And your present position is that you accept $A$ and $B$, but you *don't* accept the Hypothetical—"

"Let's call it $C$," said the Tortoise.

"—but you don't accept:

($C$) If $A$ and $B$ are true, $Z$ must be true."

"That is my present position," said the Tortoise.

"Then I must ask you to accept $C$."

"I'll do so," said the Tortoise, "as soon as you've entered it in that note-book of yours. What else have you got in it?"

"Only a few memoranda," said Achilles, nervously fluttering the leaves: "a few memoranda of—of the battles in which I have distinguished myself!"

"Plenty of blank leaves, I see!" the Tortoise cheerily remarked. "We shall need them *all*!" (Achilles shuddered.) "Now write as I dictate:

($A$) Things that are equal to the same are equal to each other.

(*B*) The two sides of this triangle are things that are equal to the same.

(*C*) If *A* and *B* are true, *Z* must be true.

(*Z*) The two sides of this Triangle are equal to each other."

"You should call it *D*, not *Z*," said Achilles. "It comes *next* to the other three. If you accept *A* and *B* and *C*, you *must* accept *Z*."

"And why *must I*?"

"Because it follows *logically* from them. If *A* and *B* and *C* are true, *Z must* be true. You don't dispute *that*, I imagine?"

"If *A* and *B* and *C* are true, *Z must* be true," the Tortoise thoughtfully repeated. "That's *another* Hypothetical, isn't it? And, if I failed to see its truth, I might accept *A* and *B* and *C*, and *still* not accept *Z*, mightn't I?"

"You might," the candid hero admitted; "though such obtuseness would certainly be phenomenal. Still, the event is *possible*. So I must ask you to grant one more Hypothetical."

"Very good. I'm quite willing to grant it, as soon as you've written it down. We will call it

(*D*) If *A* and *B* and *C* are true, *Z* must be true.

"Have you entered that in your note-book?"

"I *have*!" Achilles joyfully exclaimed, as he ran the pencil into its sheath. "And at last we've got to the end of this ideal race-course! Now that you accept *A* and *B* and *C* and *D*, *of course* you accept *Z*."

"Do I?" said the Tortoise innocently. "Let's make that quite clear. I accept *A* and *B* and *C* and D. Suppose I *still* refuse to accept *Z* "

"Then Logic would take you by the throat, and *force* you to do it!" Achilles triumphantly replied. "Logic would tell you 'You can't help yourself. Now that you've accepted *A* and *B* and *C* and *D*, you *must* accept *Z*!' So you've no choice, you see."

"Whatever *Logic* is good enough to tell me is worth *writing down*," said the Tortoise. "So enter it in your book, please. We will call it

(*E*) If *A* and *B* and *C* and *D* are true, *Z* must be true.

"Until I've granted *that*, of course, I needn't grant *Z*. So it's quite a *necessary* step, you see?"

"I see," said Achilles; and there was a touch of sadness in his tone.

Here the narrator, having pressing business at the Bank, was obliged to leave the happy pair, and did not again pass the spot until some months afterwards. When he did so, Achilles was still seated on the back of the much-enduring Tortoise, and was writing in his notebook, which appeared to be nearly full. The Tortoise was saying "Have you got that last step written down? Unless I've lost count, that makes a thousand and one. There are several millions more to come. And *would* you mind, as a personal favour— considering what a lot of instruction this colloquy of ours will provide for the Logicians of the Nineteenth Century—*would* you mind adopting a pun that my cousin the Mock-Turtle will then make, and allowing yourself to be re-named Taught-Us?"

"As you please!" replied the weary warrior, in the hollow tones of despair, as he buried his face in his hands. "Provided that *you,* for *your* part, will adopt a pun the Mock-Turtle never made, and allow yourself to be renamed A Kill-Ease!"

# THE TWO CLOCKS

Which is better, a clock that is right only once a year, or a clock that is right twice every day? "The latter," you reply, "unquestionably." Very good, now attend.

I have two clocks: one doesn't go *at all,* and the other loses a minute a day: which would you prefer? "The losing one," you answer, "without a doubt." Now observe: the one which loses a minute a day has to lose twelve hours, or seven hundred and twenty minutes before it is right again, consequently it is only right once in two years, whereas the other is evidently right as often as the time it points to come round, which happens twice a day.

So you've contradicted yourself *once.*

"Ah, but," you say, "what's the use of its being right twice a day, if I can't tell when the time comes?"

Why, suppose the clock points to eight o'clock, don't you see that the clock is right *at* eight o'clock? Consequently, when eight o'clock comes round your clock is right.

"Yes, I see *that,*" you reply.

Very good, then you've contradicted yourself *twice*: now get out of the difficulty as best you can, and don't contradict yourself again if you can help it.

You *might* go on to ask, "How am I to know when eight o'clock *does* come? My clock will not tell me." Be patient: you know that when eight o'clock comes your clock is right, very good; then your rule is this: keep your eye fixed on your clock, and *the very moment it is right* it will be eight o'clock. "But—," you say. There, that'll do; the more you argue the farther you get from the point, so it will be as well to stop.

# KNOT IX

## A SERPENT WITH CORNERS[1]

*Water, water, everywhere,*
*Nor any drop to drink.*

"It'll just take one more pebble."

"Whatever *are* you doing with those buckets?"

The speakers were Hugh and Lambert. Place, the beach of Little Mendip. Time 1:30 P.M. Hugh was floating a bucket in another a size larger, and trying how many pebbles it would carry without sinking. Lambert was lying on his back, doing nothing.

For the next minute or two Hugh was silent, evidently deep in thought. Suddenly he started. "I say, look here, Lambert!" he cried.

"If it's alive, and slimy, and with legs, I don't care to," said Lambert.

"Didn't Balbus say this morning that, if a body is immersed in liquid it displaces as much liquid as is equal to its own bulk?" said Hugh.

"He said things of that sort," Lambert vaguely replied.

"Well, just look here a minute. Here's the little bucket almost quite immersed: so the water displaced ought to be just about the same bulk. And now just look at it!" He took out the little bucket as he spoke, and handed the big one to Lambert. "Why, there's hardly a teacupful! Do you mean to say *that* water is the same bulk as the little bucket?"

"Course it is," said Lambert.

"Well, look here again!" cried Hugh, triumphantly, as he poured the water from the big bucket into the little one. "Why, it doesn't half fill it!"

"That's *its* business," said Lambert. "If Balbus says it's the same bulk, why, it *is* the same bulk, you know."

"Well, I don't believe it," said Hugh.

"You needn't," said Lambert. "Besides, it's dinner-time. Come along." They found Balbus waiting dinner for them, and to him Hugh at once propounded his difficulty.

"Let's get you helped first," said Balbus, briskly cutting away at the joint. "You know the old proverb, 'Mutton first, mechanics afterwards'?"

The boys did *not* know the proverb, but they accepted it in perfect good faith, as they did every piece of information, however startling, that came from so infallible an authority as their tutor. They ate on steadily in silence, and, when dinner was over, Hugh set out the usual array of pens, ink, and paper, while Balbus repeated to them the problem he had prepared for their afternoon's task.

"A friend of mine has a flower-garden—a very pretty one, though no great size—"

"How big is it?" said Hugh.

"That's what *you* have to find out!" Balbus gayly replied. "All *I* tell you is that it is oblong in shape—just half a yard longer than its width—and that a gravel-walk, one yard wide, begins at one corner and runs all round it."

"Joining into itself?" said Hugh.

"*Not* joining into itself, young man. Just before doing *that,* it turns a corner, and runs round the garden again, alongside of the first portion, and then inside that again, winding in and in, and each lap touching the last one, till it has used up the whole of the area."

"Like a serpent with corners?" said Lambert.

"Exactly so. And if you walk the whole length of it, to the last inch, keeping in the centre of the path, it's exactly two miles and half a furlong. Now, while you find out the length and breadth of the garden, I'll see if I can think out that sea-water puzzle."

"You said it was a flower-garden?" Hugh inquired, as Balbus was leaving the room.

"I did," said Balbus.

"Where do the flowers grow?" said Hugh. But Balbus thought it best not to hear the question. He left the boys to their problem, and, in the silence of his own room, set himself to unravel Hugh's mechanical paradox.

"To fix our thoughts," he murmured to himself, as, with hands deep-buried in his pockets, he paced up and down the room, "we will take a cylindrical glass jar, with a scale of inches marked up the side, and fill it with water up to the 10-inch mark: and we will assume that every inch depth of jar contains a pint of water. We will now take a solid cylinder, such that every inch of it is equal in bulk to *half* a pint of water, and plunge 4 inches of it into the water, so that the end of the cylinder comes down to the 6-inch mark. Well, that displaces 2 pints of water. What becomes of them? Why, if there were no more cylinder, they would lie comfortably on the top, and fill the jar up to the 12-inch mark. But unfortunately there is more cylinder, occupying half the space between the 10-inch and the 12-inch marks, so that only *one* pint of water can be accommodated there. What becomes of the other pint? Why, if there were no more cylinder, it would lie on the top, and fill the jar up to the 13-inch mark. But unfortunately—Shade of Newton!" he exclaimed, in sudden accents of terror. "When *does* the water stop rising?"

A bright idea struck him. "I'll write a little essay on it," he said.

## BALBUS'S ESSAY

"When a solid is immersed in a liquid, it is well known that it displaces a portion of the liquid equal to itself in bulk, and that the level of the liquid rises just so much as it would rise if a quantity of liquid had been added to it, equal in bulk to the solid. Lardner says precisely the same process occurs when a solid is *partially* immersed: the quantity of liquid displaced, in this case, equalling the portion of the solid which is immersed, and the rise of the level being in proportion.

"Suppose a solid held above the surface of a liquid and partially immersed: a portion of the liquid is displaced, and the level of the liquid rises. But, by this rise of level, a little bit more of the solid is of course immersed, and so there is a new displacement of a second portion of the liquid, and a consequent rise of level. Again, this second rise of level causes a yet further immersion, and by consequence another displacement of liquid and another rise. It is self-evident that this process must continue till the entire solid is immersed, and that the liquid will then begin to immerse whatever holds the solid, which, being connected with it, must for the time be considered a part of it. If you hold a stick, six feet long, with its ends in a tumbler of water, and wait long enough, you must eventually be immersed. The question as to the source from which the water is supplied—which belongs to a high branch of mathematics, and is therefore beyond our present scope—does not apply to the sea. Let us therefore take the familiar instance of a man standing at the edge of the sea, at ebb-tide, with a solid in his hand, which he partially immerses: he remains steadfast and unmoved, and we all know that he must be drowned. The multitudes who daily perish in this manner to attest a philosophical truth, and whose bodies the unreasoning wave casts sullenly upon our thankless shores, have a truer claim to be called the martyrs of science than a Galileo or a Kepler. To use Kossuth's eloquent phrase, they are the unnamed demigods of the nineteenth century."[2]

"There's a fallacy *somewhere*," he murmured drowsily, as he stretched his long legs upon the sofa. "I must think it over again." He closed his eyes, in order to concentrate his attention more perfectly, and for the next hour or so his slow and regular breathing bore witness to the careful deliberation with which he was investigating this new and perplexing view of the subject.

# ANSWERS TO KNOT IX

## § 1. THE BUCKETS

*Problem.*—Lardner states that a solid, immersed in a fluid, displaces an amount equal to itself in bulk. How can this be true of a small bucket floating in a larger one?

*Solution.*—Lardner means, by "displaces," "occupies a space which might be filled with water without any change in the surroundings." If the portion of the floating bucket, which is above the water, could be annihilated and the rest of it transformed into water, the surrounding water would not change its position: which agrees with Lardner's statement.

Five answers have been received, none of which explains the difficulty arising from the well-known fact that a floating body is the same weight as the displaced fluid. *Hecla* says that "Only that portion of the smaller bucket which descends below the original level of the water can be properly said to be immersed, and only an equal bulk of water is displaced." Hence, according to *Hecla,* a solid whose weight was equal to that of an equal bulk of water, would not float till the whole of it was below "the original level" of the water: but, as a matter of fact, it would float as soon as it was all under water. *Magpie* says the fallacy is "the assumption that one body can displace another from a place where it isn't," and that Lardner's assertion is incorrect, except when the containing vessel "was originally full to the brim." But the question of floating depends on the present state of things, not on past history. *Old King Cole* takes the same view as *Hecla*. *Tympanum* and *Vindex* assume that "displaced" means "raised above its original level," and merely explain how it comes to pass that the water, so raised, is less in bulk than the immersed portion of bucket, and thus land themselves—or rather set themselves floating—in the same boat as *Hecla*.

I regret that there is no Class List to publish for this Problem.

# §2. BALBUS'S ESSAY

*Problem.* —Balbus states that if a certain solid be immersed in a certain vessel of water, the water will rise through a series of distances, two inches, one inch, half an inch, etc., which series has no end. He concludes that the water will rise without limit. Is this true?

*Solution.* —No. This series can never reach 4 inches, since, however many terms we take, we are always short of 4 inches by an amount equal to the last term taken.

Three answers have been received—but only two seem to me worthy of honours.

*Tympanum* says that the statement about the stick "is merely a blind, to which the old answer may well be applied, *solvitur ambulando*, or rather *mergendo*." I trust *Tympanum* will not test this in his own person, by taking the place of the man in Balbus's Essay! He would infallibly be drowned.

*Old King Cole* rightly points out that the series, 2, 1, etc., is a decreasing geometrical progression: while *Vindex* rightly identifies the fallacy as that of "Achilles and the Tortoise."

## Endnotes

[1][From "A Tangled Tale," ED.]
[2]*Note by the writer.*—For the above essay I am indebted to a dear friend, now deceased.

# COMMENTARY ON CONTINUITY

The theory of continuity is an important and beautiful mathematical fiction. Mathematics could not have got along without it, yet it has been the source of formidable difficulties. In mathematics continuity is understood as being analogous to, but not identical with, the intuitive idea of continuity associated with time, space or motion. We all think of time, as did Newton, as flowing "equably" and unbrokenly; of space as smooth and without gaps or fissures; of motion as uninterrupted, unintermittent, "continuous." This is the natural, perhaps even indispensable, mode of interpreting experience, for the plain man and the philosopher.[1] The mathematician, with characteristic perversity, redefines this comfortably vague notion, thereby making it more precise, more useful and more troublesome.

I shall give two examples of mathematical continuity. The series of real numbers—composed of the rational and irrational numbers—is continuous; so is the class of points on a line segment. Each number or point is separate and possesses its own distinguishable identity; "it does not pass over by imperceptible degrees into another"; yet the series of numbers (or class of points) is what mathematicians call "everywhere dense"—which means that between any two numbers (or points), however near together, there is an infinitude of others. This is one of the essential attributes of the mathematical continuum. Another example of the concept is the continuous function, an immensely valuable tool of pure and applied mathematics. The modern definition of continuous function, upon which mathematicians have lavished exquisite care, is very precise and very technical; it will suffice here to say that a function is continuous if its graph is a smooth curve, without sudden jumps or breaks.

The problems of continuity have for centuries vexed philosophers and logicians, as well as mathematicians. One of the major problems stems from Zeno's famous puzzles. Zeno showed that the mathematical treatment of space and time required that they be broken up into infinite sets of points and instants; this, in turn, seemed to force the conclusion that motion is impossible, and also to lead to other paradoxes. Philosophers were especially displeased because they felt that resolving space and time into points and instants destroyed the intuitive property of continuity without furnishing a satisfactory substitute.[2] Imagine, they said, living in a space bristling with an infinity of points; the idea of an array of points implies discontinuity, however densely the points are packed, and the very notion of infinite number is self-contradictory.

Some, but not all of these problems have been cleared up. It is now understood that the theory of mathematical continuity is an "abstract logical scheme" which may or may not describe the structure of actual space, but whose validity is independent of such considerations. It is not yet certain that Zeno's puzzles have finally been solved, but a logically consistent theory of mathematical infinity has been devised which disposes of part of the philosophical muddle. Among the principal founders of this theory are George Cantor (see p. 1533), Richard Dedekind (see p. 519) and Bernhard Bolzano (1781–1848), an Austrian Catholic priest, whose posthumous little book, *Paradoxes of the Infinite* (1851) is a landmark of modern mathematical and logical thought. Bolzano recognized the necessity, in analyzing the paradoxes of infinity, of defining various "obvious" mathematical concepts, including that of continuity. One of the theorems in his book, on continuous function, reads as follows:

"A continuous function of a variable x which is positive for some value of $x$ and negative for some other value of $x$ in a closed interval a $\leq x \leq$ b of continuity must have the value zero for some intermediate value of $x$."[3] You will readily perceive that this is almost as self-evident a theorem as one which states that in ascending from the cellar to the roof of a building it is necessary at some point to pass the street level. Yet certain of the mathematical implications of this simple and obvious statement about continuity are utterly astonishing, as may be seen in the small problem of mechanics given below. I suspect you will not believe the solution, even after going through the proof; or that if you concede the proof is without a logical flaw, you will still have to exert yourself to *feel* the correctness of the solution. If you succeed in this effort, you may call yourself a mathematician; in any case you will appreciate the subtlety of the continuity concept and the depth of Bolzano's trivial theorem.

## ENDNOTES

[1]"They [the plain man and the philosopher] conceive continuity rather as absence of separateness, the sort of general obliteration of distinctions which characterizes a thick fog. A fog gives an impression of vastness without multiplicity or division. It is the sort of thing that a metaphysician means by 'continuity,' declaring it, very truly, to be characteristic of his mental life and of that of children and animals." Bertrand Russell, *Introduction to Mathematical Philosophy*, N.Y., 1930, p. 105.

[2]Bertrand Russell, *Our Knowledge of the External World—As a Field for Scientific Method in Philosophy*, Open Court (Chicago), 1915, pp. 129 *et seq.*

[3]As given in Courant and Robbins, *What Is Mathematics?*, New York, 1941, p. 312. See also Bernhard Bolzano, *Paradoxes of the Infinite*, translated by Dr. Fr. Přihonský, with a historical introduction by Donald A. Steele, New Haven (Yale University Press), 1950.

# 4

*He hangs between; in doubt to act or rest.*

—ALEXANDER POPE

# THE LEVER OF MAHOMET

## By Richard Courant and Herbert Robbins

Suppose a train travels in a finite time from station $A$ to station $B$ along a straight section of track. The journey need not be of uniform speed or acceleration. The train may act in any manner, speeding up, slowing down, coming to a halt, or even backing up for a while, before reaching $B$. But the exact motion of the train is supposed to be known in advance; that is, the function $s = f(t)$ is given, where $s$ is the distance of the train from station $A$, and $t$ is the time, measured from the instant of departure. On the floor of one of the cars a rod is pivoted so that it may move without friction either forward or backward until it touches the floor. (If it does touch the floor, we assume that it remains on the floor henceforth; this will be the case if the rod does not bounce.) *We ask if it is possible to place the rod in such a position that if it is released at the instant when the train starts and allowed to move solely under the influence of gravity and the motion of the train, it will not fall to the floor during the entire journey from A to B.*

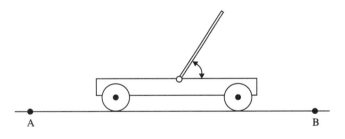

**FIGURE 1**

At first thought it might seem quite unlikely that for any given schedule of motion the interplay of gravity and reaction forces will always permit such a maintenance of balance under the single condition that the initial position of the rod is suitably chosen. But we state that such a position always exists.

Fortunately, the proof does not depend on a detailed knowledge of the laws of dynamics. (If it did, our task would be exceedingly difficult.) Only the following simple assumption of a physical nature need be granted: *The subsequent motion of the rod depends continuously on its initial position;* in particular, if for a given initial position the rod will fall down and hit the floor in one direction, then for any initial position differing sufficiently little from this, the rod will not hit the floor in the opposite direction.

Now the position of the rod in the train at any time is characterized by the angle $\alpha$ which it makes with the floor. To the angles $\alpha = 0°$ and $\alpha = 180°$ respectively correspond the two flat positions of the rod. We denote by $x$ the angle of the initial position of the rod. The proof of our statement will be given indirectly, in line with its purely existential character. We shall assume that whatever initial position $x$ we choose, the rod will *always* fall down and touch the floor, either at $\alpha = 0°$ or at $\alpha = 180°$. We may then define a function $f(x)$ whose value is to be $+1$ if the rod hits the floor at $\alpha = 0°$ and $-1$ otherwise. Since we have assumed that for each initial angle $x$ we have one of the two cases, the function $f(x)$ is defined in the whole interval $0 \le x \le 180$. Obviously $f(0) = +1$ and $f(180) = -1$, while according to the assumed continuity property of our dynamical system, $f(x)$ will be a continuous function of $x$ in the closed interval $0 \le x \le 180$. Hence, by Bolzano's theorem, it must have the value $f(x) = 0$ for some intermediate value of the initial angle $x$, contradicting the definition of $f(x)$ as only able to assume the values $+1$ or $-1$. This absurdity proves false the assumption that the rod will fall to the floor during the journey for every initial position $x$.

It is clear that this statement has an entirely theoretical character, since the proof gives no indication of how to find the desired initial position. Moreover, even if such a position could be calculated theoretically, it would probably be quite useless in practice, because of its instability. For example, in the extreme case where the train remains motionless at station $A$ during the entire journey, the solution is obviously $x = 90°$, but anyone who has attempted to balance a needle upright on a plate for any length of time will find this result of little assistance. Nevertheless, for the mathematician, the existence proof that we have given loses none of its interest.

# COMMENTARY ON
# GAMES AND PUZZLES

G ames and puzzles are the subject of a considerable mathematical literature, much of which is both difficult and tedious. The writings on magic
squares alone suffice to make a fair-sized library. Nothing more dismal can be
imagined. It is not inevitable that the mathematical treatment of games should
work a blight. Games are among the most interesting creations of the human
mind, and the analysis of their structure is full of adventure and surprises.
Unfortunately there is never a lack of mathematicians for the job of transforming delectable ingredients into a dish that tastes like a damp blanket.

A few books on the subject will delight almost any reader. This group
includes W. W. Rouse Ball's classic, *Mathematical Recreations and Essays,*
W. Ahrens' *Mathematische Unterhaltungen und Spiele,* H. E. Dudeney's
*Amusements in Mathematics,* E. Lucas' *Récréations Mathématiques,* and
the immortal Sam Loyd's helter-skelter *Cyclopaedia of Puzzles.* A recent
addition to these is *A History of Board Games Other than Chess* by H. J. R.
Murray.[1] Mr. Murray, known for his monumental *History of Chess,* presents
some 270 games, ancient and modern, of the widest provenance and diffusion.
Despite their diversity, he is able to classify the games into five basic groups:
games of "alignment and configuration" (e.g., tic-tac-toe), "war games" (e.g.,
chess and checkers), "hunt games" (e.g., fox and geese), "race games" (e.g.,
parcheesi), and "mancala games" (e.g., "dealing of beans into a series of holes
and pockets"—a diversion all but unknown in Europe but described as "the
national game of Africa"). It is surprising to find that nothing really new has
been born in the world of games during the last ten centuries. The latest
gambling device or puzzle sold by Abercrombie and Fitch is apt to have an
Egyptian ancestor of the twelfth dynasty, or to have been played for generations
in Basutoland or by Bulgarian peasants. Even where the form of a game is new,
its fundamental scheme can usually be fitted into an established classification.

Kasner and I made a similar attempt at reduction in a field broader than
Murray's. Our enumeration does not compare with his in exhaustiveness; but
many of the principal prototypes of games and puzzles are set forth. It used to
be thought that the mark of a good game was its uselessness; that is now known
to be a mistake. In recent years mathematicians have dissected these playthings
and shown that their strategy is applicable to such worth-while activities as
business and war. One of the leaders in this new branch of research is the
contemporary mathematician, John von Neumann. The theory of games has
now been brought to such a high stage of development that in the next war a
mastery of the subtler elements of draughts and rithmomachy, or even of

dominoes, may prove decisive. Our treatment of games is in the older tradition. The analysis will not, we trust, help you to win a battle or corner the market in black pepper.

The second selection, from Ball and W. E. H. Berwick, considers a special class of arithmetical recreations, some of whose problems are simple, some exceptionally taxing. The problems are known as arithmetical restorations: it is required, given an imperfect set of digits, of an exercise in addition, division, multiplication, to reconstruct the complete figures. There are no general rules, or mathematical formulas which help to solve these problems. It is a matter of pure ingenuity, reasoning power and perseverance.

## ENDNOTE

[1]Oxford University Press, 1952.

5 *Work consists of whatever a body is obliged to do, and play consists of whatever a body is not obliged to do.*

—MARK TWAIN

# PASTIMES OF PAST AND PRESENT TIMES

*By Edward Kasner and James R. Newman*

It has been said, "It is not by amusing oneself that one learns,"[1] and in reply: "It is *only* by amusing oneself that one can learn." Wherever the truth may lie, somewhere between those extremes, it is undeniable that mathematical recreations furnish a challenge to imagination and a powerful stimulus to mathematical activity. The theory of equations, of probability, the infinitesimal calculus, the theory of point sets, of topology—all are fruits grown from seeds sown in the fertile soil of creative imagination—all have grown out of problems first expressed in puzzle form.

Puzzles and paradoxes have been popular since antiquity, and in amusing themselves with these playthings men sharpened their wits and whetted their ingenuity. But it was not for amusement alone that Kepler, Pascal, Fermat, Leibniz, Euler, Lagrange, Hamilton, Cayley, and many others devoted so much time to puzzles. Researches in recreational mathematics sprang from the same desire to know, were guided by the same principles, and required the exercise of the same faculties as the researches leading to the most profound discoveries in mathematics and mathematical physics. Accordingly, no branch of intellectual activity is a more appropriate subject for discussion than puzzles and paradoxes.

The field is enormous. Puzzles have been in the making since Egyptian times and probably before. From the cryptical utterances of the oracle of Delphi, through the time of Charlemagne, down to the golden age of the crossword, paradoxes and puzzles, like the creatures of the earth, have assumed every shape and form and have multiplied. We can examine only a few of the dominating species, those which have survived in one shape or another and continue to thrive in streamlined form.

Most of the famous puzzles invented before the 17th century may be found in the first great puzzle book, *Les problèmes plaisants et délectables, qui se font par les nombres,* by Claude-Gaspard Bachet, Sieur de Meziriac. Although it appeared in 1612, two years before Napier's work on logarithms, it is still a delightful book and a quarry of information. Many collections have appeared since then,[2] Bachet's volume alone having been enlarged to almost five times its original size.

All we can hope to do is to follow the illustrious example of Mark Twain in a similar predicament. He attempted to reduce all jokes to a dozen primitive or elementary forms (mother-in-law, farmer's daughter, etc.). We shall attempt to present a few of the typical puzzles that illustrate the basic ideas from which all are evolved. . . .

Puzzles often seem difficult because they are not easy to interpret in precise terms. In attempting the solution of a problem, the method of trial and error is not only more natural, but generally easier than the mathematical attack. It is common experience that often the most formidable algebraic equations are easier to solve than problems formulated in words. Such problems must first be translated into symbols, and the symbols placed into the proper equations before the problems can be solved.

When Flaubert was a very young man, he wrote a letter to his sister, Carolyn, in which he said: "Since you are now studying geometry and trigonometry, I will give you a problem. A ship sails the ocean. It left Boston with a cargo of wool. It grosses 200 tons. It is bound for Le Havre. The mainmast is broken, the cabin boy is on deck, there are 12 passengers aboard, the wind is blowing East-North-East, the clock points to a quarter past three in the afternoon. It is the month of May—How old is the captain?" Flaubert was not only teasing, he was uttering a complaint shared by that large and respectable company "not good at puzzles," that the average puzzle both confuses and overwhelms with superfluous words.[3] For that reason, the following puzzles have been stripped of all inessential elements so as to exhibit their underlying mathematical structure. And we understand by the term "mathematical structure" not necessarily something expressed by numbers, angles, or lines, but the essential internal relationship between the component elements of the puzzle. For, at bottom, that is all that mathematical analysis can reveal, all that mathematics itself signifies.

Among the oldest problems are those which involve ferrying people and their belongings across a river under somewhat trying conditions. Alcuin, the friend of Charlemagne, suggested a problem which has since been restated and complicated in many ways. A traveler comes to a riverbank with his possessions: a wolf, a goat, and a head of cabbage. The only available boat is very small and can carry no more than the traveler and *one* of his possessions. Unfortunately, if left together, the goat will eat the cabbage and the wolf dine on the goat. How shall the traveler transport his belongings to the other side of the river, keeping his vegetables and animals intact?[4] The solution may be attempted with the aid of a match box, representing the boat, and four slips of paper for its occupants.

A more elaborate version of this problem was suggested in the sixteenth century by Tartaglia. Three beautiful brides with their jealous husbands also come to a river. The small boat which is to take them across holds only two people. To avoid any compromising situations, the crossings are to be so arranged that no woman shall be left with a man unless her husband is present. Eleven passages are required. Five passages would be required for two couples, but with four or more couples the crossing under the conditions stated would be impossible.

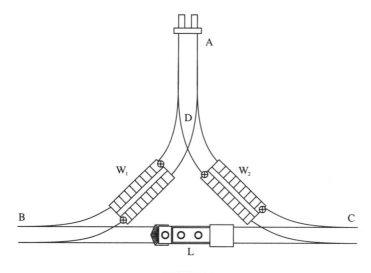

**FIGURE 1**

Similar problems involve shunting. In Figure 1 there is a locomotive, $L$, and 2 freight cars, $W_1$ and $W_2$. The common portion of the rails of the two sidings on which $W_1$ and $W_2$ are standing, $DA$, is long enough to hold $W_1$ or $W_2$, but not both, nor the locomotive $L$. Thus, a car on $DA$ can be shunted to either siding. The engineer's job is to switch the positions of $W_1$ and $W_2$. How can this be done? Although this problem presents no particular difficulties, the same theme in more complex form may demand of the engineer mathematical talents of a high order.

Simeon Poisson's family tried to make him everything from a surgeon to a lawyer, the last on the theory that he was fit for nothing better. One or two of these professions he tackled with singular ineptitude, but at last he found his métier. It was on a journey that someone posed him a problem similar to the one below. Solving it immediately, he realized his true calling and thereafter devoted himself to mathematics, becoming one of the greatest mathematicians of the nineteenth century.[5]

Two friends who have an eight-quart jug of wine wish to share it evenly. They also have two empty jars, one holding five quarts, the other three. The diagram illustrates how they were able to divide their wine into two portions of four quarts each.[6]

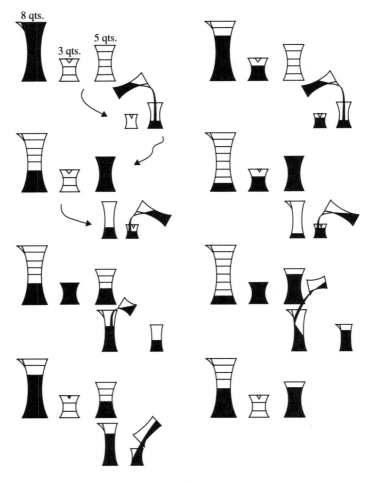

**FIGURE 2**
*Solution to the problem of the 3 jars.*

The mystifying nature of so many arithmetic tricks lies, as we have indicated, in their structure, not their content. With a strainer to sift out the essential ideas hidden between dozens of useless ones, every man could be his own magician. A silly little riddle, oft repeated among mathematicians, comes to mind. "How shall one catch the lions in the desert?" it is asked. Since there is so much sand and so few lions, simply take a strainer, strain out the sand, and there are the lions! Such a strainer, then, or perhaps a scalpel, is needed to get at the rudiments. When the verbiage has been swept away the puzzle skeletons succumb to simple arithmetic or algebra. The parlor tricks of guessing numbers which others have selected, or cards which someone has chosen seem almost as wonderful as instances of "extra sensory perception." But after we have learned to separate the lions from the sand, caging them is comparatively simple.

Card tricks are usually arithmetic puzzles in disguise. Generally, they are amenable to mathematical analysis, and are not, as is commonly believed, performed by sleight of hand. One important principle, easily overlooked, is that "cutting a pack of cards never alters the relative positions of the cards, provided that, if necessary, we regard the top card as following immediately after the bottom card in the pack."[7] Once this is understood, many tricks cease to be baffling.

Seven poker players have invested in a new deck of cards. In keeping with tradition the cards are cut, not shuffled, on the first deal. The dealer, pretending to cheat, takes his second and fourth cards from the bottom of the deck. This lapse is noticed by everyone, as intended. However, when the other players pick up their cards, they are reluctant to demand a new deal, each one finding that he has a full house. But still fearful that the dealer has fixed up a better hand for himself, they insist that he discard his 5 cards and take the next 5 from the top of the deck. Feigning indignation he acquiesces—and wins with a straight flush. Try it. Ninety-nine times out of one hundred you will succeed in cheating your friends—but then, you can't cheat an honest man.

Frequently, the arithmetic tricks of guessing a number selected by another depend on the scale of notation. When a number is expressed in the decimal system, such as 3976, what is actually meant is

$$(3 \times 10^3) + (9 \times 10^2) + (7 \times 10^1) + (6 \times 10^0) .$$

The table[8] further illustrates other numbers written to the base 10.

| Example | $10^0$ | $10^1$ | $10^2$ | $10^3$ | $10^4$ |
|---|---|---|---|---|---|
| 469 $= 9 \times 10^0$ | | $+ 6 \times 10^1$ | $+ 4 \times 10^2$ | | |
| $= 9$ | | $+ 60$ | $+ 400$ | | |
| 7901 $= 1 \times 10^0$ | | $+ 0 \times 10^1$ | $+ 9 \times 10^2$ | $+ 7 \times 10^3$ | |
| $= 1$ | | $+ 0$ | $+ 900$ | $+ 7000$ | |
| 30,000 $= 0 \times 10^0$ | | $+ 0 \times 10^1$ | $+ 0 \times 10^2$ | $+ 0 \times 10^3$ | $+ 3 \times 10^4$ |
| $= 0$ | | $+ 0$ | $+ 0$ | $+ 0$ | $+ 30,000$ |
| 21,148 $= 8 \times 10^0$ | | $+ 4 \times 10^1$ | $+ 1 \times 10^2$ | $+ 1 \times 10^3$ | $+ 2 \times 10^4$ |
| $= 8$ | | $+ 40$ | $+ 100$ | $+ 1000$ | $+ 20,000$ |

Among the wide variety of problems which arise from the use of the decimal system, the following are of some interest:

A useful device for checking multiplication goes by the copybook title of "casting out nines."

Consider $1234 \times 5678 = 7006652$. Add the digits of the multiplier, multiplicand, and product, thus obtaining 10, 26 and 26 respectively. Since each of these numbers is greater than 9, add the digits of the individual sums once more,[9] obtaining 1, 8, and 8. (If, after the first repetition a sum greater than 9 remains, the digits may be added once again.) Now, take the product of the integers corresponding to the multiplier, and the multiplicand, i.e., 1 × 8, and compare this with the integer corresponding to the sum of the digits of the product, which is also 8. Since they are the same, the result of the original multiplication is correct.

Using the same rule, let us test whether the product of 31256 and 8427 is 263395312. Again the sum of the digits of the multiplicand, multiplier, and the product are respectively 17, 21, and 34; repeating, the sum of these digits is 8, 3, and 7. The product of the first two equals 24 which has 6 for the sum of its digits. But the sum of the digits of the product equals 7. Thus, we have two different remainders, 6 and 7, whence the multiplication must be incorrect.

Closely connected with the rule of casting out nines is the following trick, which reveals a remarkable property common to all numbers.

Take any number and rearrange its digits in any order you please to form another number. The difference between the first number and the second is always divisible by 9.[10]

Another type of problem dependent on the decimal scale of notation involves finding numbers which may be obtained by multiplying their *reversals* by integers. Among such numbers with 4 digits, 8712 equals 4 times 2178, and 9801 equals 9 times 1089.

The *binary* or *dyadic* notation (using the base 2) is hardly a new concept, having been referred to in a Chinese book believed to have been written about 3000 B.C. Forty-six centuries later, Leibniz rediscovered the wonders of the binary scale and marveled at it as though it were a new invention—somewhat like the twentieth-century city dweller, who, upon seeing a sundial for the first time, and having it explained, remarked with awe: "What will they think of next?" In its use of only two symbols, Leibniz saw in the dyadic system something of great religious and mystic significance: God could be represented by unity, and nothingness by zero, and since God had created all forms out of nothingness, zero and one combined could be made to express the entire universe. Anxious to impart this gem of wisdom to the heathens, Leibniz communicated it to the Jesuit Grimaldi, president of the Tribunal of Mathematics in China, in the hope that he could thus show the Emperor of China the error of his ways in clinging to Buddhism instead of adopting a God who could create a universe out of nothing.

Whereas the decimal scale requires ten symbols: 0, 1, 2, 3, 4, . . . , 9, the binary scale uses only two: 0 and 1. On the following page, are the first 32 integers given in the binary scale. Since $2^0 = 1$, it may readily be seen that *any* number can be expressed as the *sum* of powers of 2, just as any number in the decimal system can be expressed as the *sum* of powers of 10. For example, the

| Decimal | | Binary | Decimal | | Binary |
|---|---|---|---|---|---|
| 1 | = | 1 | 17 | = | 10001 |
| 2 | = | 10 | 18 | = | 10010 |
| 3 | = | 11 | 19 | = | 10011 |
| 4 | $= 2^2 =$ | 100 | 20 | = | 10100 |
| 5 | = | 101 | 21 | = | 10101 |
| 6 | = | 110 | 22 | = | 10110 |
| 7 | = | 111 | 23 | = | 10111 |
| 8 | $= 2^3 =$ | 1000 | 24 | = | 11000 |
| 9 | = | 1001 | 25 | = | 11001 |
| 10 | = | 1010 | 26 | = | 11010 |
| 11 | = | 1011 | 27 | = | 11011 |
| 12 | = | 1100 | 28 | = | 11100 |
| 13 | = | 1101 | 29 | = | 11101 |
| 14 | = | 1110 | 30 | = | 11110 |
| 15 | = | 1111 | 31 | = | 11111 |
| 16 | $= 2^4 =$ | 10000 | 32 | $= 2^5 =$ | 100000 |

number expressed in the decimal system as 25, is expressed in the binary system, using only the two symbols 1 and 0, by 11001.

| Decimal | | Dyadic |
|---|---|---|
| 25 | = | 11001 |
| ↕ | | ↕ |
| $(2 \times 10^1) + (5 \times 10^0)$. | | $(1 \times 2^4) + (1 \times 2^3) + (0 \times 2^2)$ |
| | | $+ (0 \times 2^1) + (1 \times 2^0)$. |

Because numbers can be more briefly written in the decimal scale than in the binary, it is more convenient, although in every other respect the latter is just as accurate and efficient. Even fractions have their place in the dyadic notation. The fraction ⅓, for example, given by the nonterminating decimal, .33333 . . . , is represented in the binary notation by a nonterminating binary, .01010101 . . .[11]

The binary system easily makes understandable the solution of problems such as:

I. In many sections of Russia, the peasants employed until recently what appears to be a very strange method of multiplication. In substance, this was at one time in use in Germany, France, and England, and is similar to a method used by the Egyptians 2000 years before the Christian era.

It is best illustrated by an example: To multiply 45 by 64, form two columns. At the head of one put 45, at the head of the other, 64. Successively multiply one column by 2 and divide the other by the same number. When an odd number is divided by 2, discard the remaining fraction. The result will be:

|  | Divide | Multiply |
|---|---|---|
|  | 45 | 64 |
|  | 22 | 128 |
| (A) | 11 | 256 |
|  | 5 | 512 |
|  | 2 | 1024 |
|  | 1 | 2048 |

Take from the second column those numbers which appear opposite an odd number in the first. Add them and you obtain the desired product:

| | 45 | 64 | .......... | $64 = 2^0 \times 64$ |
|---|---|---|---|---|
| | 22 | 128 | | $= 2^1 \times 64$ |
| (B) | 11 | 256 | .......... | $256 = 2^2 \times 64$ |
| | 5 | 512 | .......... | $512 = 2^3 \times 64$ |
| | 2 | 1024 | | $= 2^4 \times 64$ |
| | 1 | 2048 | .......... | $2048 = 2^5 \times 64$ |

$$2880 = 45 \times 64$$

The relation of this method to the dyadic system may be seen upon expressing 45 in the dyadic notation.

$$45 = (1 \times 2^5) + (0 \times 2^4) + (1 \times 2^3) + (1 \times 2^2) + (0 \times 2^1) + (1 \times 2^0)$$
$$= 101101$$
$$= 32 + 0 = 8 + 4 + 0 + 1$$

Therefore,

$$45 \times 64 = (2^5 + 2^3 + 2^2 + 2^0) \times 64$$
$$= (2^5 \times 64) + (2^3 \times 64) + (2^2 \times 64) + (2^0 \times 64).$$

Since $2^4$ and $2^1$ do not appear in the dyadic expression for 45, the products $(2^4 \times 64)$ and $(2^1 \times 64)$ are not included in the numbers to be added in (B). Thus, what the peasant does in multiplying 45 by 64 is to multiply $2^5$, $2^3$, $2^2$, $2^0$, successively by 64, and then take the sum.

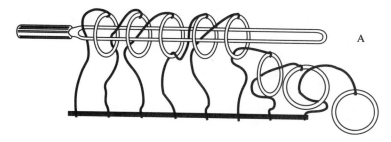

**FIGURE 3**
*The Chinese ring puzzle.*

II. Another well-known problem, already mentioned by Cardan, consists in the removal of a number of rings from a bar. The puzzle can best be analyzed by the use of the dyadic system, although the actual manipulation of the rings is at all times extremely difficult.

The rings on the bar are so connected that although the end one can be removed without difficulty, any other ring can be put on or removed only when the one next to it, toward the end (A in the figure) is on the bar, and all the rest of the rings are *off*. Thus, to remove the fifth ring, the first, second, third must be off the bar, and the fourth must be on. If the position of all the rings on or off the rack are written in the dyadic notation, 1 designating a ring which is off, and 0 designating a ring which is on, the mathematical determination of the number of moves required to remove a given number of rings is not too hard. The solution without the aid of the dyadic notation, as the rings increase in number, would be wholly beyond one's imaginative powers.

III. The problem of the Tower of Hanoi is similar in principle. The game consists of a board with three pegs, as illustrated in Figure 4.

On one of these pegs rests a number of discs of various sizes, so arranged that the largest disc is on the bottom, the next largest rests on that one, the next largest on that, and so on, up to the smallest disc which is on top. The problem is to transfer the entire set of discs to one of the other two pegs, moving only one disc at a time, and making certain that no disc is ever permitted to rest on one smaller than itself. If the removal of a disc from one peg to another constitutes one transfer, the following table shows the number of transfers required for various numbers up to $n$ discs:

TABLE FOR TRANSFERS[12]

| Discs | Transfers |
|-------|-----------|
| 1 | 1 |
| 2 | 3 |
| 3 | 7 |
| 4 | 15 |
| 5 | 31 |
| 6 | 63 |
| 7 | 127 |
| . | . |
| . | . |
| . | . |
| $n$ | $2^n - 1$ |

There is a charming story about this toy:[13]

In the great temple at Benares beneath the dome which marks the center of the world, rests a brass plate in which are fixed three diamond needles, each a cubit high and as thick as the body of a bee. On one of these needles, at the creation, God placed sixty-four discs of pure gold, the largest disc resting on the brass plate and the others getting smaller and smaller up to the top one. This is the Tower of Brahma. Day and night unceasingly, the priests transfer the discs from one diamond needle to another, according to the fixed and immutable laws of Brahma, which require that the priest on duty must not move more than one disc at a time and that he must place this disc on a needle so that there is no smaller disc below it. When the sixty-four discs shall have been thus transferred from the needle on which, at the creation, God placed them, to one of the other needles, tower, temple, and Brahmans alike will crumble into dust, and with a thunderclap, the world will vanish.

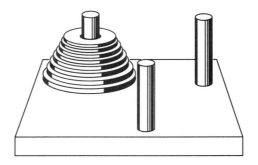

**FIGURE 4**

The number of transfers required to fulfill the prophecy is $2^{64}-1$, that is 18,446,744,073,709,551,615. If the priests were to effect one transfer every second, and work 24 hours a day for each of the 365 days in a year,[14] it would take them 58,454,204,609 centuries plus slightly more than 6 years to perform the feat, assuming they never made a mistake—for one small slip would undo all their work.

IV. One other game may be mentioned in connection with the dyadic system—Nim. In this game, two players play alternately with a number of counters placed in several heaps. At his turn, a player picks up one of the heaps, or as many of the counters from it as he pleases. The player taking the last counter loses. If the number of counters in each heap is expressed in the binary scale, the game readily lends itself to mathematical analysis. A player who can bring about a certain arrangement of the number of counters in each heap may force a win.[15]

It is interesting to note that the number $2^{64}$—18,446,744,073,709,551,-616—represented in the dyadic system by a number with 64 digits, appears in the solution of a puzzle connected with the origin of the game of chess.

According to an old tale, the Grand Vizier Sissa Ben Dahir was granted a boon for having invented chess for the Indian King, Shirhâm. Since this game is played on a board with 64 squares, Sissa addressed the king: "Majesty, give me a grain of wheat to place on the first square, and two grains of wheat to place on the second square, and four grains of wheat to place on the third, and eight grains of wheat to place on the fourth, and so, Oh, King, let me cover each of the 64 squares of the board." "And is that all you wish, Sissa, you fool?" exclaimed the astonished King. "Oh, Sire," Sissa replied, "I have asked

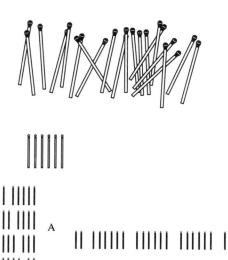

**FIGURE 5**

*The diagram illustrates how to force a win at the game of Nim. Assume each player at his turn must pick up at least one match and may pick up as many as five. The rule is that the player picking up the last match loses. Then, for example, imagine that the original heap consists of 21 matches. In that case, the one playing first can force a win by mentally dividing the matches into groups of 1, 6, 6, 6, and 2 (as in B). Since he plays first, he picks up 2 matches. Then, however many his opponent picks, the first player picks up the complement of 6. This is shown in A: If the second player takes 1, the first player takes 5; if the second player takes 2, the first player takes 4, and so on. Each of the three groups of 6 is thus exhausted, and the second player is left with the last match.*

*Had there been 47 matches, say, the grouping to force a win for the first player would have been: 1, 6, 6, 6, 6, 6, 6, and 4. Rules for any other variation of Nim can also be easily formulated.*

for more wheat than you have in your entire kingdom, nay, for more wheat than there is in the whole world, verily, for enough to cover the whole surface of the earth to the depth of the twentieth part of a cubit."[16] Now the number of grains of wheat which Sissa demanded is $2^{64}-1$, exactly the same as the number of disc transfers required to fulfill the prophecy of Benares related on pp. 2401–2402.

Another remarkable way in which $2^{64}$ arises is in computing the number of each person's ancestors from the beginning of the Christian era—just about 64 generations ago. In that length of time, assuming that each person has 2 parents, 4 grandparents, 8 great-grandparents, etc., and not allowing for incestuous combinations, everyone has at least $2^{64}$ ancestors, or a little less than eighteen and a half quintillion lineal relations alone. A most depressing thought.

The Josephus problem is one of the most famous and certainly one of the most ancient. It generally relates a story about a number of people on board a ship, some of whom must be sacrificed to prevent the ship from sinking. Depending on the time that the version of the puzzle was written, the passengers were Christians and Jews, Christians and Turks, sluggards and scholars, Negroes and whites, etc. Some ingenious soul with a knowledge of mathematics always managed to preserve the favored group. He arranged everyone in a circle, and reckoning from a certain point onward, every $n$th person was to be thrown overboard—$n$ being a specified integer. The arrangement of the circle by the mathematician was such that either the Christians, or the industrious scholars, or the whites were saved, while the rest were thrown overboard in accordance with the Golden Rule.

Originally, this tale was told of Josephus who found himself in a cave with 40 other Jews bent on self-extinction to escape a worse fate at the hands of the Romans. Josephus decided to save his own neck. He placed everyone in a circle and made them agree that each third person, counting around and around, should be killed. Placing himself and another provident soul in the 16th and the 31st position of the circle of 41, he and his companion, being the last ones left, were conveniently able to avoid the road to martyrdom.

A later version of this problem places 15 Turks and 15 Christians on board a storm-ridden ship which is certain to sink unless half the passengers are thrown overboard. After arranging everyone in a circle, the Christians, *ad majorem Dei gloriam,* proposed that every ninth person be sacrificed.

Thus, every infidel was properly disposed of, and all true Christians saved.[17]

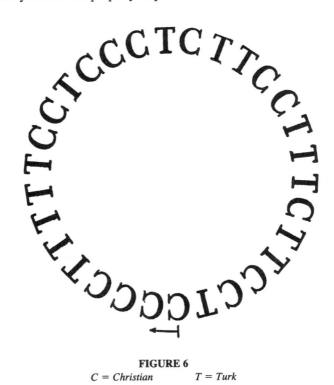

**FIGURE 6**

$C$ = *Christian*      $T$ = *Turk*

Among the Japanese, the Josephus problem assumed another form: Thirty children, 15 of the first marriage, and 15 of the second, agree that their father's estate is too small to be divided among all of them. So the second wife proposes that all the children be arranged in a circle, in order to determine her husband's heir by a process of elimination. Being a prudent mathematician, as well as the proverbially wicked stepmother, she arranges the children in such a way that one of her own is certain to be chosen. After 14 of the children of the first marriage have been eliminated, the remaining child, evidently a keener mathematician than his stepmother, proposes that the counting shall start afresh in the opposite direction. Certain of her advantage, and therefore disposed to be generous, she consents, but finds to her dismay that all 15 of her own children are eliminated, leaving the one child of the first marriage to become the heir.[18]

Elaborate mathematical solutions of more difficult and generalized versions of the Josephus problem were given by Euler, Schubert, and Tait.

**FIGURE 7**
*The Josephus problem, from Miyake Kenryu's "Shojutsu." (From Smith and Mikami, "A History of Japanese Mathematics.")*

No discussion of puzzles, however brief, can afford to omit mention of the best-known of the many puzzles invented by Sam Loyd. "15 Puzzle," "Boss Puzzle," "Jeu de Taquin," are a few of its names. For several years after its appearance in 1878, this puzzle enjoyed a popularity, particularly throughout Europe, greater than swing and contract bridge combined enjoy today. In Germany, it was played in the streets, in factories, in the royal palaces, and in the Reichstag. Employers were forced to post notices forbidding their employees to play the "15 Puzzle" during business hours under penalty of dismissal. The electorate, having no such privileges, had to watch their duly elected representatives play the "Boss Puzzle" in the Reichstag while Bismarck played the Boss. In France, the "Jeu de Taquin" was played on the boulevards of Paris and in every tiny hamlet from the Pyrenees to Normandy. A scourge of mankind was the "Jeu de Taquin," according to a contemporary French journalist,—worse than tobacco and alcohol—"responsible for untold headaches, neuralgias, and neuroses."

For a time, Europe was "15 Puzzle" mad. Tournaments were staged and huge prizes offered for the solution of apparently simple problems. But the strange thing was that no one ever won any of these prizes, and the apparently simple problems remained unsolved.

The "15 Puzzle" (figure below) consists of a square shallow box of wood or metal which holds 15 little square blocks numbered from 1 to 15. There is actually room for 16 blocks in the box so that the 15 blocks can be moved about and their places interchanged. The number of conceivable positions is 16! = 20,922,789,888,000. A problem consists of bringing about a specified arrangement of the blocks from a given initial position, which is frequently the *normal* position illustrated in Figure 8.

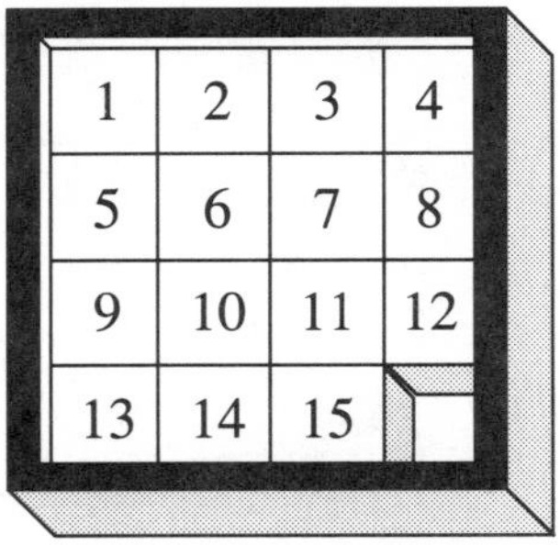

**FIGURE 8**
*The 15 Puzzle (also Boss Puzzle or Jeu de Taquin) in normal position.*

Shortly after the puzzle was invented, two American mathematicians[19] proved that from any given initial order only *half* of all the conceivable positions can actually be obtained. Thus, there are always approximately ten trillion positions which the possessor of a "15 Puzzle" can bring about, and ten trillion that he *cannot*.

The fact that there are impossible positions makes it easy to understand why such generous cash prizes were offered by Loyd and others, since the problems for which prizes were offered always entailed impossible positions. And it is heart-breaking to think of the headaches, neuralgias, and neuroses that might have been spared—to say nothing of the benefits to the Reichstag—if *The American Journal of Mathematics* had been as widely circulated as the puzzle itself. With ten trillion possible solutions there still would have been enough fun left for everyone.

In the normal position (Figure 8), the blank space is in the lower right-hand corner. When making a mathematical analysis of the puzzle, it is convenient to consider that a rearrangement of the blocks consists of nothing more than moving the blank space itself through a specific path, always making certain that it ends its journey in the lower right-hand corner of the box. To this end, the blank space must travel through the same number of boxes to the left as to the right and through the same number of boxes upwards as downwards. In other words, *the blank space must move through an even number of boxes.* If, starting from the normal position, the desired one can be attained while complying with this requirement, it is a *possible* position, otherwise it is *impossible*.

Based upon this principle, the method of determining whether a position is possible or impossible is very simple. In the normal position every numbered block appears in its proper numerical order, i.e., regarding the boxes, row by row, from left to right, no number precedes any number smaller than itself. To bring about a position different from the normal one, the numerical order of the blocks must be changed. Some numbers, perhaps all, will precede others smaller than themselves. Every instance of a number preceding another smaller than itself is called an *inversion.* For example, if the number 6 precedes the numbers 2, 4, and 5, this is an inversion to which we assign the value 3, because 6 precedes three numbers smaller than itself. If the *sum* of the *values* of *all the inversions* in a given position is even, the position is a possible one— that is, it can be brought about from the normal position. If the sum of the values of the inversions is odd, the position is impossible and cannot be brought about from the normal configuration.

The position illustrated in Figure 9 can be created from the normal position since the sum of the values of the inversions is six—an even number. But the position shown in Figure 10 is impossible, since, as may readily be seen, the sum of the value of the inversions brought about is odd.

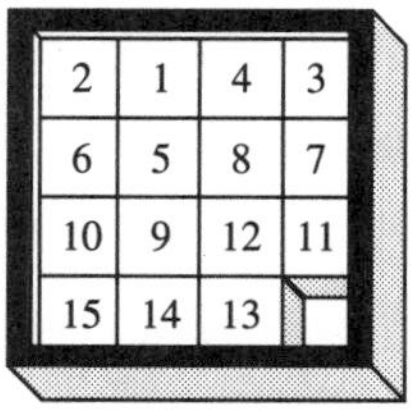

| 2 | 1 | 4 | 3 |
|---|---|---|---|
| 6 | 5 | 8 | 7 |
| 10 | 9 | 12 | 11 |
| 15 | 14 | 13 | |

**FIGURE 9**

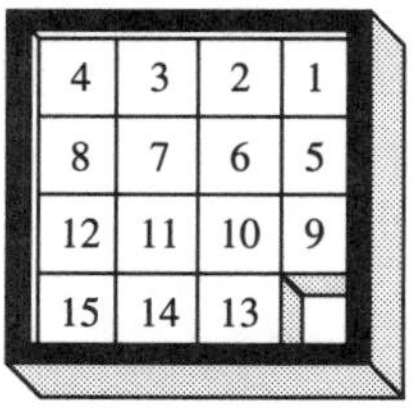

| 4 | 3 | 2 | 1 |
|---|---|---|---|
| 8 | 7 | 6 | 5 |
| 12 | 11 | 10 | 9 |
| 15 | 14 | 13 | |

**FIGURE 10**

Figures 11 a, b, c illustrate three other positions. Are they possible, or impossible to obtain from the normal order?

| 1 | 2 | 3 | 4 |
|---|---|---|---|
| 5 | 6 | 7 | 8 |
| 9 | 10 | 11 | 12 |
| 15 | 14 | 13 | |

| 11 | 7 | 4 | |
|---|---|---|---|
| 8 | 13 | 1 | 2 |
| 5 | 10 | 3 | 9 |
| 15 | 12 | 14 | 6 |

| 2 | 4 | 6 | 8 |
|---|---|---|---|
| 10 | 11 | 12 | 13 |
| 3 | 5 | 7 | 9 |
| 15 | 1 | 14 | |

**FIGURES 11 (*a, b, c*)**

## SPIDER AND FLY PROBLEM

Most of us learned that a straight line is the shortest distance between two points. If this statement is supposed to apply to the earth on which we live, it is both useless and untrue. As we have seen in the previous chapter, the nineteenth-century mathematicians Riemann and Lobachevsky knew that the statement, if true at all, applied only to special surfaces. It does not apply to a *spherical* surface on which the shortest distance between two points is the arc of a great circle. Since the shape of the earth approximates a sphere, the shortest distance between two points anywhere on the surface of the earth is *never* a straight line, but is a portion of the arc of a great circle.

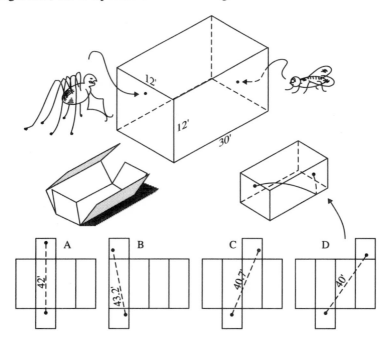

**FIGURE 12**
*The spider, his kind invitation to the fly having been rebuffed, sets out for dinner along the shortest possible route. What path is the geodesic for the hungry spider?*

Yet, for all *practical* purposes, even on the surface of the earth, the shortest distance between two points is given by a straight line. That is to say, in measuring ordinary distances with a steel tape or a yardstick, the principle is substantially correct. However, for distances beyond even a few hundred feet, allowance must be made for the curvature of the earth. When a steel rod over 600 feet in length was recently constructed in a large Detroit automobile factory, it was found that the exact measurement of its length was impossible without allowing for the earth's curvature. We indicated that the determination of a geodesic is very difficult for complicated surfaces. But we can give one puzzle showing how deceptive this problem may be for even the simplest case—the flat surface.

In a room 30 feet long, 12 feet wide, and 12 feet high, there is a spider in the center of one of the smaller walls, 1 foot from the ceiling; and there is a fly in the middle of the opposite wall, 1 foot from the floor. The spider has designs on the fly. What is the shortest possible route along which the spider may crawl to reach his prey? If he crawls straight down the wall, then in a straight line along the floor, and then straight up the other wall, or follows a similar route along the ceiling, the distance is 42 feet. Surely it is impossible to imagine a shorter route! However, by cutting a sheet of paper, which, when properly folded, will make a model of the room (see Figure 12), and then by joining the points representing the spider and the fly by a straight line, a geodesic is obtained. The length of this geodesic is only 40 feet, in other words, 2 feet shorter than the "obvious" route of following straight lines.

There are several ways of cutting the sheet of paper, and accordingly, there are several possible routes, but that of 40 feet is the shortest; and remarkably enough, as may be seen from cut *D* in Figure 12, this route requires the spider to pass over 5 of the 6 sides of the room.

This problem graphically reveals the point emphasized throughout—our intuitive notions about space almost invariably lead us astray.

## RELATIONSHIPS

Ernest Legouvé,[20] the well known French dramatist, tells in his memoirs that, while taking the baths at Plombières, he proposed a question to his fellow bathers: "Is it possible for two men, wholly unrelated to each other, to have the same sister?" "No, that's impossible," replied a notary at once. An attorney who was not quite so quick in giving his answer, decided after some deliberation, that the notary was right. Thereupon, the others quickly agreed that it was impossible. "But still it is possible," Legouvé remarked, "and I will name two such men. One of them is Eugene Sue, and I am the other." In the midst of cries of astonishment and demands that he explain, he called the bath attendant and asked for the slate on which the attendant was accustomed to mark down those who had come for their baths. On it, he wrote:

($\sim$ means married to; | means offspring of)

| Mrs. Sue ~ Mr. Sue | Mrs. Sauvais ~ Mr. Sue | Mrs. Sauvais ~ Mr. Legouvé |
|:---:|:---:|:---:|
| \| | \| | \| |
| Eugene Sue | Flore Sue | Ernest Legouvé |

"Thus, you see," he concluded, "it is quite possible for two men to have the same sister, without being related to each other."

Most of the puzzles treated hitherto required four steps for their solution:

1. Sifting out the essential facts.
2. Translating these facts into the appropriate symbols.
3. Setting up the symbols in equations.
4. Solving the equations.

To solve the problems of relationship two of these steps must be modified. A simple diagram replaces the algebraic equation; inferences from the diagram replace the algebraic solution. Without the symbols and diagrams, however, the problems may become extremely confusing.

Alexander Macfarlane, a Scotch mathematician, developed an "algebra of relationships" which was published in the proceedings of the Royal Society of Edinburgh, but the problems to which he applied his calculus were easily solvable without it. Macfarlane used the well-known jingle:

> Brothers and sisters have I none,
> But this man's father is my father's son,

as a guinea pig for his calculus, although the diagrammatic method gives the solution much more quickly.

An old Indian fairy tale creates an intricate series of relationships which would probably prove too much even for Macfarlane's algebra. A king, dethroned by his relatives, was forced to flee with his wife and daughter. During their flight they were attacked by robbers; while defending himself, the king was killed, although his wife and daughter managed to escape. Soon they came to a forest in which a prince of the neighboring country and his son were hunting. The prince (a widower) and his son (an eligible bachelor) noticing the footsteps of the mother and daughter decided to follow them. The father declared that he would marry the woman with the large footsteps—undoubtedly the older—and the son said he would marry the woman with the small footsteps who was surely the younger. But on their return to the castle, the father and his son discovered that the small feet belonged to the mother and the big feet to the daughter. Nevertheless, mastering their disappointment, they married as they had planned. After the marriage, the mother, daughter-in-law of her daughter, the daughter, mother-in-law of her mother, both had children—sons and daughters. The task of disentangling the resulting relationships we entrust to the reader, as well as the explanation of the following verse found on an old gravestone at Alencourt, near Paris:

Here lies the son; here lies the mother;
Here lies the daughter; here lies the father;
Here lies the sister; here lies the brother;
Here lie the wife and the husband.
Still, there are only three people here.

In Albrecht Dürer's famous painting, "Melancholia," there appears a device about which more has been written than any other form of mathematical amusement. The device is a *magic square*.

A magic square consists of an array of integers in a square which, when added up by rows, diagonals, or columns, yield the same total. Magic squares date back at least to the Arabs. Great mathematicians like Euler and Cayley found them amusing and worth studying. Benjamin Franklin admitted somewhat apologetically that he had spent some time in his youth on these "trifles"—time "which," he hastened to add, "I might have employed more usefully." Mathematicians have never pretended that magic squares were anything more than amusement, however much time they spent on them, although the continual study devoted to this puzzle form may incidentally have cast some light on relations between numbers. Their chief appeal is still mystical and recreational.

There are other puzzles of considerable interest not discussed here[21] because we treat them more fully in their proper place.[22] Among these are problems connected with the theory of probability, map-coloring, and the one-sided surfaces of Möbius.

Only one extensive group of problems remains—those connected with the theory of numbers. The modern theory of numbers, represented by a vast literature, engages the attention of every serious mathematician. It is a branch of study, many theorems of which, though exceedingly difficult to prove, can be simply stated and readily understood by everyone. Such theorems are therefore more widely known among educated laymen than theorems of far greater importance in other branches of mathematics, theorems which require technical knowledge to be understood. Every book on mathematical recreations is filled with simple or ingenious, cunning or marvelous, easy or difficult puzzles based on the behavior and properties of numbers. Space permits us to mention only one or two of those significant theorems about numbers which, in spite of their profundity, can be easily grasped.

Ever since Euclid proved[23] that the number of primes is infinite, mathematicians have been seeking for a test which would determine whether or not a given number is a prime. But no test applicable to all numbers has been found. Curiously enough, there is reason to believe that certain mathematicians of the seventeenth century, who spent a great deal of time on number theory, had means of recognizing primes unknown to us. The French mathematician Mersenne and his much greater contemporary, Fermat, had an uncanny way of determining the values of $p$, for which $2^p - 1$ is a prime. It has not yet been clearly determined how completely they had developed their method, or indeed, exactly what method they employed. Accordingly, it is still a source of wonder that Fermat replied without a moment's hesitation to a letter which asked whether 100895598169 was a prime, that it was the product of 898423 and 112303, and that each of these numbers was prime.[24] Without a general formula for all primes, a mathematician, even today, might spend years hunting for the correct answer.

One of the most interesting theorems of number theory is Goldbach's, which states that every even number is the sum of two primes. It is easy to understand; and there is every reason to believe that it is true, no even number having ever been found which is *not* the sum of two primes; yet, no one has succeeded in finding a proof valid for all even numbers.

But perhaps the most famous of all such propositions, believed to be true, but never proved, is "Fermat's Last Theorem." In the margin of his copy of Diophantus, Fermat wrote: "If $n$ is a number greater than two, there are no whole numbers, $a$, $b$, $c$ such that $a^n + b^n = c^n$. I have found a truly wonderful proof which this margin is too small to contain." What a pity! Assuming Fermat actually had a proof, and his mathematical talents were of such a high order that it is certainly possible, he would have saved succeeding generations of mathematicians unending hours of labor if he had found room for it on the margin. Almost every great mathematician since Fermat attempted a proof, but none has ever succeeded.

Many pairs of integers are known, the sum of whose squares is also a square, thus:

$$3^2 + 4^2 = 5^2; \text{ or, } 6^2 + 8^2 = 10^2.$$

But no three integers have ever been found where the sum of the cubes of two of them is equal to the cube of the third. It was Fermat's contention that this would be true for all integers when the power to which they were raised was greater than 2. By extended calculations, it has been shown that Fermat's theorem is true for values of $n$ up to 617. But Fermat meant it for *every* $n$ greater than 2. Of all his great contributions to mathematics, Fermat's most celebrated legacy is a puzzle which three centuries of mathematical investigation have not solved and which skeptics believe Fermat, himself, never solved.

Somewhat reluctantly we must take our leave of puzzles. Reluctantly, because we have been able to catch only a glimpse of a rich and entertaining subject, and because puzzles in one sense, better than any other single branch of mathematics, reflect its always youthful, unspoiled, and inquiring spirit. When a man stops wondering and asking and playing, he is through. Puzzles are made of the things that the mathematician, no less than the child, plays with, and dreams and wonders about, for they are made of the things and circumstances of the world he lives in.

# ENDNOTES

[1]Anatole France, *The Crime of Sylvestre Bonnard.*

[2]W. W. R. Ball, *Mathematical Recreations and Essays,* 11th ed. New York: Macmillan, 1939.

W. Lietzmann, *Lustiges und Merkwürdiges von Zahlen und Formen,* Breslau: Hirt, 1930.

Helen Abbot Merrill, *Mathematical Excursions,* Boston: Bruce Humphries, 1934.

W. Ahrens, *Mathematische Unterhaltungen und Spiele,* Leipzig: B. G. Teubner, 1921, vols. I and II.

H. E. Dudeney, *Amusements in Mathematics,* London: Thomas Nelson, 1919.

E. Lucas, *Récréations Mathématiques,* Paris: Gautier-Villars. 1883–1894, vols. I, II, III and IV.

[3]Here is an example of a type of puzzle quite fashionable of late, which, though apparently wordy, contains no unessential facts.

## THE ARTISANS

There are three men, John, Jack and Joe, each of whom is engaged in two occupations. Their occupations classify each of them as two of the following: chauffeur, bootlegger, musician, painter, gardener, and barber.

From the following facts find in what two occupations each man is engaged:

1. The chauffeur offended the musician by laughing at his long hair.
2. Both the musician and the gardener used to go fishing with John.
3. The painter bought a quart of gin from the bootlegger.
4. The chauffeur courted the painter's sister.
5. Jack owed the gardener $5.
6. Joe beat both Jack and the painter at quoits.

[4]There are two different ways, both of which are symbolized in the following table.

|  | *First Solution* |  |  |  | *Second Solution* |  |
|---|---|---|---|---|---|---|
|  |  |  | W = Wolf   C = Cabbage |  |  |  |
|  |  |  | G =  Goat →  = Crossing |  |  |  |
| 1. | WGC |  |  | 1. | WGC |  |
| 2. | WC | G→ | G | 2. | WC | G→ | G |
| 3. | WC | ← | G | 3. | WC | ← | G |
| 4. | C | W→ | WG | 4. | W | C→ | GC |
| 5. | GC | ←G | W | 5. | WG | ←G | C |
| 6. | G | →C | WC | 6. | G | W→ | WC |
| 7. | G | ← | WC | 7. | G | ← | WC |
| 8. |  | G→ | WGC | 8. |  | G→ | WGC |

[5]At least so says his biographer, Arago. Not only was the quality of Poisson's work extremely high, but the output was enormous. Besides occupying several important official positions, he turned out over 300 works in a comparatively short lifetime (1781–1840). "La vie, c'est le travail," said this erstwhile shadow on the Poisson household, though oddly enough, a puzzle brought him to a life dedicated to unceasing labor.

[6]Fill the 5 quart jar from the 8 quart jar and pour 3 quarts from the 5 quart jar into the 3 quart jar. Then pour the 3 quarts back into the 8 quart jar. Pour the remaining 2 quarts from the 5 quart jar into the 3 quart jar. Now fill the 5 quart jar again. Since there are 2 quarts in the 3 quart jar, one additional quart will fill this jar. Pour enough wine from the 5 quart jar to fill the 3 quart jar. The 5 quart jar will then have 4 quarts remaining in it. Now pour the 3 quarts from the 3 quart jar into the 8 quart jar. This, together with the 1 quart remaining in the 8 quart jar, will make 4 quarts.

[7]W. W. R. Ball. *op. cit.*

[8]Other bases have been suggested. There is reason to believe that the Babylonians employed the base 60, and in more recent times, the use of the base 12 has been urged rather strongly.

[9]Thus 10 = 1 + 0 = 1
        26 = 2 + 6 = 8, etc.

[10]Hall and Knight, *Higher Algebra.*

[11]Arnold Dresden, *An Invitation to Mathematics.* New York : Henry Holt & Co., 1936.

[12]W. Ahrens, *op. cit.*

[13]W. W. R. Ball, *op. cit.*

[14](Making allowance for leap years.—ED.).

[15]See W. Ahrens, *op. cit.,* and Bouton, *Annals of Mathematics,* series 2, vol. III (1901–1902), pp. 35–39. for the mathematical proof of Nim.

[16]One-twentieth of a cubit is about one inch.

[17]The general rule of solution of all such problems may be found in P. G. Tait, *Collected Scientific Papers,* 1900.

[18]Smith & Mikami, *A History of Japanese Mathematics,* p. 83.

[19]Johnson & Story, *American Journal of Mathematics,* vol. 2 (1879).

[20]Ahrens, *op. cit.,* volume 2.

[21]See, for example, Kasner and Newman's discussion of paradoxes, pp. 1907–1925.

[22]There are also puzzles which though very amusing and deceptive, present no mathematical idea which has not been already considered—and such puzzles have, therefore, been omitted. We may, nevertheless, give three examples, chosen because they are so often solved incorrectly:

(a) A glass is half-filled with wine, and another glass half-filled with water. From the first glass remove a teaspoonful of wine and pour it into the water. From the *mixture* take a teaspoonful and pour it into the wine. Is the quantity of wine in the water glass now greater or less than the quantity of water in the wine glass? To end all quarrels—they are the same.

(b) The following puzzle troubled the delegates to a distinguished gathering of puzzle experts not long ago. A monkey hangs on one end of a rope which passes through a pulley and is balanced by a weight attached to the other end. The monkey decides to climb the rope. What happens? The astute puzzlers engaged in all sorts of futile conjectures and speculations, ranging from doubts as to whether the monkey could climb the rope, to rigorous "mathematical demonstrations" that he couldn't. (We yield to a shameful and probably superfluous urge to point out the solution—the weight rises, like the monkey!)

(c) Imagine we have a piece of string 25,000 miles long, just long enough to exactly encircle the globe at the equator. We take the string and fit it snugly around, over oceans, deserts, and jungles. Unfortunately, when we have completed our task we find that in manufacturing the string there has been a slight mistake, for it is just a yard too long.

To overcome the error, we decide to tie the ends together and to distribute this 36 inches evenly over the entire 25,000 miles. Naturally (we imagine) this will never be noticed. How far do you think that the string will stand off from the ground at each point, merely by virtue of the fact that it is 36 inches too long?

The correct answer seems incredible, for the string will stand 6 inches from the earth over the entire 25,000 miles.

To make this seem more sensible you might ask yourself: In walking around the surface of the earth, how much further does your head travel than your feet?

[23]Euclid's proof that there is an infinite number of primes is an elegant and concise demonstration. If $P$ is any prime, a prime greater than $P$ can always be found. Construct $P! + 1$. This number, obviously greater than $P$, is not divisible by $P$ or any number less than $P$. There are only two alternatives: (1) It is not divisible at all; (2) It is divisible by a prime lying between $P$ and $P! + 1$. But both of these alternatives prove the existence of a prime greater than $P$. Q.E.D.

[24]Ball, *op. cit.*

**6**

*I am ill at these numbers.*

—Shakespeare *(Hamlet)*

*You mentioned your name as if I should recognize it, but beyond the obvious facts that you are a bachelor, a solicitor, a Freemason, and an asthmatic, I know nothing whatever about you.*

—Sir Arthur Conan Doyle
*(The Memoirs of Sherlock Holmes, The Norwood Builder)*

# ARITHMETICAL RESTORATIONS

## By W. W. Rouse Ball

I take next a class of problems dealing with the reconstruction of arithmetical sums from which various digits have been erased. Some of these questions are easy, some difficult. This kind of exercise has attracted a good deal of attention in recent years. I give examples of three kinds of restoration.

*Class A.* The solutions of one group of these restoration questions depend on the well-known propositions that every number $a + 10b + 10^2c + 10^3d + \ldots$ is equal to any of certain expressions such as

$$M(9) + a + b + c + d + \ldots$$
$$M(11) + a - b + c - d + \ldots$$
$$M(33) + (a + 10b) + (c + 10d) + (e + 10f) + \ldots$$
$$M(101) + (a + 10b) - (c + 10d) + (e + 10f) - \ldots$$
$$M(m) + (a + 10b + 10^2c) + (d + 10e + 10^2f) + \ldots$$
$$M(n) + (a + 10b + 10^2c) - (d + 10e + 10^2f) + \ldots$$

where, in the penultimate line, $m = 27$, or 37, or 111, and in the last line, $n = 7$, or 13, or 77, or 91, or 143.

Examples, depending on such propositions, are not uncommon. Here are four easy instances of this class of questions.

(i) The product of 417 and .1. . . is 9. . .057. Find the missing digits, each of which is represented by a dot. If the undetermined digits in the multiplier are denoted in order by $a$, $b$, $c$, $d$, and we take the steps of the multiplication in their reverse order, we obtain successively $d = 1$, $c = 2$, $b = 9$. Also the product has 7 digits, therefore $a = 2$. Hence the product is 9,141,057.

(ii) The seven-digit number 70. .34. is exactly divisible by 792. Find the missing digits, each of which is represented by a dot. Since 792 is $8 \times 9 \times 11$, we can easily show that the number is 7,054,344.

(iii) The five-digit number 4.18. is divisible by 101. Find the missing digits.[1]

Denote the two missing digits, from right to left, by $x$ and $y$. Applying the theorem for 101, noting that each of the unknowns cannot exceed 9, and for convenience putting $y = 10 - z$, this equation gives $z = 1$, $x = 7$, $y = 9$. Hence the number is 49,187.

(iv) The four-digit number .8. . is divisible by 1287. Find the missing digits.[2]

Denote these digits, from right to left, by $x$, $y$, $z$. We have $1287 = 9 \times 11 \times 13$. Applying the suitable propositions, and noting that each of the unknowns cannot exceed 9, we get $x = 1$, $y = 6$, $z = 3$. Hence the number is 3861.

(v) As a slightly harder example of this type, suppose we know that 6. 80. 8. .51 is exactly divisible by 73 and 137. Find the missing digits.[3] The data suffice to determine the number, which is 6,780,187,951.

*Class B.* Another and more difficult class of restoration problems is illustrated by the following examples. Their solutions involve analytical skill which cannot be reduced to rules.

(i) I begin with an easy instance, said to be of Hindoo origin, in which the problem is to restore the missing digits in the annexed division sum where a certain six-digit number when divided by a three-digit number gives a three-digit result.[4]

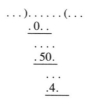

The solution involves no difficulty. The answer is that the divisor is 215, and the quotient 573; the solution is unique.

......)........(.............
        .......
        ......
        .......
        ......
        .......
        ......
        .......
        ......
        .......
        ......
        .......
        ......
        .......
        ......
        .......
            ......
            ......
            ......

(ii) As a more difficult specimen, on the previous page, I give the following problem, proposed in 1921 by Prof. Schuh of Delft. A certain seven-digit integer when divided by a certain six-digit integer gives a result whose integral part is a two-digit number and whose fractional part is a ten-digit expression of which the last nine digits form a repeating decimal, as indicated in the following work, where a bar has been put above the repeating digits. It is required to restore the working.[5] This problem is remarkable from the fact that not a single digit is given explicitly. The answer is that the divisor is 667334 and the dividend is 7752341.

Here are three additional examples of arithmetical restorations.[6] The solutions are lengthy and involve much empirical work.

(iii) The first of these Berwick questions is as follows. In the following division sum all the digits, except the seven "7's" shown, have been erased: each missing digit may be 1, 2, 3, 4, 5, 6, 7, 8, 9, or (except in the first digit of a line) 0. Observe that every step in the working consists of two lines each of which contains an equal number of digits. The problem is to restore the whole working of the sum. The solution is unique and gives a divisor of 125473 and a quotient of 58781.

$$\ldots\ldots 7.\,)\ldots 7.\ldots\ldots(\ldots 7\ldots$$

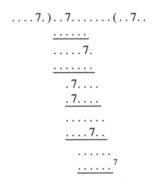

(iv) The second problem is similar and requires the restoration of the digits in the following division sum, where the position of four "4's" is given,

$$\ldots)\ldots\ldots 4(\,.\,4\ldots$$

To this problem there are four solutions, the divisors being 846, 848, 943, 949; and the respective quotients 1419, 1418, 1418, 1416.

If we propound the problem (using five "4's") thus:

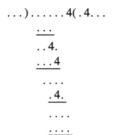

$$\dots)\dots\dots 4(.4\dots$$

there is only one solution, and some will think this is a better form in which to enunciate it.

(v) In the last of these Berwick examples, it is required to restore the working of the following division sum where all the digits, except five "5's," have been erased.

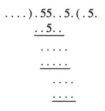

$$\dots).55..5.(.5.$$

To this problem there is only one solution, the divisor being 3926 and the quotient 652.

*Class C.* A third class of digit problems depends on finding the values of certain symbols which represent specified numbers. Two examples will suffice.

(i) Here is a very simple illustrative specimen. The result of multiplying $bc$ by $bc$ is $abc$, where the letters stand for certain numbers. What are the numbers? A brief examination shows that $bc$ stands for 25, and therefore $a$ stands for 6.

(ii) Here is another example. The object is to find the digits represented by letters in the following sum:[8]

$$
\begin{array}{r}
a\,b\,)\,c\,d\,e\,e\,b\,(\,b\,f\,b \\
c\,e\,b \\
\hline
g\,g\,e \\
g\,c\,h \\
\hline
c\,e\,b \\
c\,e\,b \\
\hline
\end{array}
$$

A solution may be obtained thus: Since the product of $b$ by $b$ is a number which ends in $b$, $b$ must be 1, 5, or 6. Since the product of $ab$ by $b$ is a number of three digits, $b$ cannot be 1. The result of the subtraction of $h$ from $e$ is $e$, hence $h = 0$, and therefore if $b = 5$ we have $f$ even, and if $b = 6$ we have $f = 5$. Also the result of the subtraction of $c$ from $g$ is $c$, hence $g = 2c$, and therefore $c$ cannot be greater than 4: from which it follows that $b$ cannot be 6. A few trials now show that the question arose from the division of 19,775 by 35.

It is possible to frame digit restoration examples of a mixed character involving the difficulties of all the examples given above, and to increase the difficulty by expressing them in a non-denary scale of notation. But such elaborations do not add to the interest of the questions.

## ENDNOTES

[1] P. Delens, *Problèmes d'Arithmétique Amusante*, Paris, 1914, p. 55.

[2] *Ibid.*, p. 57.

[3] *Ibid.*, p. 60.

[4] *American Mathematical Monthly*, 1921, vol. XXVIII, p. 37.

[5] *Ibid.*, p. 278.

[6] All are due to W. E. H. Berwick. The "7" problem appeared in the *School World*, July and October 1906, vol. VIII, pp. 280, 320; the "4" problem appeared in the *Mathematical Gazette*, 1920, vol. X, pp. 43, 359–360; the "5" problem in the same paper, vol. X, p. 361, vol. XI, p. 8.

[7] This problem, fully worked out, appears on page 2421.

[8] *Strand Magazine*, September–October, 1921.

7

*It is quite a three-pipe problem.*

—Sir Arthur Conan Doyle (*The Red-
Headed League*)

# THE SEVEN SEVEN'S

## By W. E. H. Berwick

The following original problem may interest some readers of *The School World*. The problem, as will be seen, consists of an ordinary long-division sum in which some of the figures are replaced by dots. There is one, and only one, solution possible.

W. E. H. BERWICK.

The Grammar School, Bradford.

```
....7.).. 7.......)..7..
        ......
        ....7.
        ......
          .7....
          .7....
        ......
        ....7..
        ......
        ......
```

## SOLUTION OF THE PROBLEM OF THE SEVEN SEVEN'S[1]

I send you herewith a solution of the interesting problem which appeared in last month's *School World,* and I have given with it, in some detail, the process of reasoning by which the result was reached.

(1) Since multiplication of the divisor by 7 gives 6 figures in the product, while the 2nd and 4th multiplications give 7 figures, the divisor must begin with 11, 12, 13, or 14, and the 2nd and 4th figures of the quotient are 8 or 9.

(2) Since the product of the divisor multiplied by 7 has its 2nd digit 7, it is found by trial that the divisor must begin with 111, 124, 125, 138, or 139.

(3) The 3rd remainder must evidently begin with 10, and therefore also the 4th product begins with 10. Hence either divisor begins with 111, and 4th figure of quotient is 9, or divisor begins 125, and 4th figure of quotient is 8.

(4) But if 9 were the 4th figure of the quotient, then, since the product has its 3rd digit from the right 7, the divisor would require to begin with 11197, which would make the product by 7 have an 8 in the 2nd place instead of a 7. Hence

the 4th figure in the quotient is 8, and the divisor begins with 12547, the 6th figure being less than 5.

(5) Since the 3rd product (by 7) begins 878, and the line above cannot be greater than 979 . . ., therefore the remainder begins 10I; and since the 4th product begins 100, therefore the last remainder begins with a 1, and the last figure of the quotient must be 1.

(6) On trying 4, 3, 2, 1 in succession as 6th digit in the divisor, and working the division backwards, it is found that the 2nd of these alone satisfies the conditions not yet considered, and that the following is the full division.

```
1 2 5 4 7 3 ) 7 3 7 5 4 2 8 4 1 3 ( 5 8 7 8 1
              6 2 7 3 6 5
              1 1 0 1 7 7 8
              1 0 0 3 7 8 4
                  9 7 9 9 4 4
                  8 7 8 3 1 1
                  1 0 1 6 3 3 1
                  1 0 0 3 7 8 4
                      1 2 5 4 7 3
                      1 2 5 4 7 3
```

J. W. REID.

Drummond Villa, Inverness.

## APPENDIX ON BERWICK'S PROBLEM OF THE SEVEN SEVEN'S

Berwick's Division Problem, as given by Callandreau (*Célèbres Problèmes Mathématiques,* Edition Albin Michel, Paris, 1949).

A division—dividend $\Delta$, divisor $\Delta'$, quotient $\Omega$—is set out below, the points denoting numbers which have to be determined.

```
    . . . . 7.         . . 7. . . . . .         . . 7. .
                       . . . . . .
                       . . . . . 7.
                       . . . . . .
                        . 7. . . .
                        . 7. . . .
                       . . . . . .
                       . . . . 7. .
                       . . . . . .
                       . . . . . .
```

This odd problem, which illustrates well the power of quite elementary arithmetic, was proposed and solved by Berwick in I906, in the review *The School World.*

Denoting the missing figures by letters, the operation may be written

$$\alpha\,\beta\,\gamma\,\delta\,7\,\epsilon\,|\,\text{A B 7 C D E L Q W Z}'\,|\,\iota\,\lambda\,7\,\mu\,\nu \qquad \text{(1st Line)}$$
$$\underline{\text{a b x c d e}} \qquad\qquad\qquad\qquad\qquad\qquad \text{(2nd Line)}$$
$$\text{F G H J K 7 L} \qquad\qquad\qquad\qquad\qquad \text{(3rd Line)}$$
$$\underline{\text{f g h j k y l}}$$
$$\text{M 7 N O P Q}$$
$$\underline{\text{m 7 n o p q}}$$
$$\text{R S T U z V W}$$
$$\underline{\text{r s t u 7 v w}}$$
$$\text{X Y Z X}'\text{Y}'\text{Z}'$$
$$\text{X Y Z X}'\text{Y}'\text{Z}'$$

The first figure $\alpha$ of the divisor must be 1 for the product $7\Delta$ in the sixth line has only six figures; even with $\alpha = 2$ it would have to have seven.

Considering now the partial remainders in the third and seventh lines, since they each have six figures F and R must be equal to unity, for FGHJK7 and RSTUZV are each less than $\alpha\beta\gamma\delta7\epsilon$, and $\alpha$ is equal to unity. So F = R = 1 and therefore also f = r = 1.

The divisor $\Delta'$ therefore cannot be greater than 199,979 and since $\mu$ is not greater than 9 the partial product in the eighth line is at most 1,799,811, so s is less than 8. But since S is the difference of the two figures 7 in the column immediately above it S must be either 9 or 0. Now in the ninth line the remainders (R − r) and (S − s) are zero, and R = r = 1 therefore S = s = 0. Further, with S = 0 and R = 1 it follows that M = m + 1, therefore m ≤ 8. So the partial product in the sixth line is at most 87nopq.

The second figure $\beta$ of the divisor can only be equal to 0, 1 or 2; for if it were 3 or more the divisor would be at least 130,000, and multiplication of this by 7 gives 910,000, which is greater than 87nopq. Suppose $\beta = 0$: the divisor is then at most 109,979, the product of which with the highest possible figure, 9, of the quotient still gives a partial product of six figures; but at the eighth line the partial product has seven figures. $\beta$ must therefore be greater than 0. Suppose $\beta = 1$: then $\gamma$ can only be equal to 0 or 1, for if $\gamma \geq 2$ the partial product $7\Delta$ will give for the second figure of the sixth line a figure greater than 7; but $\gamma$ cannot be 0, for even with a figure 9 in the quotient the partial product $9 \times 110,979$ would not have seven figures, as it must have at the eighth line. Assuming that $\gamma = 1$, we must determine $\delta$, $\epsilon$, $\mu$ in such a way that the partial product $\mu \times 111,\!\delta7\epsilon$ gives a seven-figure number in which the third figure from the right is 7. $\mu$ must be equal to 9, for $\mu = 8$ would give a six-figure number; it follows that $\delta$ must be equal to 0 or 9. In the first case the divisor would be at most 110,979 and its multiple by the highest figure 9 of the quotient would not give a seven-figure number as required. In the second case the divisor would be 111.97$\epsilon$ which on multiplication by 7 would give for the sixth line 783. . ., which is impossible since the second figure from the left must be a 7. Therefore $\beta$ cannot be equal to 1. Therefore it must be equal to 2, from which it follows that m = 8 and M = 9.

We have then that the partial product in the sixth line is $7 \times 12\gamma\delta7\epsilon$ and this must be equal to 87nopq; it follows that $\gamma$ must be 4 or 5, for $7 \times 126\delta7\epsilon$ would be greater than 87nopq, and $7 \times 123{,}979$ would be less than this number. Now $7 \times 126{,}979$ would still give a partial product of six figures, instead of the seven required for the eighth line (10tu7vw, say), and $9 \times 123{,}979$ would be greater than it; so $\mu = 8$, and since for $\gamma = 4$ the maximum value of the partial product in the eighth line would be $8 \times 124{,}979$ and thus less than 10tu7vw, $\gamma$ must be equal to 5.

But the partial product $8 \times 125\delta7\epsilon$, that is 10tu7vw, must have the third figure from the right equal to 7, which is only possible for $\delta = 4$ or 9; but $7 \times 125{,}97\epsilon$ would give a partial product greater than 87nopq in the sixth line, so $\delta = 4$, and it will be seen also that $\epsilon$ cannot be greater than 4. From this we conclude that $n = 8$; and the partial product 10tu7vw in the eighth line is $8 \times 125{,}47\epsilon$, so that $t = 0$ and $u = 3$.

X, not being zero, is $\geq 1$, so that $T \geq 1$; but since $n = 8$ and $N \leq 9$ we have $T \leq 1$, so that $T$ must be equal to 1. Consequently $N = 9$ and $X = 1$. The partial product $\nu \times 125{,}47\epsilon$ in the tenth line requires $\nu = 1$ since $X = 1$. From this it follows that $Y = 2$, $Z = 5$, $X' = 4$, $Y' = 7$, and $Z' = \epsilon$.

The partial product $\lambda \times 125{,}47\epsilon$ gives at the fourth line a seven-figure number, from which it follows that $\lambda$ must be equal to 8 or 9.

Suppose $\epsilon = 0, 1, 2, 3, 4$ in turn, with $\lambda = 8$ or 9: the partial products in the fourth, sixth and eighth lines are determined, and on reconstructing the division we must have a 7 as the second figure from the right on the third line. Trial and error shows that this is satisfied only in the case $\lambda = 8$ and $\epsilon = 3$. The partial products may now be obtained, giving vw = 84, opq = 311, ghjkyl = 003,784; from which it follows that UzVW = 6,331, OPQ = 944, and GHJK7L = 101,778.

It only remains to determine the first and second lines, and i. The first partial product in the second line, $i \times 125{,}473$, must give a number which added to the remainder 110,177 gives a 7 in the third place from the left in the dividend. Only $i = 5$ will do this. The quotient and the divisor are thus completely determined; it follows that abxcde = 627,365 and finally that the dividend AB7CDE = 737,542.

## ENDNOTE

[1](The reader will find in an appendix, p. 2422, a much more detailed, meticulously reasoned solution of Berwick's problem, by Callandreau, *Célèbres Problèmes Mathématiques*, Edition Albin Michel, Paris, 1949. I add this material on the theory there are readers as intrigued as I am by this sort of logical exercise.*Ed.*)

# COMMENTARY ON THOMAS JOHN I'ANSON BROMWICH

T homas John I'Anson Bromwich (1875–1929) was an English mathematician and, for all I know, a tennis player—but not *the* tennis player.
Bromwich was noted for his work in algebra, his encyclopedic treatise *The
Theory of Infinite Series,* and certain classic researches in mathematical
physics. He lectured for some years at St. John's College, Cambridge, and was
a Fellow of the Royal Society. Bromwich was an enthusiastic exponent of the
"elementary dynamics of the lawn tennis court," a subject which, in its mathematical form, at least, does not appear to have a wide following. Those who
regard tennis as too important to be left to the mercy of crude empirics will wish
to study carefully the analysis presented below.

You will note that Bromwich says the mathematics of his essay is easy. This
opinion, I have learned, is debatable among tennis players. Most of those whom I
consulted implied they would rather play a dozen sets of singles in Death Valley
than sweat through Bromwich's algebra. I do not guarantee that the tactics here
outlined actually work; it may be that their effectiveness can be realized only in
the form of gamesmanship. A few words about $\frac{1}{2}gt^2$ and parabolic trajectories
might well be helpful if spoken when your opponent is about to serve or smash.
Bromwich's reminder that "few ladies will volley with any real confidence"
evidently refers to a sort of lady that used to volley with Cambridge dons. The
tactic based on this assumption is no longer recommended.

8   *You know my methods. Apply them.*

—SIR ARTHUR CONAN DOYLE (*The Sign of Four*)

# EASY MATHEMATICS AND LAWN TENNIS

*By T. J. I'A. Bromwich*

In view of the popular appeal of Lawn Tennis, and of the fact that a large number of professional mathematicians depend upon this game for much of their recreation, it seems strange that no one has pointed out that a good deal may be learned (as to suitable tactics) by applying a few quite simple principles of dynamics and geometry to the game.

The following remarks refer to the game as played in a men's double:[1] after the age of thirty-five the players usually cease to take very great interest in a strenuous men's single, and (at about the same age) less anxiety to play in mixed doubles will be found. And once the ideas have been suggested which govern these tactics in a double, any one really interested will have little difficulty in working out similar rules for singles, or for mixed doubles, where it must be borne in mind that few ladies will volley with any real confidence.

In the first place, any student of geometry will have no difficulty in seeing that for a pair to cover the court (with as little trouble to themselves as possible) *they should adopt a formation in which the line joining the players is parallel to the net* (or nearly so); and again it is easy to see that *the most really vulnerable area is* between *the two players, rather than the areas which are left outside the players* (that is, between each player and the nearer side-line).

These are, of course, general principles; *they cannot be of universal application,* and they must be modified in the light of the known methods of play and the special personality of individual opponents. For this reason closeness to the net has the further advantage (apart from the mathematical reasons given below) that the probable play of the attackers can be studied more easily, and their strokes (to some extent) anticipated in consequence of the knowledge so gained.

The court is taken as a rectangle $80^2$ feet by 32 feet, and the net as 3 feet high.

## ELEMENTARY DYNAMICS OF THE LAWN TENNIS COURT

Although the actual flight of the ball must be greatly affected by air resistance, yet we can get an idea of the general character of the problem by using the familiar parabolic trajectory.[3] Take then the height of the net as $h$, and suppose that a ball is to be hit at a height $h_0$ so as to skim the net and just fall into the court; we take the distances from the net to be $x$, $l$ respectively.

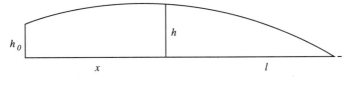

**FIGURE 1**

Then if $u$ is the horizontal component of velocity we easily find that

$$\frac{h - h_0}{x} + \frac{h}{l} = \frac{g}{2u^2}(x + l) = \frac{\frac{1}{2}gt^2}{x + l}$$

where $t$ is the whole time of flight.

From the point of view of the driver, it is of very great importance to make $t$ *small* (so as to give his opponent little time to anticipate the stroke); and accordingly, $h_0$ being partly at his disposal, we have the FIRST WORKING RULE: *Take the ball as high above the ground as possible.*

To consider the effect of position, we may write the formula as

$$\tfrac{1}{2}gt^2 = (h - h_0)\left(1 + \frac{l}{x}\right) + h\left(1 + \frac{x}{l}\right),$$

and in the subsequent discussion there is one fundamental difference, according as $h_0$ is *greater* or *less* than $h$: for definiteness we may distinguish these alternatives as the *smash* and the *drive*. The latter may be considered first.

*The Drive.*

Here $h_0 < h$; and thus $(h - h_0)/h$ is *positive,* but it will be supposed *fairly small,* say about $\frac{1}{4}$; that is, *the height $h_0$ will be $\frac{3}{4}h$ (= 2 ft. 3 in.).* Thus:

$$\frac{\frac{1}{2}gt^2}{h} = \frac{1}{4}\left(1 + \frac{l}{x}\right) + \left(1 + \frac{x}{l}\right),$$

and so the value of $t$ is *least* for $x = \frac{1}{2}l$, and then is $t_0$ where

$$t_0^2 = \tfrac{3}{16}(\tfrac{9}{4}) = \tfrac{27}{64}, \text{ or } t_0 = \tfrac{1}{8}(5.2) = .65 \text{ sec.}$$

This corresponds to a distance of some 20 feet from the net, if the driver aims at the back-line; but in practice it is usual to come up to, say, 10 or 12 feet, and to aim at a point perhaps 30 feet beyond the net. The value is then

$$t_1^2 = \tfrac{3}{16}(\tfrac{7}{3}), \text{ or } t_1 = .66 \text{ sec.},$$

but the slight increase in time for *this* stroke is more than compensated by the gain to be found in smashes. The speed $u$ is about 60 ft./sec. in the second position and nearly 90 ft./sec. in the former.

*The Smash.*[4]

Here $h_0 > h$; to have a clear idea of the magnitudes involved, we shall suppose $h_0$ to be from 4 to 5 feet; and so an average value for $(h - h_0)/h$ would be $-\tfrac{1}{2}$. It is now evident that $x$ *must be as small as possible* (to reduce $t$ as far as may be). But as $x$ is decreased, it is evident that the negative term (the first) in the formula will overpower the positive term: this simply indicates the fact (familiar in all *net-play*) that *the net is no longer any real obstacle*. Then the racket usually directs the ball slightly *downwards*: but to make an easier calculation, let us suppose that the stroke is *horizontal,* and then we have the two simple formulae:

$$\tfrac{1}{2}gt'^2 = h_0, \; ut' = x + l \,.$$

These give

$$t' = .5 (\text{for } h_0 = 4 \text{ feet})$$

and

$$u = 80 \text{ ft./sec. (taking } x + l = 40 \text{ feet).}$$

*The Base-line Drive.*

Here we shall have $x = l$, at least approximately, and so $\tfrac{1}{2}gt^2 = 2 (2h - h_0)$; nowadays it is usual to take the ball at a height of perhaps 2 feet, so that again $t$ is about .7, but *the necessary speed is considerably greater,* and would seem to be at least 100 ft./sec. Since the energy varies as the square of the initial speed, *the amount of energy used in a base-line drive is about three times as much as from a point* 10 *feet from the net.*

# GEOMETRY APPLIED TO THE QUESTION OF POSITION IN THE COURT

If the opponent is driving from the base-line (unless he possesses exceptional skill) he is almost forced to aim at the base-line (of course a *lob* can be dropped on the side-lines, but lobs must be treated on other principles). Thus, at a distance of 10 feet from the net, a length of $(^{40+10}\!/_{80})$ 32 $=$ 20 feet is exposed to fire: thus each player is responsible for 10 feet to be covered. It is a matter of common experience that few players can aim quite accurately at the corners of the court, so that probably not more than 9 feet needs to be covered by each player, but there is more need to cover the inside than the outside. The length to be covered by the *left-hand* player has been thickened in the figure. It is to be carefully noticed that, *in order to provide the best cover, the players* (*represented by M, N*) *should move to their* RIGHT, *following the position of the driver at P.*

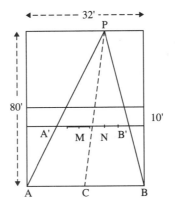

**FIGURE 2**

The players should, therefore, try to form mental pictures of the area into which driving is possible, so as to place themselves to the best possible advantage along $A'B'$. Once this idea is grasped, the mind quickly indicates suitable points for defence; of course, in trying to defend against a smash, the problem is far more complicated, since *then* cross-shots are comparatively easy. Still the same principles can be used as a guide.

It is not difficult to apply similar considerations to *the position of the server's partner*: he must first decide, after observing the pitch of the service, whether there is any real danger of being passed at the side-line. When the pitch is near the centre-line, a passing shot is so difficult that this risk is almost negligible; then the player will be well-advised to move towards the centre-line of his court (so as to help the server to cover the danger zone). But if the pitch is near the side-line, a passing-shot is quite likely, and the player must remain as close to the side-line as is necessary to cover it; for the server will be fully engaged in trying to cover the rest of the court. As to the server himself, in either case his (further) side-line will be in little danger, and his best policy is to come up towards the centre-line, keeping a watch on his partner, so as to cover up any possible gap between them.

## POSTSCRIPT

About three weeks after putting down the above formulae I was asked by a lady player if there was an explanation (by mathematics) of a statement made by Señorita Alvarez to the effect that *if a ball is* ABOVE *the shoulder it may be hit at* ANY *speed, without being driven out of court, provided that the ball* GRAZES THE NET.

This statement can easily be justified by the same formulae; but it is a little easier to write down first some general formulae (from which our earlier formulae were found in the first place). The parabolic trajectory is used (as already explained); then suppose that $u$, $v$ are the horizontal and vertical components of the initial velocity, and that $t_0$ is the time to the net. Then we have

$$x = ut_0, \qquad x + a = ut,$$
$$h_0 - h = vt_0 + \tfrac{1}{2}gt_0^2, \qquad h_0 = vt + \tfrac{1}{2}gt^2.$$

Thus

$$\frac{h_0 - h}{x} = \frac{v}{u} + \tfrac{1}{2}gt_0, \qquad \frac{h}{a} = \frac{v}{u} + \tfrac{1}{2}g(t_0 + t),$$

and so

$$\frac{h}{a} - \frac{h_0 - h}{x} = \tfrac{1}{2}gt = \frac{g}{2u}(x + a).$$

From this formula those used previously will follow instantly.

For our immediate purpose we note that $(h_0 - h)/x$ and also $\tfrac{1}{2}gt$ are both positive.

Thus: $$\frac{h}{a} > \frac{h_0 - h}{x} \quad \text{or} \quad \frac{x}{a} > \frac{h_0 - h}{h}.$$

Accordingly, then *if $h_0 \geqq 2h$ (i.e. 6 ft.), no ball can possibly be driven out.*

In the play of a person of moderate height it is perhaps reasonable to reckon *"shoulder-height"* at about 5 ft. 6 in. Then we have

$$\frac{h_0 - h}{h} = \frac{66 - 36}{36} = \frac{5}{6},$$

and so $(a - x) < \tfrac{1}{6}a$.

This means that, *unless the driver is within 6 feet (say) of the base-line, a ball at the level of the head cannot be driven out, however hard the ball is hit (keeping it down to the net).*

# ENDNOTES

[1] The players are supposed also to be *right-handed*; the necessary modifications due to one or two left-handed players can be easily introduced, but no new ideas are involved.

[2] Accurately 78 feet; but exact figures are not very valuable here.

[3] In actual play, the flight is modified a good deal by "cut" and "top-spin." The former is a very important feature in Court-Tennis, and consequently figured largely in the early days of lawn-tennis—in modern good fast play, "cut" is almost nonexistent. "Top-spin" on the other hand may be seen often, but its effect is too complicated to be considered here.

[4] In what is usually called *"smashing"* the ball will be hit at a height of 6 or 7 feet: here we have used the phrase *smash* as a conveniently short phrase to distinguish the two types of stroke.

# COMMENTARY ON
## STEPHEN BUTLER LEACOCK

S tephen Butler Leacock, who died in 1944, was a political economist—
head of department at McGill University—and the author of a number of
very funny short stories and essays. I have never understood why it was
considered so remarkable that an economist should also be a humorist. Economics is apt to parch those who practice it and it is not surprising to find them
turning to humor for relief. Leacock has been described as a "genial satirist."
He was not genial. It is true that he was inclined more to use the flat of his
sword than the point or edge; but he was much too aware of society's follies to
be genial. He had a sharp eye for every sort of idiot's delight and a delightful
way of arraigning the world's ample stupidity. Much of his earlier writing—he
was most prolific from 1910 to 1925—has gone flat, but there are pieces in
*Nonsense Novels, Behind the Beyond, Frenzied Fiction* and *Winnowed Wisdom*
that still sparkle. Burlesque was Leacock's greatest skill. He knew about
science and enjoyed deflating its pretensions. The essay *Mathematics for
Golfers* uses arithmetic to dissolve a few of the solemn principles of this
pompous and snobbish game that was obviously invented to kill off overweight
executives on Sundays. *Common Sense and the Universe* discusses irreverently
some of the holiest mathematical concepts in this book, from probability and
quantum theory to relativity and entropy. I wish that Bishop Berkeley could
have lived to take on these theories as he took on the calculus; Leacock's
treatment is less pungent—he was not an Irishman—but no less gratifying.

*When ye come to play golf ye maun hae a heid.*
—Charles Baird Macdonald (*1928*)

# Mathematics for Golfers

*By Stephen Leacock*

It is only quite recently that I have taken up golf. In fact I have played for only three or four years, and seldom more than ten games in a week, or at most four in a day. I have had a proper golf vest for only two years. I bought a "spoon" only this year and I am not going to get Scotch socks till next year.

In short, I am still a beginner. I have once, it is true, had the distinction "of making a hole in one," in other words, of hitting the ball into the pot, or can, or receptacle, in one shot. That is to say, after I had hit, a ball was found in the can, and my ball was not found. It is what we call circumstantial evidence—the same thing that people are hanged for.

Under such circumstances I should have little to teach to anybody about golf. But it has occurred to me that from a certain angle my opinions may be of value. I at least bring to bear on the game all the resources of a trained mind and all the equipment of a complete education.

In particular I may be able to help the ordinary golfer, or "goofer"—others prefer "gopher"—by showing him something of the application of mathematics to golf.

Many a player is perhaps needlessly discouraged by not being able to calculate properly the chances and probabilities of progress in the game. Take for example the simple problem of "going round in bogey." The ordinary average player, such as I am now becoming—something between a beginner and an expert—necessarily wonders to himself, "Shall I ever be able to go around in bogey; will the time ever come when I shall make not one hole in bogey, but all the holes?"

To this, according to my calculations, the answer is overwhelmingly "yes." The thing is a mere matter of time and patience.

Let me explain for the few people who never play golf (such as night watchmen, night clerks in hotels, night operators, and astronomers) that "bogey" is an imaginary player who does each hole at golf in the fewest strokes that a first-class player with ordinary luck ought to need for that hole.

Now an ordinary player finds it quite usual to do one hole out of the nine "in bogey"—as we golfers, or rather, "us goofers" call it; but he wonders whether it will ever be his fate to do all the nine holes of the course in bogey. To which we answer again with absolute assurance, he will.

The thing is a simple instance of what is called the mathematical theory of probability. If a player usually and generally makes one hole in bogey, or comes close to it, his chance of making any one particular hole in bogey is one in nine. Let us say, for easier calculation, that it is one in ten. When he makes it, his chance of doing the same with the next hole is also one in ten; therefore, taken from the start, his chance of making the two holes successively in bogey is one-tenth of a tenth chance. In other words, it is one in a hundred.

The reader sees already how encouraging the calculation is. Here is at last something definite about his progress. Let us carry it farther. His chance of making three holes in bogey one after the other will be one in 1000, his chance of four one in 10,000, and his chance of making the whole round in bogey will be exactly one in 1,000,000,000—that is one in a billion games.

In other words, all he has to do is to keep right on. But for how long? he asks. How long will it take, playing the ordinary number of games in a month, to play a billion? Will it take several years? Yes, it will.

An ordinary player plays about 100 games in a year, and will, therefore, play a billion games in exactly 10,000,000 years. That gives us precisely the time it will need for persons like the reader and myself to go round in bogey.

Even this calculation needs a little revision. We have to allow for the fact that in 10,000,000 years the shrinking of the earth's crust, the diminishing heat of the sun, and the general slackening down of the whole solar system, together with the passing of eclipses, comets, and showers of meteors, may put us off our game.

In fact, I doubt if we shall ever get around in bogey. Let us try something else. Here is a very interesting calculation in regard to "allowing for the wind."

I have noticed that a great many golf players of my own particular class are always preoccupied with the question of "allowing for the wind." My friend Amphibius Jones, for example, just before driving always murmurs something, as if in prayer, about "allowing for the wind." After driving he says with a sigh, "I didn't allow for the wind." In fact, all through my class there is a general feeling that our game is practically ruined by the wind. We ought really to play in the middle of the Desert of Sahara where there isn't any.

It occurred to me that it might be interesting to reduce to a formula the effect exercised by the resistance of the wind on a moving golf ball. For example, in our game of last Wednesday, Jones in his drive struck the ball with what, he assures me, was his full force, hitting it with absolute accuracy, as he himself admits, fair in the center, and he himself feeling, on his own assertion, absolutely fit, his eye being (a very necessary thing with Jones) absolutely "in," and he also having on his proper sweater—a further necessary condition of first-class play. Under all the favorable circumstances the ball advanced only 50 yards! It was evident at once that it was a simple matter of the wind: the wind, which was of that treacherous character that blows over the links unnoticed, had impinged full upon the ball, pressed it backward, and forced it to the earth.

Here, then, is a neat subject of calculation. Granted that Jones—as measured on a hitting machine the week the circus was here—can hit 2 tons, and that this whole force was pressed against a golf ball only one inch and a quarter in diameter. What happens? My reader will remember that the superficial area of a golf ball is $\pi r^3$, that is $3.141567 \times (⅝$ inches$)^3$.[1] And all of this driven forward with the power of 4000 pounds to the inch!

In short, taking Jones's statements at their face value, the ball would have traveled, had it not been for the wind, no less than 6½ miles.

I give next a calculation of even more acute current interest. It is in regard to "moving the head." How often is an admirable stroke at golf spoiled by moving the head! I have seen members of our golf club sit silent and glum all evening, murmuring from time to time, "I moved my head." When Jones and I play together I often hit the ball sideways into the vegetable garden from which no ball returns (they have one of these on every links; it is a Scottish invention). And whenever I do so Jones always says, "You moved your head." In return when *he* drives his ball away up into the air and down again ten yards in front of him, I always retaliate by saying, "You moved your head, old man."

In short, if absolute immobility of the head could be achieved, the major problem of golf would be solved.

Let us put the theory mathematically. The head, poised on the neck, has a circumferential sweep or orbit of about 2 inches, not counting the rolling of the eyes. The circumferential sweep of a golf ball is based on a radius of 250 yards, or a circumference of about 1600 yards, which is very nearly equal to a mile. Inside this circumference is an area of 27,878,400 square feet, the whole of which is controlled by a tiny movement of the human neck. In other words, if a player were to wiggle his neck even ¹⁄₁₉₀ of an inch the amount of ground on which the ball might falsely alight would be half a million square feet. If at the same time he multiplies the effect by rolling his eyes, the ball might alight anywhere.

I feel certain that after reading this any sensible player will keep his head still.

A further calculation remains—and one perhaps of even greater practical interest than the ones above.

Everybody who plays golf is well aware that on some days he plays better than on others. Question—How often does a man really play his game?

I take the case of Amphibius Jones. There are certain days, when he is, as he admits himself, *"put off his game"* by not having on his proper golf vest. On other days the light puts him off his game; at other times the dark; so, too, the heat; or again the cold. He is often put off his game because he has been up late the night before; or similarly because he has been to bed too early the night before; the barking of a dog always puts him off his game; so do children; or adults; or women. Bad news disturbs his game; so does good; so also does the absence of news.

All of this may be expressed mathematically by a very simple application of the theory of permutations and probability; let us say that there are altogether 50 forms of disturbance any one of which puts Jones off his game. Each one of these disturbance happens, say, once in ten days. What chance is there that a day will come when *not a single one of them occurs?* The formula is a little complicated, but mathematicians will recognize the answer at once as $x/1 + x^2/1 + \ldots . x^n/1$. In fact, that is exactly how often Jones plays at his best; $x/1 + x^2/1 \ldots x^n/1$ worked out in time and reckoning four games to the week and allowing for leap years and solar eclipses, comes to about once in 2,930,000 years.

And from watching Jones play I think that this is about right.

## ENDNOTE

[1][I hope it is not too heavy handed to point out that the formulae are wrong and the mathematics egregious. ED.]

# 10

*In the space of one hundred and seventy-six years the Lower Mississippi has shortened itself two hundred and forty-two miles. That is an average of a trifle over one mile and a third per year. Therefore, any calm person, who is not blind or idiotic, can see that in the old Oölitic Silurian Period, just a million years ago next November, the Lower Mississippi River was upward of one million three hundred thousand miles long, and stuck out over the Gulf of Mexico like a fishing-rod. And by the same token any person can see that seven hundred and forty-two years from now the Lower Mississippi will be only a mile and three-quarters long, and Cairo and New Orleans will have joined their streets together, and be plodding comfortably along under a single mayor and a mutual board of aldermen. There is something fascinating about science. One gets such wholesale returns of conjecture out of such a trifling investment of fact.*

—MARK TWAIN (*Life on the Mississippi*)

# COMMON SENSE AND THE UNIVERSE

## By Stephen Leacock

### I

Speaking last December [1941] at the annual convention of the American Association for the Advancement of Science, and speaking, as it were, in the name of the great 100-inch telescope under his control, Professor Edwin Hubble, of the Mount Wilson Observatory, California, made the glad announcement that the universe is not expanding. This was good news indeed, if not to the general public who had no reason to suspect that it was expanding, at least to those of us who humbly attempt to 'follow science.' For some twenty-five years past, indeed ever since the promulgation of this terrific idea in a paper published by Professor W. de Sitter in 1917, we had lived as best we could in an expanding universe, one in which everything, at terrific speed kept getting further away from everything else. It suggested to us the disappointed lover in the romance who leaped on his horse and rode madly off in all directions. The idea was majestic in its sheer size, but it somehow gave an uncomfortable sensation.

Yet we had to believe it. Thus, for example, we had it on the authority of Dr. Spencer Jones, the British Astronomer Royal, as recently as in his new and fascinating book of 1940, *Life on Other Worlds,* that 'a distant universe in the constellation of Boötes has been found to be receding with a velocity of 24,300 miles a second. We can infer that this nebula is at a distance of 230,000,000 light-years.' I may perhaps remind my fellow followers of science that a light-year means the distance traveled in one year by light, moving at 186,000 miles a second. In other words, this 'distant universe' is now 1,049,970,-980,000,000,000,000 miles away.

Some distance! as Mr. Churchill would say.

But now it appears that that distant universe has *not* been receding at all; in fact, it isn't away out there. Heaven knows where it is. Bring it back. Yet not only did the astronomers assert the expansion but they proved it, from the behavior of the red band in the spectrum, which blushed a deeper red at the revelation of it, like the conscious water that 'saw its God and blushed' at Cana in Galilee long ago. One of the most distinguished and intelligible of our astronomers, Sir Arthur Eddington, had written a book about it, *The Expanding Universe,* to bring it down to our level. Astronomers at large accepted this universal explosion in all directions as calmly as they once accepted the universal fall of gravitation, or the universal death in the cold under Carnot's Second Law of Thermodynamics.

But the relief brought by Professor Hubble is tempered on reflection by certain doubts and afterthoughts. It is not that I venture any disbelief or disrespect toward science, for that is as atrocious in our day as disbelief in the Trinity in the days of Isaac Newton. But we begin to doubt whether science can quite keep on believing in and respecting itself. If we expand today and contract tomorrow; if we undergo all the doubled-up agonies of the curvature of space only to have the kink called off, as it has been; if we get reconciled to dying a martyr's death at one general, distributed temperature of 459 degrees below zero, the same for all, only to find that the world is perhaps unexpectedly warming up again—then we ask, where are we? To which, of course, Einstein answers 'Nowhere,' since there is no place to be. So we must pick up our little book again, follow science, and wait for the next astronomical convention.

Let us take this case of the famous Second Law of Thermodynamics, that inexorable scroll of fate which condemned the universe—or at least all life in it—to die of cold. I look back now with regret to the needless tears I have wasted over that, the generous sympathy for the last little band of survivors, dying at 459 degrees below our zero ($-273°$ centigrade), the absolute zero of cold when the molecules cease to move and heat ends. No stove will light at that, for the wood is as cold as the stove, and the match is as cold as both, and the dead fingers motionless.

I remember meeting this inexorable law for the first time in reading, as a little boy, a piece of 'popular science' entitled *Our Great Timepiece Running Down.* It was by Richard Proctor, whose science-bogeys were as terrifying as Mrs. Crow's *Night Thoughts,* only slower in action. The sun, it appeared, was cooling; soon it would be all over. Lord Kelvin presently ratified this. Being

Scotch, he didn't mind damnation and he gave the sun and whole solar system only ninety million years more to live.

This famous law was first clearly enunciated in 1824 by the great French physicist, Nicolas Carnot. It showed that all bodies in the universe kept changing their temperature—hot things heated cold, and cold things chilled hot. Thus they pooled their temperature. Like the division of a rich estate among a flock of poor relations, it meant poverty for all. We must all share ultimately the cold of absolute space.

It is true that a gleam of hope came when Ernest Rutherford and others, working on radioactivity, discovered that there might be a contrary process of 'stoking up.' Atoms exploding into radioactivity would keep the home fires burning in the sun for a long time. This glad news meant that the sun was both much older and much younger than Lord Kelvin had ever thought it was. But even at that it was only a respite. The best they could offer was 1,500,000,000 years. After that we freeze.

And now what do you think! Here comes the new physics of the Quantum Theory and shatters the Second Law of Thermodynamics into gas—a word that is Dutch for chaos. The world may go on forever. All of this because of the final promulgation of the Law of the *Quantum,* —or, shall we say, the Law of the Just So Much,—of which we shall presently speak. These physical people do not handle their Latin with the neat touch of those of us who knew our declensions as they know their dimensions. Of course they mean *Tantum*—but let it go at that. *Quantum* is drugstore Latin, *quantum sufficit. Tantum* is the real thing—*Virgilium vidi tantum* ('I saw something of Virgil').

At this point I may perhaps pause to explain that the purpose of this article is not to make fun of science, nor to express disbelief in it, but only to suggest its limits. What I want to say is that when the scientist steps out from recording phenomena and offers a general statement of the nature of what is called 'reality,' the ultimate nature of space, of time, of the beginning of things, of life, of a universe, then he stands exactly where you and I do, and the three of us stand where Plato did—and long before him, Rodin's primitive thinker.

Consider this. Professor Hubble, like Joshua, has called upon the universe to be still. All is quiet. The universe rests, motionless, in the night sky. The mad rush is over. Every star in every galaxy, every island universe, is at least right where it is. But the old difficulty remains: Does it go forever, this world in the sky, or does it stop? Such an alternative has posed itself as a problem for every one of us, somewhere about the age of twelve. We cannot imagine that the stars go on forever. It's unthinkable. But we equally cannot imagine that they come to a stop and that beyond them is nothing, and then more nothing. Unending nothing is as incomprehensible as unending something. This alternative I cannot fathom, nor can Professor Hubble, nor can anyone ever hope to.

Let me turn back in order to make my point of view a little clearer. I propose to traverse again the path along which modern science has dragged those who have tried to follow it for about a century past. It was at first a path singularly easy to tread, provided that one could throw aside the inherited burden of superstition, false belief, and prejudice. For the direction seemed verified and assured all along by the corroboration of science by actual physical results. Who could doubt electricity after the telegraph? Or doubt the theory of light after photography? Or the theory of electricity when read under electric light? At every turn, each new advance of science unveiled new power, new mechanism of life—and of death. To 'doubt science' was to be like the farmer at the circus who doubted the giraffe. Science, of course, had somehow to tuck into the same bed as Theology, but it was the theologian who protested. Science just said, 'Lie over.'

Let us follow then this path.

## II

When mediæval superstition was replaced by the new learning, mathematics, astronomy, and physics were the first sciences to get organized and definite. By the opening of the nineteenth century they were well set; the solar system was humming away so drowsily that Laplace was able to assure Napoleon that he didn't need God to watch over it. Gravitation worked like clockwork and clockwork worked like gravitation. Chemistry, which, like electricity, was nothing but a set of experiments in Benjamin Franklin's time, turned into a science after Lavoisier had discovered that fire was not a thing but a process, something happening to things—an idea so far above the common thought that they guillotined him for it in 1794. Dalton followed and showed that all things could be broken up into a set of very, very small atoms, grouped into molecules all acting according to plan. With Faraday and Maxwell, electricity, which turned out to be the same as magnetism, or interchangeable with it, fell into its place in the new order of science.

By about 1880 it seemed as if the world of science was fairly well explained. Metaphysics still talked in its sleep. Theology still preached sermons. It took issue with much of the new science, especially with geology and the new evolutionary science of life that went with the new physical world. But science paid little attention.

For the whole thing was so amazingly simple. There you had your space and time, two things too obvious to explain. Here you had your matter, made up of solid little atoms, infinitely small but really just like birdseed. All this was set going by and with the Law of Gravitation. Once started, the nebulous world condensed into suns, the suns threw off planets, the planets cooled, life resulted and presently became conscious, conscious life got higher up and higher up till you had apes, then Bishop Wilberforce, and then Professor Huxley.

A few little mysteries remained, such as the question of what space and matter and time and life and consciousness really were. But all this was conveniently called by Herbert Spencer the *Unknowable,* and then locked in a cupboard and left there.

Everything was thus reduced to a sort of Dead Certainty. Just one awkward skeleton remained in the cupboard. And that was the peculiar, mysterious aspect of electricity, which was not exactly a thing and yet more than an idea. There was also, and electricity only helped to make it worse, the old puzzle about 'action at a distance.' How does gravitation pull all the way from here to the sun? And if there is *nothing* in space, how does light get across from the sun in eight minutes, and even all the way from Sirius in eight years?

Even the invention of 'ether' as a sort of universal jelly that could have ripples shaken across it proved a little unconvincing.

Then, just at the turn of the century the whole structure began to crumble.

The first note of warning that something was going wrong came with the discovery of X-rays. Sir William Crookes, accidentally leaving round tubes of rarefied gas, stumbled on 'radiant matter,' or 'matter in the fourth state,' as accidentally as Columbus discovered America. The British Government knighted him at once (1897) but it was too late. The thing had started. Then came Guglielmo Marconi with the revelation of more waves, and universal at that. Light, the world had learned to accept, because we can see it, but this was fun in the dark.

There followed the researches of the radioactivity school and, above all, those of Ernest Rutherford which revolutionized the theory of matter. I knew Rutherford well as we were colleagues at McGill for seven years. I am quite sure that he had no original intention of upsetting the foundations of the universe. Yet that is what he did and he was in due course very properly raised to the peerage for it.

When Rutherford was done with the atom all the solidity was pretty well knocked out of it.

Till these researches began, people commonly thought of atoms as something like birdseed, little round solid particles, ever so little, billions to an inch. They were small. But they were there. You could weigh them. You could apply to them all the laws of Isaac Newton about weight and velocity and mass and gravitation—in other words, the whole of first-year physics.

Let us try to show what Rutherford did to the atom. Imagine to yourself an Irishman whirling a shillelagh round his head with the rapidity and dexterity known only in Tipperary or Donegal. If you come anywhere near you'll get hit with the shillelagh. Now make it go faster; faster still; get it going so fast that you can't tell which is Irishman and which is shillelagh. The whole combination has turned into a green blur. If you shoot a bullet at it, it will probably go through, as there is mostly nothing there. Yet if you go up against it, it won't hit you now, because the shillelagh is going so fast that you will seem to come against a solid surface. Now make the Irishman smaller and the shillelagh longer. In fact you don't need the Irishman at all; just his force, his Irish determination, so to speak. Just keep that, the *disturbance*. And you don't need the shillelagh either, just the *field* of force that it sweeps. There! Now put in two Irishmen and two shillelaghs and reduce them in the same way to one solid body—at least it seems solid but you can shoot bullets through it anywhere now. What you have now is a hydrogen atom—one proton and one electron flying round as a *disturbance* in space. Put in more Irishmen and more shillelaghs—or, rather, more protons and electrons—and you get other kinds of atoms. Put in a whole lot—eleven protons, eleven electrons; that is a sodium atom. Bunch the atoms together into combinations called molecules, themselves flying round—and there you are! That's solid matter, and nothing in it at all except disturbance. You're standing on it right now: the molecules are beating against your feet. But there is nothing there, and nothing in your feet. This may help you to understand how 'waves,' ripples of disturbance,—for instance, the disturbance you call radio,—go right through all matter, indeed right through *you,* as if you weren't there. You see, you aren't.

The peculiar thing about this atomic theory was that whatever the atoms were, birdseed or disturbance, it made no difference to the way they acted. They followed all the laws of mechanics and motion, or they seemed to. There was no need to change any idea of space or time because of them. Matter was their 'fort,' like wax figures with Artemus Ward.

One must not confuse Rutherford's work on atoms with Einstein's theories of space and time. Rutherford worked all his life without reference to Einstein. Even in his later days at the Cavendish Laboratory at Cambridge when he began, ungratefully, to smash up the atom that had made him, he needed nothing from Einstein. I once asked Rutherford—it was at the height of the popular interest in Einstein, in 1923—what he thought of Einstein's relativity. 'Oh, that stuff!' he said. 'We never bother with that in our work!' His admirable biographer, Professor A. S. Eve, tells us that when the German physicist Wien told Rutherford that no Anglo-Saxon could understand relativity Rutherford answered, 'No, they have too much sense.'

But it was Einstein who made the real trouble. He announced in 1905 that there was no such thing as absolute rest. After that there never was. But it was not till just after the Great War that the reading public caught on to Einstein and little books on 'Relativity' covered the bookstalls.

Einstein knocked out space and time as Rutherford knocked out matter. The general viewpoint of relativity towards space is very simple. Einstein explains that there is no such place as *here.* 'But,' you answer, 'I'm here; here is where I am right now.' But you're moving, you're spinning round as the earth spins; and you and the earth are both spinning round the sun, and the sun is rushing through space towards a distant galaxy, and the galaxy itself is beating it away at 26,000 miles a second. Now where is that spot that is here! How did you mark it? You remember the story of the two idiots who were out fishing, and one said, 'We should have marked that place where we got all the fish,' and the other said, 'I did, I marked it on the boat.' Well, that's it. That's *here.*

You can see it better still if you imagine the universe swept absolutely empty: nothing in it, not even *you.* Now put a *point* in it, just one point. Where is it? Why, obviously it's nowhere. If you say it's right there, where do you mean by there? In which direction is there? In that direction? Oh! hold on, you're sticking yourself in to make a direction. It's in *no* direction; there aren't any directions. Now put in another point. Which is which? You can't tell. They *both* are. One is on the right, you say, and one on the left. You keep out of that space! There's no right and no left.

The discovery by Einstein of the curvature of space was greeted by the physicists with the burst of applause that greets a winning home-run at baseball. That brilliant writer just mentioned, Sir Arthur Eddington, who can handle space and time with the imagery of a poet, and even infiltrate humor into gravitation, as when he says that a man in an elevator falling twenty stories has an ideal opportunity to study gravitation—Sir Arthur Eddington is loud in his acclaim. Without this curve, it appears, things won't fit into their place. The fly on the globe, as long as he thinks it flat (like Mercator's map), finds things shifted as by some unaccountable demon to all sorts of wrong distances. Once he gets the idea of a sphere everything comes straight. So with our space. The mystery of gravitation puzzles us, except those who have the luck to fall in an elevator, and even for them knowledge comes too late. They weren't falling at all: just curving. 'Admit a curvature of the world,' wrote Eddington in his Gifford Lectures of 1927, 'and the mysterious agency disappears. Einstein has exorcised this demon.'

But it appears now, fourteen years later, that Einstein doesn't care if space is curved or not. He can take it either way. A prominent physicist of today, head of the department in one of the greatest universities of the world, wrote me on this point: 'Einstein had stronger hopes that a general theory which involved the assumption of a property of space, akin to what is ordinarily called curvature, would be more useful than he now believes to be the case.' Plain talk for a professor. Most people just say Einstein has given up curved space. It's as if Sir Isaac Newton years after had said, with a yawn, 'Oh, about that apple— perhaps it wasn't falling.'

### III

But unhappily we can't get away from the new physics quite as simply as that. Even if we beat them out on space and time, there is far worse to come. That's only the start of it, for now, as the fat boy in *Pickwick* said, 'I'm going to make your flesh creep.' The next thing to go is cause and effect. You may think that one thing causes another. It appears that it doesn't. And of course, when cause and effect go, the bottom is out of the universe, since you can't tell, literally can't, what's going to happen next. This is the consequence of the famous Quantum Theory, first hinted at by Professor Max Planck about forty years ago and since then scrambled for by the physicists like dogs after a bone. It changes so fast that when Sir Arthur Eddington gave the Gifford Lectures referred to, he said to his students that it might not be the same when they met next autumn.

But we cannot understand the full impact of the Quantum Theory, in shattering the world we lived in, without turning back again to discuss time in a new relation, namely, the forward-and-backwardness of it, and to connect it up again with the Second Law of Thermodynamics—the law, it will be recalled, that condemns us to die of cold. Only we will now call it by its true name, which we had avoided before, as the Law of Entropy. All physicists sooner or later say, 'Let us call it Entropy,' just as a man says, when you get to know him, 'Call me Charlie.'

So we make a new start.

I recall, as some other people still may, a thrilling melodrama called *The Silver King*. In this the hero, who thinks he has committed a murder (of course, he hasn't really), falls on his knees and cries, 'Oh, God, turn back the universe and give me yesterday.' The supposed reaction of the audience was 'Alas, you *can't* turn back the universe!'

But nowadays it would be very different. At the call the Spirit of Time would appear—not Father Time, who is all wrong, being made old, but a young, radiant spirit in a silver frock made the same back and front. 'Look,' says the Spirit, 'I'm going to turn back the universe. You see this wheel turning round. Presto! It's going the other way! You see this elastic ball falling to the floor. Presto! It's bouncing back. You see out of the window that star moving west. Presto! It's going east. Hence accordingly,' continues the Spirit, now speaking like a professor, so that the Silver King looks up in apprehension, 'time as evidenced by any primary motion is entirely reversible so that we cannot distinguish between future time and past time: indeed if they move in a circle both are one.'

The Silver King leaps up, shouts 'Innocent! Innocent!' and dashes off, thus anticipating Act V and spoiling the whole play. The musing Spirit, musing of course backwards, says, 'Poor fellow, I hadn't the heart to tell him that this only applies to primary motion and not to Entropy. And murder of course is a plain case of Entropy.'

And now let us try to explain. Entropy means the introduction into things that happen of a random element, as opposed to things that happen and 'unhappen,' like a turning wheel, good either way, or a ball falling and bouncing as high as it falls, or the earth going around the sun. These primary motions are 'reversible.' As far as they are concerned, time could just as well go back as forward. But now consider a pack of cards fresh from the maker, all in suits, all in order. Shuffle them. Will they ever come all in order again? They might, but they won't. Entropy.

Here then is Entropy, the smashing down of our world by random forces that don't reverse. The heat and cold of Carnot's Second Law are just one case of it. This is the only way by which we can distinguish which of two events came first. It's our only clue as to which way time is going. If procrastination is the thief of time, Entropy is the detective.

The Quantum Theory begins with the idea that the quantities of disturbance in the atom, of which we spoke, are done up, at least they act that way, in little fixed quantities (each a Quantum—no more, no less), as if sugar only existed by the pound. The smallness of the Quantum is beyond comprehension. A Quantum is also peculiar. A Quantum in an atom flies round in an orbit. This orbit may be a smaller ring or a bigger ring. But when the Quantum shifts from orbit to orbit it does not pass or drift or move *from one to the other*. No, sir. First, it's here and then it's there. Believe it or not, it has just shifted. Its change of place is random, and *not because of anything*. Now the things that we think of as matter and movements and events (things happening) are all based, infinitely far down, on this random dance of Quantums. Hence, since you can't ever tell what a Quantum will do, you can't ever say what will happen next. Cause and effect are all gone.

But as usual in this bright, new world of the new physics, the statement is no sooner made than it is taken back again. There are such a lot of Quantums that we can feel sure that one at least will turn up in the right place—by chance, not by cause.

The only difficulty about the Quantum Theory has been that to make the atomic 'orbits' operate properly, and to put the Quantum *into two places at once*, it is necessary to have 'more dimensions' in space. If they are not in one they are in another. You ask next door. What this means I have no idea.

Nor does it tell us any ultimate truth about the real nature of things to keep on making equations about them. Suppose I wish to take a holiday trip and am selecting a place to go. I ask, How far is it? how long does it take? what does it cost? These things all come into it. If I like I can call them 'dimensions.' It does no harm. If I like I can add other dimensions—how hot it is, how much gold it has, and what sort of women. I can say, if I wish, that the women are therefore found out to be the seventh dimension of locality. But I doubt if I can find anything sillier to say than the physicists' talk of ten and twelve dimensions added to space.

Let it be realized, I say, that making equations and functions about a thing does not tell us anything about its real nature. Suppose that I sometimes wonder just what sort of man Chipman, my fellow club member is. While I am wondering another fellow member, a mathematician, comes in. 'Wondering about Chipman, were you?' he says. 'Well, I can tell you all about him, as I have completed his dimensions. I have here the statistics of the number of times he comes ($t$), the number of steps he takes before he sits down ($s$), his orbit in moving around ($o$), aberrations as affected by other bodies ($ab$), velocity ($v$), specific gravity (*sp*), and his saturation (*S*). He is therefore a function of these things, or shall we say quite simply:—

$$F \int \frac{s \cdot v \cdot o \cdot sp \cdot S}{t \cdot ab}$$

Now this would be mathematically useful. With it I can calculate the likelihood of my friend being at the club at any particular time, and whether available for billiards. In other words, I've got him in what is called a 'frame' in space-time. But just as all this tells me nothing of ultimate reality, neither do the super-dimensions of the new physics.

People who know nothing about the subject, or just less than I do, will tell you that science and philosophy and theology have nowadays all come together. So they have, in a sense. But the statement, like those above, is just a 'statistical' one. They have come together as three people may come together in a picture theatre, or three people happen to take apartments in the same building, or, to apply the metaphor that really fits, as three people come together at a funeral. The funeral is that of Dead Certainty. The interment is over and the three turn away together.

'Incomprehensible,' murmurs Theology reverently.

'What was that word?' asks Science.

'Incomprehensible; I often use it in my litanies.'

'Ah yes,' murmurs Science, with almost equal reverence, 'incomprehensible!'

'The comprehensibility of comprehension,' begins Philosophy, staring straight in front of him.

'Poor fellow,' says Theology, 'he's wandering again; better lead him home.'

'I haven't the least idea where he lives,' says Science.

'Just below me,' says Theology. 'We're both above you.'

# INDEX

## A

# S

Book pages from the first printing of Simon and Schuster's 1956 edition of
The World of Mathematics were scanned on a Kurzweil 4000 Intelligent
Scanning System. Electronic pagination using the Linoscreen 7000 and math
and text composition in Times Roman were by Douglas Vogt, Nova Type-
setting, Inc., Bellevue, Washington. Output was done on a Linotron 202/N.


Project editor: Evan Konecky
Cover and interior text design: Darcie S. Furlan
Cover artwork: Richard Kehl
Illustrations redrawn using Adobe Illustrator: Rick Bourgoin